Lecture Notes in Physics

Springer-Verlag Berlin Heidelberg GmbH

Editorial Policy

The series *Lecture Notes in Physics* (LNP), founded in 1969, reports new developments in physics research and teaching -- quickly, informally but with a high quality. Manuscripts to be considered for publication are topical volumes consisting of a limited number of contributions, carefully edited and closely related to each other. Each contribution should contain at least partly original and previously unpublished material, be written in a clear, pedagogical style and aimed at a broader readership, especially graduate students and nonspecialist researchers wishing to familiarize themselves with the topic concerned. For this reason, traditional proceedings cannot be considered for this series, though volumes to appear in this series are often based on material presented at conferences, workshops and schools (in exceptional cases the original papers and/or those not included in the printed book may be added on an accompanying CD-ROM, together with the abstracts of posters and other material suitable for publication, e.g. large tables, colour pictures, program codes, etc.).

Acceptance

A project can only be accepted tentatively for publication, by both the editorial board and the publisher, following thorough examination of the material submitted. The book proposal sent to the publisher should consist of at least a preliminary table of contents outlining the structure of the book together with abstracts of all contributions to be included.
Final acceptance is issued by the series editor in charge, in consultation with the publisher, only after receiving the complete manuscript. Final acceptance, possibly requiring minor corrections, usually follows the tentative acceptance unless the final manuscript differs significantly from expectations (project outline). In particular, the series editors are entitled to reject individual contributions if they do not meet the high quality standards of this series. The final manuscript must be camera-ready, and should include both an informative introduction and a sufficiently detailed subject index.

Contractual Aspects

Publication in LNP is free of charge. There is no formal contract, no royalties are paid, and no bulk orders are required, although special discounts are offered in this case. The volume editors receive jointly 30 free copies for their personal use and are entitled, as are the contributing authors, to purchase Springer books at a reduced rate. The publisher secures the copyright for each volume. As a rule, no reprints of individual contributions can be supplied.

Manuscript Submission

The manuscript in its final and approved version must be submitted in camera-ready form. The corresponding electronic source files are also required for the production process, in particular the online version. Technical assistance in compiling the final manuscript can be provided by the publisher's production editor(s), especially with regard to the publisher's own LaTeX macro package which has been specially designed for this series.

Online Version/ LNP Homepage

LNP homepage (list of available titles, aims and scope, editorial contacts etc.):
http://www.springer.de/phys/books/lnpp/

LNP online (abstracts, full-texts, subscriptions etc.):
http://link.springer.de/series/lnpp/

D. Shi B. Aktaş L. Pust F. Mikailov (Eds.)

Nanostructured Magnetic Materials and Their Applications

Springer

Editors

Prof. Donglu Shi
Dept. of Materials Science
and Engineering
University of Cincinnati
Cincinnati, OH 45221-0012
USA

Dr. Ladislav Pust
Seagate Technology
Bloomington
Minnesota 55378
USA

Prof. Bekir Aktaş
Department of Physics
Gebze Institute of Technology
P.O. 141, Gebze 41400
Kocaeli, Turkey

Dr. Faik Mikailov
Department of Physics
Gebze Institute of Technology
P.O. 141, Gebze 41400
Kocaeli, Turkey

Cover picture: See Fig. 10.1a for details

Die Deutsche Bibliothek - CIP-Einheitsaufnahme

Nanostructured magnetic materials and their applications / D. Shi ... (ed.).
(Lecture notes in physics ; Vol. 593)
(Physics and astronomy online library)

DOI 10.1007/978-3-540-36872-4

ISSN 0075-8450

http://www.springer.de

Originally published by Springer-Verlag Berlin Heidelberg New York in 2002
MyCopy version of the original edition 2002

Typesetting: Camera-ready by the authors/editor
Camera-data conversion by Steingraeber Satztechnik GmbH Heidelberg
Cover design: *design & production*, Heidelberg

Printed on acid-free paper
SPIN: 10885630 57/3141/du - 5 4 3 2 1 0
www.springer.com/mycopy

Preface

Today the role of magnetism in human life is more important than ever. The growing need for even higher data-recording densities has driven the size of particles used in recording media down into the nanometer range while the rapidly increasing power of computers has made it feasible to perform simulations of the dynamic properties of realistic model systems of sizes comparable to experimental ones. The quality and the resolution of the devices have been increased in parallel with the developments in scientific research. It is thus possible to fabricate and study these materials at very small, almost atomic scales. On the other hand the physical properties of the materials at such a small scale are observed to change dramatically compared to those of their macroscopic counterparts. That is, the nanoscale materials exhibit new or crossover phenomena connected with quantum size effects.

There is now intensive interest in the magnetic properties of nanostructured materials. The scientific research has concentrated on the design, synthesis and characterization of nanostructured materials during the past years. A number of methods based on physical principles have been developed for manufacturing and characterizing nanostructured materials.

There is a great expectation in industry that the results of these exciting scientific researches can be exploited in a new generation of high-technology devices. Many of the unique properties of these materials have high potential for technical applications in diverse areas such as information technology, magnetosensors, electronics, data storage, magnetic heads of computer hard disks, single electron devices, microwave electronic devices, etc. In fact, for instance, the GMR materials have already found applications as sensors of low magnetic fields, computer hard disk heads, magnetoresistive RAM chips etc. Even new terminologies, for example magnetoelectronics, spintronics, spin valve etc. have recently been introduced to refer to aspects of the field involving magnetic phenomena. Thus this generic nanotechnology will inevitably have great impact on a wide range of industrial sectors and on the everyday lives of humans. In other words, one of the pillars of industry in the 21st century may involve the field of magnetic and electrical nanoscale materials.

The idea for this book was born at the International Workshop on Nanostructured Magnetic Materials and their Application (NMMA2001), held in Gebze, Istanbul (Turkey) from September 3-7, 2001. The contributions are concentrated on magnetic properties of nanoscale magnetic materials, especially on fabrication, characterization and the physics behind the behavior of these structures.

We would like to thank all the authors for their contributions. We should also acknowledge the great efforts of the people, especially Prof. Ahmet Ayhan, Prof. Alinur Buyukaksoy, Dr. Bulat Rameev and others who made major contributions to the organization of the meeting and made this publication possible.

Gebze Institute of Technology, Turkey,
April 2002

Donglu Shi
Bekir Aktaş
Ladislav Pust
Faik Mikailov

Contents

List of Contributors

B. Aktaş
Gebze Institute of Technology
P.O. Box 141,
Gebze, 41400, Kocaeli, Turkey
aktas@penta.gyte.edu.tr

A. K. M. Akther Hossain
Bangladesh University of Engineering & Technology, Dhaka 1000,
Bangladesh
ahossain@phy.buet.edu

M. Alper
Uludag University,
16059, Görükle,
Bursa, Turkey
malper@uludag.edu.tr

S. Bhattacharya
TU Darmstadt,
Petersenstrasse 23,
64287 Darmstadt, Germany
sarbari@hrz2.hrz.tu-darmstadt.de

F. Briones
Instituto de Microelectrónica de Madrid
Isaac Newton 8 (PTM) Tres Cantos,
Madrid 28760, Spain
briones@imm.cnm.csic.es

G. Brown
Florida State University,
Tallahassee, FL 32306-4120, USA
browngrg@csit.fsu.edu

H. Brückl
Universität Bielefeld,
Universität str. 25,
33501 Bielefeld, Germany
brueckl@physik.uni-bielefeld.de

A. Cebollada
Instituto de Microelectrónica de Madrid
Isaac Newton 8 (PTM) Tres Cantos,
Madrid 28760, Spain
alfonso@imm.cnm.csic.es

P. Crespo
Instituto de Magnetismo Aplicado,
U. C. M.-RENFE, P.O.Box 155,
28230 Madrid, Spain
pcrespo@renfe.es

M. S. Flores
Instituto de Magnetismo Aplicado,
U. C. M.-RENFE, P.O.Box 155,
28230 Madrid, Spain
scexu90@renfe.es

R. Ganguly
Bhabha Atomic Research Centre,
Mumbai - 400 085, India
rajibg@magnum.barc.ernet.in

W. Gawalek
Institut für Physikalische Hochtechnologie e.V., POB 100239,
D-07702 Jena, Germany
gawalek@ipht-jena.de

M. Ghafari
TU Darmstadt,
Petersenstrasse 23,
64287 Darmstadt, Germany
ghafari@tu-darmstadt.de

A. Ghazali
Université Paris 7,
75251 Paris Cedex 05, France
ghazali@gps.jussieu.fr

I. K. Gopalakrishnan
Bhabha Atomic Research Centre,
Mumbai - 400 085, India
ikgopal@magnum.barc.ernet.in

T. Habisreuther
Institut für Physikalische
Hochtechnologie e.V., POB 100239,
D-07702 Jena, Germany
habisreuter@ipht-jena.de

H. Hahn
TU Darmstadt,
Petersenstrasse 23,
64287 Darmstadt, Germany
hhahn@hrzpub.tu-darmstadt.de

S. Heitmann
Universität Bielefeld,
Universität str. 25,
33501 Bielefeld, Germany
heitmann@physik.uni-bielefeld.de

A. Hernando
Instituto de Magnetismo Aplicado,
U. C. M.-RENFE, P.O.Box 155,
28230 Madrid, Spain
ahernando@renfe.es

R. Hiergeist
Institut für Physikalische
Hochtechnologie e.V., POB 100239,
D-07702 Jena, Germany
hiergeist@ipht-jena.de

A. Hütten
Universität Bielefeld,
Universität str. 25,
33501 Bielefeld, Germany
huetten@physik.uni-bielefeld.de

H. Kubota
Tohoku University,
05 Aoba, Aramaki, Aoba-ku,
Sendai 980-8579, Japan
kubota@mlab.apph.tohoku.ac.jp

J.-C. S. Lévy
Université Paris 7,
75251 Paris Cedex 05, France
levy@ccr.jussieu.fr

D. Litzkendorf
Institut für Physikalische
Hochtechnologie e.V., POB 100239,
D-07702 Jena, Germany
litzgendorf@ipht-jena.de

A. V. Lukashin
Moscow State University, Vorobievy
Gory, 119899, Moscow, Russia
alex@inorg.chem.msu.ru

H. O. Lutz
Universität Bielefeld,
Universität str. 25,
33501 Bielefeld, Germany
lutz@physik.uni-bielefeld.de

V. V. Matveev
St.Petersburg State University,
1 Ulianovskaya, Petrodvorets,
198504, Saint-Petersburg, Russia
vmatveev@esr.phys.spbu.ru

J. L. Menéndez
Instituto de Microelectrónica de
Madrid
Isaac Newton 8 (PTM) Tres Cantos,
Madrid 28760, Spain
menendez@imm.cnm.csic.es

S. J. Mitchell
Florida State University,
Tallahassee, FL 32306-4120, USA
mitchell@csit.fsu.edu

R. Müller
Institut für Physikalische
Hochtechnologie e.V., POB 100239,
D-07702 Jena, Germany
robert.mueller@adam.ipht-jena.de

G. A. Nikolaychuk
St.Petersburg State University,
1 Ulianovskaya, Petrodvorets,
198504, Saint-Petersburg, Russia
sergio nicola@mail.ru

M. A. Novotny
Mississippi State University,
Mississippi State, MS 39762, USA
man40@ra.msstate.edu

Y. Öner
Istanbul Technical University,
Maslak, Istanbul, Turkey
oner@itu.edu.tr

M. Özdemir
Marmara University,
Gožtepe, Istanbul, Turkey
mustafaozdemir@marmara.edu.tr

I. V. Pleshakov
A. F. Ioffe Physical-Technical
Institute, 21 Politechnicheskaya,
194021, Saint-Petersburg, Russia
vp shuv@pop.ioffe.rssi.ru

Z. Qian
NVE Corporation,
11409 Valley View Road,
Eden Prairie, MN 55344, USA
zqian@nve.com

A. G. Ravlik
National Technical University-
-Kharkiv Polytechnical Institute
21 Frunze st., Kharkiv,
61002, Ukraine
ship@kpi.kharkov.ua

G. Reiss
Universität Bielefeld,
Universität str. 25,
33501 Bielefeld, Germany
reiss@physik.uni-bielefeld.de

P. A. Rikvold
Florida State University,
Tallahassee, FL 32306-4351, USA
rikvold@scri.fsu.edu

J. Schmalhorst
Universität Bielefeld,
Universität str. 25,
33501 Bielefeld, Germany
jschmalh@physik.uni-bielefeld.de

W. Schepper
Universität Bielefeld,
Universität str. 25,
33501 Bielefeld, Germany
schepper@physik.uni-bielefeld.de

H. Schmitt
TU Darmstadt,
Petersenstrasse 23,
64287 Darmstadt, Germany
holsmi@web.de

Z. Sefrioui
Instituto de Microelectrónica de
Madrid
Isaac Newton 8 (PTM) Tres Cantos,
Madrid 28760, Spain
sefrioni@imm.cnm.csic.es

V. Siruguri
Bhabha Atomic Research Centre,
Mumbai - 400 085, India
siruguri@apsara.barc.ernet.in

B. Stahl
TU Darmstadt,
Petersenstrasse 23,
64287 Darmstadt, Germany
stahl@hrzpub.tu-darmstadt.de

H. Steinmetz
Institut für Physikalische
Hochtechnologie e.V., POB 100239,
D-07702 Jena, Germany
steinmetz@ipht-jena.de

A. Thomas
Universität Bielefeld,
Universität str. 25,
33501 Bielefeld, Germany
athomas@physik.uni-bielefeld.de

M. Tondra
NVE Corporation,
11409 Valley View Road,
Eden Prairie, MN 55344, USA
markt@nve.com

A. V. Vedyayev
CEA, SP2M/NM,
38054 Grenoble, France
vedy@magn.ru

D. Wang
NVE Corporation,
11409 Valley View Road,
Eden Prairie, MN 55344, USA
dexinw@nve.com

J. V. Yakhmi
Bhabha Atomic Research Centre,
Mumbai - 400 085, India
yakhmi@magnum.barc.ernet.in

R. Yilgin
Gebze Institute of Technology
P.O. Box 141,
Gebze, 41400, Kocaeli, Turkey
yilgin@penta.gyte.edu.tr

M. Zeisberger
Institut für Physikalische
Hochtechnologie e.V., POB 100239,
D-07702 Jena, Germany
zeisberger@ipht-jena.de

M. Ye. Zhuravlev
Universität Bielefeld,
Universität str. 25,
33501 Bielefeld, Germany
zhur@Physik.Uni-Bielefeld.de

1 A Survey of Nanomagnetism

Bekir Aktaş

Gebze Institute of Technology, Dept. of Physics,
P.O. Box 141, 41400 Gebze, Kocaeli, Turkey

Abstract. This review is mainly concerned with size effects on the magnetic properties of low-dimensional magnetic systems. Firstly the basic physics behind the magnetism-related parameters and newly appeared phenomena such as spin polarized tunneling, oscillatory exchange coupling between magnetic-nonmagnetic multilayers, magnetoresistivyty etc., which take place in low dimensional materials are briefly addressed. Then the dependence of these parameters on the sizes of the magnetic systems are discussed as the sample dimensions are lowered from the three dinesional bulk materials through two (films, quantum wells), one (wires) and zero (quantum dots) dimensions. The critical sizes for these phenomena to be observable are also considered. Then the experimental techniques for both preparation and physical characterisation of nano-sized especially magnetic materials are summarized. The recent progresses in methods and devices and their advantage and/or disadvantages are briefly discussed as well. Lastly present situation of the technical applications of the low dimensional magnetic systems and the trends for future developments are mentioned.

1.1 Introduction

The physics and chemistry of nano-scale systems has advanced rapidly over the last few years. Like in semiconductors, the structure of magnetic materials has been shown to become controllable in very low dimensions as well. As their dimensions decrease down to the nanometer scale, these materials start to exhibit new and very rich interesting physical properties mainly due to quantum size effects. For instance magnetic behavior is changed, the magnetization of noble metals is considerably induced, electron interference patterns is observed, oscillatory exchange interactions occur between adjacent layers separated by a non-magnetic spacer, magnetoresistivity is enhanced many orders of magnitude, etc. Even the intrinsic physical characteristics of nano- structured material are observed to change drastically compared to those of their macroscopic counterparts. This means that the intrinsic properties of the materials become extrinsic (size dependent) as they are tailored in nanometer size. Obviously a particular material on a macroscopic scale has a limited number of physical parameters having certain values. However, the same substance shows additional phenomena and most of these parameters can be adjusted by the size in nano-scale allowing virtually continuous spectra for them. Various combinations of nano-structured elements in a single electronic component further enrich the physics allowing the devices virtually having any characteristic. These phenomena introduce very good oportunities to scientists to exploit these new properties to manufacture new and high-technology devices, which could not be obtained by using only macroscopic materials.

However, almost all of the physical parameters in these new phenomena are affected by the imperfections of the nano-structures. The homogeneity and purities in chemical composition, crystalline structures, external morphology, etc determine the physical properties. Therefore very high quality materials should be designed and implemented in nano-scale systems in order to get a device with desired characteristics. That is one encounters many problems in the preparation of magnetic materials in such a low dimension. Developments in nano-structured magnetic materials have been summarized in the literatures [1- 8]. The most recent and detailed review was given by Himpsel et al. [9]. Our intent is not to give detailed information on recent progress of magnetic nano- structures. Rather, in this introductory section only some of main subjects will be addressed very briefly.

1.2 Physical Properties of Magnetic Nanostructures

1.2.1 Substrate Effects on Structures and Related Properties

Low dimensional behaviors start to manifest themselves a certain film thickness. The electrical and magnetic properties are the most important ones in magnetic films as far as application is concerned. The magnetic data for some of the films have been given by Gradmann [10]. As well known the most famous, oldest and perhaps most important and well-characterized ferromagnetic bulk materials have been made from Fe so far. However, the magnetic phases are dramatically changed and new magnetic phases arise in ultra thin films of epitaxially-grown metallic Fe, for instance, on a single crystal Cu(100) substrate. The magnetic properties, phase boundaries, and critical temperatures were shown to depend on the film thickness [11-15]. A very strong perpendicular magnetic anisotropy is induced as well to magnetize the film perpendicular to the film for thickness below 6 nm. Qualitatively similar behaviors were also observed for other over-layer-substrate combinations [16]. These dramatic changes are caused by a different crystal structure of the film (fcc) that is imposed by substrate through adhesion at the interface. Band structure and exchange splitting are varied by crystalline symmetry as well [17,18]. It should be remembered that the most important parameter determining the magnetic structure is exchange overlap integral of electronic wave functions on neighboring atoms. Obviously this overlap depends on neighboring atomic distances determined by crystal lattice parameters. In some cases a ferromagnetic substrate can magnetize a thin over layer films made of paramagnetic materials in certain temperature ranges [19-22].

On the other hand the major contribution to the magnetism comes from electronic spins while the magnetic anisotropies originate from the interaction between spins and orbit, which is strongly linked to the lattice. Since the orbital wave functions of local electrons at the surface are affected by the discontinuity of crystalline field, a surface anisotropy is induced. Also exchange interactions are changed because of lack of neighbours. A significant magneto-elastic energy due to lattice mismatch is induced as well. In most cases the easy axis of this induced (thickness dependent) anisotropy becomes parallel to the film normal. For example, the critical thickness for bcc Fe/Cu(100) and bcc Fe/Ag(100) is around five atomic layers [23,24]. This implies that up to the next nearest neighbor interactions between surface and the central layer atoms play major role in ferromagnetic ordering in these films. These anisotropies can overcome the magneto static demagnetizing energy to give perpendicular magnetization for ultra thin film cases. The thickness-dependence of the crystal structure of Fe on various substrates was theoretically explained by Nordström and Singh [25].

1.2.2 Oscillatory Exchange Coupling

One of the most important phenomena is observed in ferromagnetic thin films separated by a nonmagnetic spacer, namely oscillatory exchange coupling, J, [26-30]. In this case one of the layer polarizes the conduction electrons of nonmagnetic metallic spacer. If the life-time of the polarization is long enough, these polarized electrons carry this information to the layer across the spacer. Thus second layer is coupled with the first one through polarized conduction electrons (Ruderman-Kittel-Kasuya-Yoshida (RKKY) [31] interactions). Since the density of the conduction electrons are modulated by electronic wave function at the Fermi level, this indirect interaction is connected to the Fermi wavelength and oscillates along spacer thickness with $2k_F$ (k_F: Fermi wave vector). This effect is very important in magnetoresistivity of multilayer films and even fine particles embedded in a non-magnetic matrix [32].

1.2.3 Spin Polarized Tunneling

Spin polarized tunneling is one of the most promising phenomena for applications. In this case, two ferromagnetic materials are separated by a very thin insulator. The tunneling current from one layer to the other depends on the relative orientation of magnetization of the layers and the potential barrier, which in turn, depends on the insulator thickness and the type of the layer materials. Parallel orientation of the adjacent layer corresponds to the lower resistivity. Indeed tunneling rate depends on the junction quality. The interface of the junction should be very sharp and well characterized. The temperature and the voltage are also very effective on the tunneling current. The tunneling current is significantly increased in sharp metal insulator junction [33,34].

1.2.4 Magnetoresistivity

Another very important phenomenon in nano-structures is the magnetoresistivity, which is observed as a change of electrical resistivity when an external magnetic field is applied to the substance under investigation. The direct connection between magnetism and electrical properties makes this phenomenon ideal for magnetic sensors. This effect has been well known for a long time and produced by many different mechanisms as will be explained in Chapter 2. In some cases this effect is dramatically enhanced and it is called giant (GMR) [35-38], colossal (CMR) magnetoresistance, etc. This effect arises from spin dependent scattering of the charge carriers. Usually, the resistance in normal metals is increased in the presence of an external magnetic field and depends on the relative orientation of the current with respect to the field (anisotropic magnetoresistance, AMR). However the resistivity decreases with applied field for magnetic-nonmagnetic composite systems when the size of nonmagnetic metallic component separating magnetic components becomes smaller than the mean free path of charge carriers. This is called giant magnetoresistance (GMR) and can be much larger than AMR. This negative magnetoresistance is enhanced further even in the bulk of some manganite perovskites (CMR), [39]. The effect is believed to be due to electron hopping between spin-polarized Mn^{3+} and Mn^{4+} ions. A detailed explanation for this effect is given in Chapter 2 .

The magnetoresistivity depends also on the direction of the current with respect to the film normal. The effect is increased for current perpendicular to the plane (CPP, see [40,41]). However the CPP case is more difficult to realize experimentally than the configuration with the current parallel to the plane (CIP). Therefore, most studies have been carried out for the CIP case. The GMR effect arises from spin-dependent scattering of conduction electrons by spins in ferromagnetic layers. When the spin of ferromagnetic layers is untiparallel to the conduction electron then the scattering is increased. Therefore the magnetoresistivity will be varied by the relative orientation of magnetization in neighboring ferromagnetic layers separated by a non-magnetic metallic thin layer. In the ideal case this three-layer system can behave like a spin valve for electron transport from one ferromagnetic layer to the other. The relative direction of the ferromagnetic layer magnetization can easily be manipulated by exchange interaction between the layers, exchange biasing by magnetically hard layer, dipolar interaction between two ferromagnetic layers having different coercive fields, external magnetic fields, etc.

As can be expected, there are many parameters affecting the magnetoresistivity for multilayer films: Spin scattering parameters for outer surfaces and interfaces, effective masses, inner potentials and relaxation times for magnetic and non-magnetic films. It is believed that interface effects have a dominant role on the spin dependent scattering [42]. All of these parameters are strongly influenced by imperfections of the samples. Many sophisticated techniques have been developed to prepare high quality samples, as will be mentioned next section.

1.2.5 Two Dimensional Systems

Thin Films

A transition from three- to two-dimensional behavior is observed as the film thickness decreases down to some certain thickness [43,44]. The magnetization of the sample is obtained by using usual spin wave theory. However, for a two dimensional system, as the films becomes thinner, the spin wave vector perpendicular to the film plane is quantized, analogous to lattice vibrations (sound waves). Consequently spin wave energy spacing between quantized states can far exceed kT at finite temperatures for a sufficiently thin film. In this case only low energy spin waves with wave vector in two dimensions (film plane) should be considered in the calculation of the magnetization. Thus the effective number of degrees of freedom is decreased from three to two for spin waves. This effect is expected to reflect itself in critical exponents during magnetic phase transition near critical temperature T_c. As a results magnetization in two dimensions falls off more rapidly than in three dimension with increasing temperature around T_c. This effects is observed in a number of materials, for instance in Ni(111) films on W(110) substrate [45,46].

The Curie temperature T_c is determined mainly by the number and coordination symmetry of exchange coupled neighboring magnetic atoms and strength of J. However, all these parameters are different in the case of very thin films. While the number of neighboring atoms is determined by crystal symmetry, the exchange interaction parameters, J, is changed by both the lattice parameters (which is controlled by the substrate, as will be explained later) and symmetry. In the case of very thin films, absence of layer on the vacuum side is felt by the deeper layers since the ferromagnetism is cooperative effect. Consequently, the Curie temperature drastically changes in thin film case [47-49]. T_c is found to depend on the number of layers. T_c even vanishes when we go to a one-dimensional system [50].

While the Curie temperature, T_c, generally decreases with decreasing film thickness, the magnetization can be enhanced for the ultra-thin films, especially for monolayers. The enhancement in the surface magnetization for some ferromagnets was explained by Freeman and Wu [51]. This increment is believed to originate from the increase of individual atomic moments at the surfaces. As well known, the atoms try to maximize their spins as much as possible in accordance with the first Hund's rule, while trying to obey Pauli exclusion principle. However, if the spin quantum number is the same for the electrons, then their orbital states have to be different to satisfy Pauli principle. The electrons want to be far apart from each other to minimize the Coulomb energy. However, they are confined a small volume to share same orbit in a solids (cf: maximum orbital quantum number compatible with maximum spin: Second Hund's rule). But electrons at the vacuum side of the surface atoms are relatively free to have different spatial orbital wave function with the same spins. Thus the contribution to the surface magnetism and to the exchange coupling from individual atoms is expected to increase.

On the other hand, according the Wigner-Eckart theorem [52], the electron spatial coordinates appearing in crystalline field Hamiltonian can be replaced by the corresponding angular momentum operator in spin Hamiltonian. For instance z coordinate perpendicular to the film is replaced either by J_z, L_z or S_z. Therefore as the electrons go toward vacuum side (larger z) of the surface the Coulomb energy decreases corresponding to larger Sz and therefore larger atomic moments and easy perpendicular anisotropy for magnetization result. Heavy interface atoms such as Ir, Pt, and Au with large spin-orbit interaction make an additional contribution to perpendicular anisotropy. Conceptually similar effects are expected for interface electrons as well [53,54].

Monolayers

For monolayers, this effect is much more enhanced. Local density calculation [55] showed that the energy band of monolayer of 3d transition metals on a Ag(100) substrate becomes narrower and the density of states for majority spin electrons increases at the Fermi level, giving larger magnetic moments. Even a non-magnetic metal in its bulk form can become spontaneously ferromagnetic in its monolayer form [56,57]. Experimental proof is not satisfactory yet although there are some indications supporting these theoretical results [58]. The lack of experimental evidence is due to difficulties in growth of ideal monolayers. However for a high quality Cr monolayer on Fe(100) the atomic magnetic moment was observed to rise up to $3\,\mu_B$ instead of $0.4\,\mu_B$ for antiferromagnetic Cr in its bulk form [59,60] (μ_B: Bohr mahneton).

Quantum Wells

As the films thickness is decreased more and more, the film can behave like a quantum well for spin carriers. Thus the carriers are reflected by the walls of the wells and the wave functions interferes to form a standing wave with discrete energy levels. The reflection coefficient indeed depends on surface roughness and the energy of the carriers, which in turn, depend on the spin directions. That is, the reflection coefficient is spin-dependent (different for minority and majority spins). In order to observe these phenomena, the surface roughness must be smaller than the electron wavelength at the Fermi level. This effect influences almost all the physical properties, for example inverse photoemission and photoemission [61,62], magnetic anisotropy [63], magneto-optic response [64-66], electrical conductivity [67], Hall effect [67], superconductivity [69], etc. Even at an early stage of growth, the film- growth is influenced by quantum well effect via its surface free energy [70,71]. The shape of the well is indeed varied by the surface conditions and the type of material under investigation. In multilayers the spin carriers are partially reflected and partially transmitted into the neighboring layer. However periodic boundary conditions can be applied and discrete energy levels result again.

1.2.6 One Dimensional Systems

Quasi-one-dimensional magnetic system on an atomic level naturally occur as a chain of magnetic atoms even in bulk form of some crystalline materials. These magnetic chains are magnetically decoupled from neighboring chains by non- magnetic intermediated atoms. For instance, some inorganic manganites perovskite-derived transition metal-oxide structures exhibit one-dimensional magnetic chains.

Recently artificial one-dimensional magnetic structures are becoming more attractive for their magnetoresistance [72]. As mentioned above, the maximum magnetoresistance is for perpendicular transport from one magnetic layer to the other (Current Perpendicular to Plane - CPP) case. Although the geometry is experimentally difficult for continuous thin film case it can be realized by growing parallel stripes separated by non-magnetic very narrow spacer wires on a substrate [73]. In this case the current in the substrate plane can be applied perpendicular to the wires, giving high magnetoresistance.

A stepped surface of a substrate (either insulating or conducting) may contain many parallel magnetic stripes. In this structure the parallel stripes cause a discontinuity for the electrons to be partially reflected back to give a standing wave [74]. Consequently, many properties of the interface are changed. For instance, continuous spectra split into discrete energy levels [75] corresponding to the wave vector perpendicular wires, new step states arise and in-plane anisotropy is induced, etc. From the competition of this anisotropy with the shape anisotropy of the magnetic stripes, the easy magnetization direction in the sample plane may be switched [76,77] between perpendicular and parallel to the stripes. By manipulating the density of steps and the wire thickness one can obtain various magnetic states.

1.2.7 "Zero Dimensional" Systems

Since the surface to volume ratio is highest for "zero dimensional" systems most remarkable changes are observed in ultra-fine magnetic particles. The surface magnetism increases or decreases, or even can be disordered [78] and/or dead for some surface treated (e.g. Ni) particles [79] at low temperatures. An increment in T_c has been observed for $MnFe_2O_4$ particles [80]. The smallest magnetic particles are magnetic molecules and clusters [81]. The size of artificial magnetic particles may be reduced down to even one nanometer. The fine particles behave as a monodomain magnet because the particle diameter would be even smaller than domain wall thickness. When the particle size shrinks further the superparamagnetic limit is reached. Sometimes the system enters spin glass regime due to dipolar interactions between the particles. The spin flip can take place either via thermal excitation or quantum tunneling [82]. For fine particles one needs to use suitable micro-magnetic modeling to get the magnetic behavior. The fine particles system on a substrate can be in distincly different phase from the bulk, depending on the particle density and size in nano-meter range [83]. The ultimate goal is to achieve particle array for magnetic recording [84,85]. Thus both signal to noise ratio and storage capacities would be increased a few orders of magnitude.

1.3 Experimental Techniques for Nano-scaled Magnetic Materials

As mentioned above the physical properties of nano-structured materials can differ substantially from those of their macroscopic counterparts. The parameters characterizing these new features are deeply affected by sample geometry and defects, such as surface roughness, purity, structural homogeneity, etc. The misfit between substrate and the sample can have enormous influence on very thin films. Quantum effects become measurable and more clearly observable as the systems are prepared as small and perfect as possible. Therefore as the sizes of physical systems to be studied become smaller, the need for more sophisticated experimental devices increases. A number of methods have been developed.

1.3.1 Recent Progress in Nano-sized Sample Preparation

Up to now, there have been many methods and devices developed for small sized sample preparation. In principle, various growth techniques used for many non-magnetic nano-structured materials can be used also for magnetic materials. Of these, epitaxial layer-by-layer growths in ultra high vacuum systems are the most useful and common, since lattice symmetry and size of the crystal strongly influence the magnetic properties. However, due to their relatively higher surface free energy, the magnetic materials are difficult to grow on any substrate. Nevertheless, it was demonstrated that any particular ferromagnetic material can be deposited in more than one crystal structure by using different substrates, even different lattice planes of the same substrate, as an interface between the substrate and desired film [86]. Hence some artificial solids are possible to grow, exhibiting new physical properties.

As well known, the most critical parameters in magnetic multilayer growth are lattice and relative surface free energies mismatch between substrate and overlayer film (even interface) [87]. The interdiffusion, chemical reactions and alloying between substrate and overlayer atoms give additional problems ([88]). In order to solve both interdiffusion and lattice mismatch problems, sometime inert substances are used as a surfactant [89-91]. Some surface passivation method have also been used to suppress interdiffusion between layers ([92]). One of the major issues is to develop suitable buffer layers to prevent undesirable interface effects for each combination of magnetic multilayers [93-95].

For epitaxial multilayer film preparations, noble metals, semiconductors and some oxides are used as single crystal substrate [86,96]. Epitaxial growth of films in thickness even up to some micrometers can easily be achieved for sufficient matching between substrate and over layer. Usually when the mismatch exceeds a few per cent, the growing film after some nano-meter thickness eventually breaks up into islands on the substrate in order to relax progressively growing total misfit strain energy. Generally the overlayer imitates the substrate structure for the case of sufficient lattice and surface energy matching. But when the surface energy of the substrate is larger than those of over-layer and interface, then after the first wetting layer, the molecules at subsequent layers form islands to avoid the strain energy due to lattice mismatch (equilibrium mode). In order to overcome this problem, many tricks have been developed. One of these is low temperature and high growth rate [97,98]. In this mode the thermal energy for the atoms is not large enough to move around at the surface of the growing films at sufficiently low temperatures. Therefore each molecule has to stick at a nearest place with an energy minimum of the previous layer, and thus, growth symmetry follows the substrate structure. After deposition the film is annealed at suitable temperature for (non-equilibrium growth mode). Even at lowest temperature a very rough surface is obtained due to nucleation of arriving atoms on top of each other. By applying some chemical reactions or, heat and/or light pulse to the surface the thickness can be controlled layer-by-layer [99,100]. Well- characterized sharp surfaces [101] were shown to be obtainable by electrochemical deposition technique.

One-dimensional nanostructures (nanowires with a diameter of nanometer range) can be obtained in various ways. The well-known lithography method is used to produce a template on a suitable substrate and filling this step by evaporated magnetic materials [102]. Self-organization is another useful and most promising method. In particular, after a few steps of chemical and electrical processes on a substrate layer, different kinds of polymer chains in a mixture are regularly ordered like a lattice (usually hexagonal) and oriented perpendicular to the substrate. Then one type of these molecules are removed from the matrix, leaving regular pores as templates. Finally magnetic materials are deposited on these cylindrical pores to get nanowires perpendicular to the surface of the substrate. The beam of magnetic materials can be focused by interference field of an intense laser onto the substrate to form parallel wires [103]. The steps on a surface of the substrate can be used as template for magnetic nano-wires as well. The step edge guides the atoms to form wires or stripes [104]. A row of anisotropic surface atoms can also attract the magnetic atoms to be ordered parallel to themselves ([105]).

Some biological organisms contain magnetic (Fe) particles naturally [106]. There are many ways to grow fine particles. For instance ([107]) spinel ferrite particles with uniform size distribution can be obtained from the precipitation of required constituents solution by adjusting physical parameters such as pH, temperature etc. Also misfit dislocations can serve as templates for fine particles [108]. Even ultra fine particles, consisting of a few atoms, are obtained by laser ablation of target materials onto carrier gas [109]. Optical lithography can be used to get fine particles on a surface of suitable substrate materials as well. But the particle size and density are limited by the wavelength of light. A standard way to get ultra-fine structures is to use electron beam lithography [110,111]. Even scanning tunneling microscopy (STM) method can be used to carry an individual atom and put it at a desired position in order to create any desired pattern with atomic resolution [112].

1.3.2 Measurements Techniques

Magnetic Measurements Techniques

The magnetic measurements can be done by using one of the experimental methods summarized in [1, 113]. Some of the traditional bulk techniques (neutron, x-ray and Brillouin scattering) are not sensitive enough to get signal from nano-sized materials. The neutron scattering could still be used to obtain information about magnetic structure after some improvements [114]. The magnetic roughness of the surface can be characterized by using circularly polarized light [115]. Also the magnon dispersion curve contains magnetization, exchange coupling, relaxation and various magnetic anisotropies parameters. Two examples for determination of magnon energies are Brillouin scattering [116,117] and magnetic resonance methods [6,118]. Today a modern ESR spectrometer is sensitive to 10^9 paramagnetic spins, as the line width is smaller than one Oersted at room temperature. Since the magnetization is much higher and exchange narrowing prevents further line broadening in ferromagnetic substances, the ESR signal, even from very few ferromagnetic layers, can be detected. The accuracy of the magnetic parameters does not depend on the amount of the sample. Other sensitive techniques for magnetic properties for nano-sized materials are alternating-gradient magnetometer [119], usual superconducting quantum interference device (SQUID), and torsion magnetometer [47]. Conversion electron Mössbauer spectroscopy extracts indirect information about magnetization by measuring hyperfine shift [120]. Another sensitive method is magneto-optic Kerr effect [2, 121,122]. Also spin polarized secondary-electron emission can be used particularly for surface magnetism [7].

Microscopy Techniques

The magnetic maps can be obtained by using very high spatial resolution magnetic microscopy techniques. Today a large variety of magnetic microscopes are available. These techniques can be grouped regarding the physical effects they depend on. One of the microscopy methods uses visible light. Since the polarization direction of the light is influenced by both direction and magnitude of the magnetization, a laser beam of optical Kerr effect devices can reveal the map of magnetic domains [123,124] by scanning the surface of the sample. However the resolution is limited by wavelength. The resolution is further increased by near field scanning optical microscopy using a very narrow capillary for the light beam [125,126].

The magnetic force microscopy has also been developed to scan the surface and to get the magnetic map of the sample. In this technique the magnetic force between a magnetic tip on the device and the local magnetization on the sample under investigation is used to obtain information about the magnetization [110,127]. A very high resolution (10-50 nm, even atomic resolution with some new non-contact techniques) can be reached in this case.

In addition, a very sophisticated magnetic resonance method has been been developed for magnetic microscopy. In this case, the resolution and sensitivity of the magnetic force microscopy is combined with the dynamic character of high frequency ferromagnetic resonance [128]. Thus dynamic properties of the magnetic materials can be profiled by the scanning procedure.

SQUID is very sensitive to a magnetic field, even to tiny fractions of a single flux quantum, and it is used for magnetization measurements on very small samples. The same effects can also be used to get magnetic domain profile at sample surface with some minor modifications. In this case very small SQUID detector elements are integrated on a silicon chip to measure the magnetic field emanating from a single magnetic domain. Thus the magnetic profile can be obtained by scanning the surface by this chip [129]. The resolution is limited by the size of an individual SQUID element on the chip.

As well known, the resolution of electron microscope is much higher than optical microscopes due to shorter de Broglie wave length (determined by the kinetic electron's energy) compared to optical wavelength. The basic principle is to scan and measure the state of the electrons coming from the sample surface in various conditions. A number of electron microscopy techniques have been developed for magnetic characterization. For instance, the spins of secondary electrons are polarized as they are emitted from a magnetized sample and give useful information about the polarization of the sample (secondary electron microscopy with polarization analysis, SEMPA) and even 20 nm resolution could be reached [7,130]. In some cases, low-energy electrons emitted from magnetic surface (LEEM) are accelerated up to suitable energy to image the surface [131] and a few tens nanometer resolution can be achieved. Even circularly polarized light can be used to obtain magnetic contrast in law-energy photoemission electron microscopy (PEEM), but the resolution decrease at least by one order of magnitude due to broad energy spectrum [132]. Another electron microscopy technique uses the Lorentz force on the electrons from the magnetic field of the magnetic domains. Since the force increases linearly with to both magnetic field and speed of the electron, by increasing the electronic energy one might, principally, get stronger deflection of the electrons and higher sensitivity and even higher resolution [111,133].

In order to achieve atomic resolution one needs to go to scanning tunneling microscopy (STM) for magnetic characterization. Again the current of the tunneling electrons from a very sharp magnetic tip through a very narrow vacuum gap (or an insulating spacer) to the magnetic surface is partly determined by polarization of the electrons with respect to local magnetization of the sample (spin dependent tunneling). The contribution of polarized electrons to the tunneling currents is increased for narrower spacers and thus a magnetic map can be obtained [33,134].

1.4 Applications

Some of these exciting scientific developments have already been used in various products. Current high-density magnetic recording is an example of such widely used application. The small-scale magnetic materials are used both in new recording media and, in particular, in the most critical elements of recording systems, in recording heads [135,136].

Magnetic layers serving as recording medium in a storage media are becoming increasingly complex with increasing areal density of recording. The size of a single recorded bit in most recent hard disks is now of nanometer scale. Magnetic recording density is increasing by nearly a factor of 2 every year. Todays recording technology has already demonstrated recording density of over 100 Gbit per square inch. However, even if all the technological problems could be solved there is still a theoretical limit for recording density. With decreasing siza of recorded bits, these magnetic particles are already now entering the superparamagnetic regime. The thermal agitation energy, kT, becomes comparable to the effective magnetic anisotropy energy (including the shape anisotropy) of a single particle. Then, magnetic moments start to flip within a finite time and therefore the recorded data are partly washed out. The stability of the bit depends on barrier energy (anisotropy energy) and in turn particle volume exponentially.

The superparamagnetic limit for some typical materials is of the order of 10 nm, which corresponds to a recording density of the order of Terabits per square inch [137]. The important parameters are anisotropy, coercivity, size, preferential orientation and the density of magnetic particles. The density can still be improved further by using perpendicular recording. In this case very thin and long enough magnetic wires are ordered perpendicular to the surface of the media. Even an individual wire or particle can be used as a bit unit in near future. Production of magnetic fine particles with uniform physical properties can be achieved by lithography techniques [85,110,124, 138].

The dimensions of magnetic components used as sensor and writer elements in recording heads are smaller than the limit of optical lithography. Dimensions of the sensors detecting magnetic fields of recorded data bits are on the nanometer scale. The sensitivity and the resolution of reading is going to be increase drastically as the size of the particles is decreased down to nanometer scale. In a recording head, magneto-resistive elements based on GMR or spin tunneling effect are used. The sensitivity of magneto-restive device is defined as relative change of resistivity per Oersted. By using the spin tunneling effect [33], the sensitivity can be an order of magnitude higher compared to a conventional magneto-resistive Ni-Fe sensor. High sensitivity has been achieved by GMR [140] or CMR [141] materials. In order to enhance the sensitivity further, the current perpendicular to the surface mode is going to be implemented.

Recently magnetic and semiconducting elements are being combined into a chip (MRAM) in order to avoid disadvantages of the mechanical part [142]. (Speed of data transfer in HD is now close to 1 Gbit/s.) Even logic devices based on nanomagnetism are under investigation. Magnetic transistors have recently been demonstrated [143,144]. Basic principle of this device depends on the injection of spin-polarized current between magnetic-nonmagnetic layers. Also, some researchers are trying to use magneto-optic, magneto-elastic etc. properties to develop novel devices.

1.5 Outline of the Book

The present book consists of 16 chapters. Chapter 1 is devoted to the general survey on recent developments in the nanomagnetism. A comprehensive review on historical developments of magnetoresistivity and magnetoresistive field elements is given in Chap.2. Classical (AMR) and giant (GMR) magnetoresistivity occurring in normal bulk conductors and in some low-dimensional magnetic-nonmagnetic heterostructures are summarized. In this chapter, the authors describe the AMR in monolayer magnetic alloys and GMR in multilayer magnetic-nonmagnetic thin films studied experimentally and they propose some possible structures for magnetic field sensors. They also address the basic principles of the effects and make a detailed comparison of the characteristics of the devices based on AMR and GMR effects with those of other magnetic field sensors and demonstrate the superiority of the former over the latter.

A very nice theoretical approach for giant magnetoresistivity (GMR) in magnetic/nonmagnetic multilayer systems is described in Chap.3. After summarizing a variety of approaches to the problem the authors introduce their own method based on the single-electron Green function (GF). Finally, the theoretical model is successfully applied to decribe the GMR in the CIP (current-in-plane) geometry in magnetron-sputtered "combination multilayers" (CML) of type Py1.8nm/Cu1.8nm/Py1.6nm/Cu0.9nm/Py1.6nm (with Py=$Ni_{81}Fe_{19}$). This chapter is very useful for researchers wishing to analyse their experimental data and extract relevant physical parameters.

In Chap.4, the authors give a very brief survey on the tunneling magnetoresistive (TMR) devices that are being developed for non-volatile memory chips (MRAM), spin electronic components and sensor applications such as read heads for hard disks. In order to elucidate the effects of imperfections on the tunnelling phenomena, they studied the structural, electronic and magnetic imperfections leading to additional scattering of the conduction electrons at the inteface in multilayers like CoFe(1.5)/Ru(0.9)/CoFe(2.2)/Al2O3(1.5)NiFe(6). The subsequently emitted conversion and Auger electrons in backscattering geometry were used in the non-destructive depth selective Mössbauer technique with ^{56}Fe and ^{57}Fe isotopes in different atomic layers of a thin film structure, allowing one to probe the hyperfine interaction at selected sites while other locations are suppressed.

The effects of stray fields on the electronic parameters of magnetic tunnel junctions (MTJs) are studied in Chapter 5. The authors consider very small physical area with different lateral geometry consisting of magnetic-nonmagnetic multilayers like CoFe(1.5)/Ru(0.9)/CoFe(2.2)/Al_2O_3(1.5) /NiFe(6) and obtained very nice pictures for the spin configurations. Thus, considering the other interactions as a source of effective fields in addition to the stray fields the authors successfully obtain the magnetic behavior as a function of the size of the multilayer elements in great detail. They also discuss how to improve the performance of MRAM cells by changing the size of the cells.

The effects of heat, and of magnetic and electric field treatments on tunneling of conduction electrons from a hard antiferromagnetic three-layer to the soft ferromagnetic one through an insulating dielectric layer at magnetic tunnel junctions (MTJ) are presented in Chap.6. In this study, a magnetically hard electrode of MTJ consisting of an artificial antiferromagnet (AAF)(either Co/Cu/Co or CoFe/Ru/CoFe) was combined with magnetically soft ferromagnetic bilayers (Fe/Co or $Ni_{81}Fe_{19}$/Co) having no intentionally induced magnetic anisotropy. They found that a variety of aspects of the stability of magnetic tunnel junctions are related to subtle physical mechanisms of diffusion, breakdown and magnetic coupling phenomena. This subject is still open to further investigation in order to improve the physical parameters for MRAM applications.

A simple and still quite effective film preparation technique is introduced in Chap.7. The technique is based on the well-known electrodeposition of the constitutent elements of a physical system to be investigated. It is shown that the method can be applied to prepare a sample of multilayer thin films as thin as a few nanometers of each magnetic/nonmagnetic metallic layer. It can also be applied for fine magnetic particles and nanowires. The method is very cheap, practical and quite efficient for speeding up sample production. Also the sample quality is unexpectedly high as far as magnetoresitivity is concerned (GMR up to 50% was observed in some of these films).

Chapter 8 describes the use of the ferromagnetic resonance (FMR) technique to study magnetic properties of disordered magnetic thin films. Epitaxial Fe_3O_4 films on MgO(100) substrate and polycrystalline $Ni_{77}Mn_{23}$ films of various thickness (50, 100, 300, 600, 1000, 3700 Å) on quartz substrate are chosen as examples. The FMR spectra are analyzed and the FMR parameters deduced for magnetic energy density, including Zeeman, effective magnetocrystalline and unidirectional bulk and surface anisotropies, spin-wave energy (exchange energy) and magnetic damping terms. The films also show easy-axis (perpendicular to the film plane) bulk and easy-plane surface anisotropy energies. The bulk anisotropy was found to increase strongly with decreasing temperature and film thickness, while linearly temperature dependent surface anisotropy shows very weak thickness dependence.

In Chap.9 authors analyse new, unusual and anomalous magnetic behavior of fine iron particles deposited in non-magnetic or magnetic host materials. The authors outline in detail and give a comprehensive review of the literature on the general behavior of this kind of materials. It was pointed out that, as the structural coherence length becomes comparable to the magnetic exchange coherence length, the magnetic properties of the particles changed dramatically from the bulk properies of the same materials. The magnetic structure of Fe atoms at the interface of the magnetic grains is shown to differ from that in deeper regions of these grains. As the temperature and/or particle size are decreased, some frustration and freezing of the spins occur and the magnetic anisotropy and magnetoresistivity change significantly. The authors also point out that the origin of these phenomena is common for quite different iron-based magnetic systems and related to the superparamagnetic behaviour of lattice regions at which competing exchange interactions take place. This chapter serves as a very useful tutorial for the beginners in the field of nanostructured magnetic particles.

Chapter 10 presents a theoretical investigation of magnetization dynamics in nano-particles and ultrathin films. The authors summarize recent developments of nanostructured magnetic materials (nano-sized particles randomly distributed or in an ordered array) especially applications in magnetic recording technologies. It is shown that, as the volume of the magnetic particles decrease to reach recording densities of the order of 100Gb/in.^{2} or even more, the particles enter the superparamagnetic regime due to temperature fluctuations (switching) of monodomain throughout a single particle and this degrades the magnetic recording quality of the media. In order to overcome this difficulty one needs to use materials with high enough magnetocrystalline anisotropy and still soft enough to permit the data to be written easily. Taking into account the magnetic anisotropy, the authors use kinetic Ising and classical Heisenberg models for magnetic nanoparticles and ultrathin films to explain the some aspects of the theory of magnetization switching. They also discuss hysteresis in nanoparticles and ultrathin films, in particular the frequency dependence of hysteresis loops and a dynamic order-disorder phase transition.

In Chap.11 a new theoretical method is given to examine in detail the magnetic spin configuration in nanostructures. The authors include long-ranged dipolar interaction in addition to the short range exchange interactions and crystalline anisotropy terms in magnetic Hamiltonians in order to obtain the magnetic ground state of two-dimensional nanostructures. They properly treate two-dimensional dipolar interactions for an infinite lattice and for dots. This approach is based on the derivation of a local Landau-like Hamiltonian. They find that the competition between long (dipolar) and short (exchange and crystalline anisotropy) range interactions gives rise to the frustration of the spins and causes both stable and metastable topological defects as well as uniform domains in nanostructured magnetic materials.

A very brief and still general survey on magnetic-nonmagnetic and magnetic-superconducting nanostructured composite systems is given in Chap.12. As one or more dimensions of this composite system decrease down to the nanometer scale, new unusual physical properties (magnetic anisotropy peculiarities, and giant magnetoresistance, etc.) arise due to quantum size effects. The authors have studied two special class of composite systems; ferrimagnetic/superconductor structures and cobalt nanocomposites dispersed in porous matrices. They used x- ray, electron microscopy and nuclear magnetic resonance (NMR) methods to get information about structural and magnetic properties. The ^{57}Co isotope in ferromagnetic Co particles (or films) is shown to be a very convenient probe to sense local magnetic field. They also show that the effective magnetic moment of fine Co particles is significantly enhanced due to the interactions between ferromagnetic (metallic Co) and antiferromagnetic (CoO) occurring at the surface of metalic cobalt particles.

Another magnetic/superconducting composite system is studied in Chap.13. The authors present the results of their experimental studies on the influence of nanocrystalline particles on properties of magnetic and superconducting systems. They mainly investigated the change of the critical temperature (T_c) of high-

T_c superconductors by the inclusion of ferrite particuls of nanometer size. The hysteresis parameters, losses and initial susceptibility parameters were measured. The sample preparation techniques are explained in detail. It is observed that the critical current density of the high- T_c superconductor (YBCO) is increased considerably by the inclusion of superparamagnetic pinning centers.

A magnetic system that has been rediscovered in the last decade is perovskite manganite, $Ln_{1-x}A_xMnO_3$ (Ln=La, rare earth; A= divalent ions). The system exhibits magnetoresistivity and has attracted much attention due to its possible application in magnetic recording technology. In thin films of this kind of materials the absolute value of the magnetoresistivity change far exceeds even GMR and is hence called collosal magnetoresistance (CMR); it is an intrinsic property of the materials. The connection between the conductivity and external magnetic field comes from the spin-dependent double exchange of electrons of Mn ions separated by O atoms through which exchange occurs. If the spins of the neighbouring Mn ions are parallel to each other, then the exchange probablity increases giving smaller resistivity. The system undergoes a ferromagnetic phase transition at a critical temperature which depends on both the composition and external field. Therefore the phase transition from magnetic to ferromagnetic configuration is accompanied by a simultaneous insulator-metal phase transition. Thus, the electrical conductivity drastically increases as the external field orients the neighbouring Mn electron spins parallel to each other. Very systematic and comprehensive studies in the bulk perovskite manganite, $Ln_{1-x}A_xMnO_3$, as well as in the layered manganite, $(La,Sr)_3Mn_2O_7$ are presented in Chaps. 14 and 15. The authors have tried to evaluate the phase stability of some of these layered and perovskite manganites through systematic selective substitutions. Their physical characteristics point to the existence of a fine interplay between spin, charge and lattice interactions. This in turn affects the doping state (electron or hole) and thus give a variety of physical aspects for the system.

Finally the current knowledge of magnetoresistivity of nanostructured materials and its applications is summarized in Chap. 16. The authors do not cover all possible applications of the magnetoresistivity. They concentrate instead on some of the materials and devices under active development at Non-Volatile Electronics (NVE) and give a brief discussion of newly developed magnetoresistive materials and their possible applications. One of these applications of nano-size structures is connected to magnetoresistive insulators with enhanced speed by nanocrystalline ferromagnetic layers, while other ones focus mainly on the detection and manipulation of nanomagnetic particles using magnetoresistive sensors and integrated straps. Although the discussion mainly concerns on GMR effects, this section can serve as a very good reference for researchers interested in magnetic field sensors.

References

1. L. M. Falicov, D.T. Pierce, S. D. Bader, R. Gronsky, K. B. Hathaway, H. J. Hopster, D. N. Lambeth, S. S. P. Parkin, G. Prinz, M. Salamon, I. K. Schuller, and R. H. Victora, J. Mater. Res. **5**, 1299 (1990)

2. S. D. Bader, Proc. Inst. elect. electron. Engrs. **78**, 909 (1990)
3. U. Gradmann, J. Magn. Magn. Mater. **100**, 481 (1991)
4. T. Shinjo, Surf. Sci. Rep. **12**, 49 (1991)
5. H. C. Siegmann, J. Phys. condens. Matter **4**, 8395 (1992)
6. B. Heinrich, and J. F. Cochran, Adv. Phys. **42**, 523 (1993)
7. R. Allenspach, J. Magn. magn. Mater. **129**, 160 (1994)
8. G. A. Prinz, and K. Hathaway, Phys. Today, April 24 (1995)
9. F. J. Himpsel, E. J. Ortega, G. J. Mankey, R. F. Willis R.F, Advances in Physics **47**, 551 (1998)
10. U. Gradmann, Landolt-Börnstein New Series, Group III, Vol. 19g, edited by H. P. J. Wijn (Berlin: Springer-Verlag) p. 1 (1988)
11. D.-Q. Li, M. Freitag, J. Pearson, Z. Q. Qiu, and S. D. Bader, Phys. Rev. Lett. **72**, 3112 (1994)
12. M. Wuttig, B. Feldmann, J. Thomassen, F. May, H. Zillgen, A. Brodde, H. Hannemann, and H. Neddermeyer, Surf. Sci. **291**, 14 (1993)
13. M. A. Mueller, K. Mukasa, K. Sueoka, H. Hasegawa, Y. Tazuke, and K. Hayakawa, Mater. Sci. Eng. **B31**, 69 (1995)
14. H. Landskron, G. Schmidt, K. Heinz, K. Muller, C. Stuhlmann, U. Beckers, M. Wuttig, and H. Ibach, Surf. Sci. **256**, 115 (1991)
15. P. Schmailz, K. Schmidt, P. Bayer, R. Doll, and K. Heinz, Surf. Sci. **312**, 73 (1994)
16. W. L. O' Brien, and B. P. Tonner, Phys. Rev. **B52**, 15332 (1995)
17. F. J. Himpsel, Phys. Rev. **B44**, 5966 (1991)
18. H. Glatzel, R. Schneider, T. Fauster, and V. Dose, Z. Phys. **B88**, 53 (1992)
19. W. L. O' Brien, B. P. Tonner, G. R. Harp, and S. S. P. Parkin, J. appl. Phys. **76**, 6462 (1994),
20. M. G. Samant, J. Stohr, S. S. P. Parkin, G. A. Held, B. D. Hermsmeier, and F. Herman, Phys. Rev. Lett. **72**, 1112 (1994)
21. M. A. Tomaz, D. C. Ingram, G. R. Harp, D. Lederman, E. Mayo, and W. L. O' Brien, Phys. Rev. **B56**, 5474 (1997)
22. E. E. Fullerton, D. Stoeffler, K. Ounadjela, B. Heinrich, Z. Celinski, and J. A. C. Bland, Phys. Rev. **B51**, 6364 (1995)
23. B. N. Engel, C. D. England, R. A. Van Leeuwen, M. H. Wiedmann, and C. M. Falco, Phys. Rev. Lett. **67**, 1910 (1991)
24. G. Güntherodt, B. Hillerbrands, P. Krams, J. V. Harzer, F. Lauks, R. L. Stamps, W. Weber, D. Hartmann, D. A. Wesner, A. Rampe, U. A. Effner, H. P. Oepen, D. Weller, R. F. C. Farrow, B. N. Engel, and C. M. Falco, Phil. Mag. **B70**, 767 (1994)
25. L. Nordstrom, and D. Singh, J. vac. Sci. Technol. **B14**, 3160 (1996)
26. J. E. Ortega, and F. J. Himpsel, Phys. Rev. **B47**, 16441 (1993)
27. S. S. P. Parkin, C. Chappert, and F. Herman, Europhys. Lett. **24**, 71 (1993)
28. P. Bruno, Phys. Rev. **B52**, 411 (1995)
29. M. S. Ferreira, J. D' Albuquerque e Castro, D. M. Edwards, and J. Mathon, J. Magn. magn. Mater. **154**, L1 (1996)
30. Gh. Würsch, Ch. Stamm, S. Egger, and D. Pescia, 1997
31. I. Dzialoshinski, J. Phys. Chem. Solids **4**, 241 (1985); T. Moria, Phys. Rev. **120**, 91 (1960)
32. A. E. Berkowitz, J. R. Mitchell, M. J. Carey, A. P. Young, S. Zhang, F. E. Spada, F. T. Parker, A. Hutten, and G. Thomas, Phys. Rev. Lett. **68**, 3745 (1992)
33. J. S. Moodera, and L. R. Kinder, J. appl. Phys. **79**, 4724 (1996)

34. W. J. Gallagher, S. S. P. Parkin, Y. Lu, X. P. Bian, A. Marley, K. P. Roche, R. P. Altman, S. A. Rishton, C. Jahnes, T. M. Shaw, and G. Xiao, J. appl. Phys. **81**, 3741 (1997)
35. M. N. Baibich, J. M. Broto, A. Fert, F. Nguyen-Van-Dan, F. Petroff, P. Etienne, G. Creuzet, A. Friedrich, and J. Chazelas, Phys. Rev. Lett. **61**, 2472 (1988)
36. S. S. Parkin, N. More, and K. P. Roche, Phys. Rev. Lett. **64**, 2304 (1990)
37. S. S. P. Parkin, R. Bhadra, and K. P. Roche, Phys. Rev. Lett. **66**, 2152 (1991)
38. R. Schad, C. D. Potter, P. Belien, G. Verbanck, V. V. Moshchalkov, and Y. Bruynseraede, Appl. Phys. Lett. **64**, 3500 (1994)
39. G. A. Prinz, 1991, J. Magn. magn. Mater. **100**, 469 (1991, Phys. Today, April, 58.
40. H. E. Camblong, S. Zhang, and P. M. Levy, Phys. Rev. **B47**, 4735 (1993)
41. A. Fert, T. Valet, and J. Barnas, J. appl. Phys. **75**, 6693 (1994)
42. W. P. Jr. Pratt, S.-F. Lee, Q. Yang, P. Holody, R. Loloee, P. A. Schroeder, and J. Bass, J. appl. Phys. **73**, 54326 (1993)
43. M. E. Fisher, Rev. mod. Phys. **46**, 597 (1974)
44. D. L. Mills, J. Magn. magn. mater. **100**, 515 (1991)
45. Y. Li, and K. Baberschke, Phys. Rev. Lett. **68**, 1208 (1992)
46. F. Huang, M. T. Kief, G. J. Mankey, and R. F. Willis, Phys. Rev. **B49**, 3962 (1994)
47. U. Gradmann, and M. Przybylski, Thin Film Growth Techniques for L ow- Dimensional Structures, edited by R. F. C. Farrow, S. S. P. Parkin, P. J. Dobson, J. H. Neave and A. S. Arrott, NATO Advanced Study Institute Series, Series B: Physics **163**, 261 (1987)
48. B. N. Engel, M. H. Wiedmann, and C. M. Falco, J. appl. Phys. **75**, 6401 (1994)
49. R. Wu, and A. J. Freeman, J. appl. Phys. **79**, 6500 (1966)
50. H. J. Elmers, J. Hauschild, H. Fritzsche, G. Liu, U. Gradmann, and U. Kohler, Phys. Rev. Lett. **75**, 2031 (1995)
51. A. J. Freeman, and R.-Q. Wu, J. Magn. magn. Mater. **100**, 497 (1991)
52. A. Abragam and B. Bleaney, Electron Paramagnetic Resonanse of Transition Ions, Clarendon Press, Oxford, U.K., 1970
53. W. Weber, D. A. Wesner, D. Hartmann, and G. Güntherodt, Phys. Rev. **B46**, 6199 (1992)
54. O. Rader, E. Vescovo, J. Redinger, S. Blugel, C. Carbone, W. Eberhardt, and W. Gudat, Phys. Rev. Lett. **72**, 2247 (1994)
55. S. Blugel, B. Drittler, R. Zeller, and P. H. Dederichs, Appl. Phys. **A49**, 547 (1989)
56. C. L. Fu, A. J. Freeman, and T. Oguchi, Phys. Rev. Lett. **54**, 2700 (1985)
57. M. J. Zhu, D. M. Bylander, and L. Kleinman, Phys. Rev. **B43**, 4007 (1991)
58. R. Pfandzelter, G. Steierl, and C. Rau, Phys. Rev. Lett. **74**, 3467 (1995)
59. H. Hopster, Surf. Rev. Lett. **1**, 89 (1994)
60. L. Pizzagalli, M. Freyss, G. Moraitis, D. Stoeffler, C. Demangeat, H. Dreysse, A. Vega, S. Miethauer, and G. Bayreuther, J. Appl. Phys. **81**, 4347 (1997)
61. J. E. Ortega, and F. J. Himpsel, Phys. Rev. Lett. **69**, 844 (1992)
62. J. E. Ortega, F. J. Himpsel, G. J. Mankey, and R. F. Willis, Phys. Rev. **B47**, 1540 (1993)
63. W. Weber, A. Bischof, R. Allenspach, Ch. Wursch, C. H. Back, and D. Pescia, Phys. Rev. Lett. **76**, 3424 (1996)
64. M. Hayashi, T. Katayama, Y. Suzuki, M. Taninaka, A. Thiaville, and W. Geerts, J. Magn. magn. Mater. **126**, 54 (1993)

65. A. Carl, and D. Weller, Phys. Rev. Lett. **74**, 190 (1995)
66. A. Kirilyuk, Th. Rasing, R. Megy, and P. Beauvillain, J. Appl. Phys. **79**, 5632 (1996)
67. M. Jalochowski, E. Bauer, H. Knoppe, and G. Lilienkamp, Phys. Rev. **B45**, 13607 (1992)
68. M. Jalochowski, M. Hoffman, and E. Bauer, Phys. Rev. Lett. **76**, 4227 (1996)
69. B. G. Orr, H. H. Jaeger, and A. M. Goldman, Phys. Rev. Lett. **53**, 2046 (1984)
70. A. R. Smith, K.-J. Chao, Q. Niu, and C. K. Shih, Science **273**, 226 (1996)
71. J. Braun, and J. P. Toennies, Surf. Sci. Lett. (1997, to be published).
72. W. Hansen, J. P. Kotthaus, and U. Merkt, Semicond. Semimetals **35**, 279 (1992)
73. W. Oepts, M. A. M. Gijs, A. Reinders, R. M. Jungblut, R. M. J. van Gansewinkel, and W. J. M. de Jonge, Phys. Rev. **B53**, 14024 (1996)
74. J. Li, W.-D. Schneider, and R. Berndt, Phys. Rev. **B56**, 7656 (1997)
75. Ph. Avouris, L.-W. Lyo, R. E. Walkup, and Y. Hasegawa, J. Vac. Sci. Technol. **B12**, 1447 (1994)
76. B. Hillebrands, P. Baumgart, and G. Guntherodt, Phys. Rev. **B36**, 2450 (1987)
77. P. Krams, B. Hillebrands, G. Guntherodt, and H. P. Oepen, Phys. Rev. **B49**, 3633 (1992)
78. R. H. Kodama, A. E. Berkowitz, E. J. McNiff, and S. Foner, Phys. Rev. Lett. **77**, 394 (1996),
79. D. A. Van Leeuwen, J. M. van Ruitenbeek, L. J. de Jongh, A. Ceriotti, G. Pcchioni, O. D. Häberlen, and N. Rösch, Phys. Rev. Lett. **73**, 1432 (1994)
80. Z. X. Tang, C. M. Sorensen, K. J. Klabunde, and Hadjipanayis, Phys. Rev. Lett. **67**, 3602 (1991)
81. D. Gatteschi, A. Canesci, L. Pardi, and R. Sessoli, Science **265**, 1054 (1994)
82. D. D. Awschalom, and D. P. DiVincenzo, Phys. Today, April, 43 (1995)
83. M. R. Scheinfein, K. E. Schmidt, K. R. Heim, and G. G. Hembree, Phys. Rev. Lett. **76**, 1541 (1996)
84. R. M. H. New, R. F. W. Pease, and R. L. White, J. vac. Sci. Technol. **B12**, 3916 (1995); **B13**, 1089 (1995)
85. R. O' Barr, S. Y. Yamamoto, S. Schultz, W. Xu, and Scherer, J. Appl. Phys. **81**, 4730 (1997)
86. G. A. Prinz, J. Magn. magn. Mater. **100**, 469 (1991)
87. E. Bauer, Z. Kristallogr. **110**, 372 (1958)
88. J. H. Weaver, Phys. Today, January, 24 (1986)
89. H. A. van der Vegt, H. M. van Pinxteren, M. Lohmeier, and E. Vlieg, Phys. Rev. Lett. **68**, 3335 (1992)
90. S. D. Bader, and E. R. Moog, J. Appl. Phys. **61**, 3729 (1987)
91. D. A. Steigerwald, I. Jacob, and W. F. Jr. Egelhoff, Surf. Sci. **202**, 472 (1988)
92. G. W. Anderson, M. C. Hanf, and P. R. Norton, Phys. Rev. Lett. **74**, 2764 (1995)
93. R. F. C. Farrow, S. S. P. Parkin, and V. S. Speriosu, J. appl. Phys. **64**, 5315 (1988)
94. K. I. Sano, and T. Miyagawa, Jpn. J. appl. Phys, **30**, 1434 (1991)
95. F. Nguyen-Van-Dan, M. Sussian, A. Schuhl, and P. Galtier, J. Appl. Phys. **81**, 4482 (1997)
96. G. A. Prinz, J. Magn. magn. Mater. **100**, 469 (1991); Phys. Today, April, 58. (1995)
97. W. K. Burton, N. Cabrera, and F. C. Frank, Phil. Trans. E. Soc. A. **243**, 299 (1951)
98. E. Bauer, and J. H. van der Merwe, Phys. Rev. **33**, 3657 (1986)

99. E. Bauer, and J. H. van der Merwe, Phys. Rev. **33**, 3657 (1986)
100. D. Welipitiya, Y. L. He, J. Zhang, P. I. Oden, T. Thundat, R. J. Warmack, I. Gobulukoglu, S. Z. Shan, D. J. Sellmyer, and P. A. Dowben, J. appl. Phys. **80**, 1867 (1996)
101. W. Schindler, O. Schneider, and J. Kirschner, J. appl. Phys. **81**, 3915 (1997)
102. K. Liu, S. M. Baker, M. Tuominen, T. P. Russell, and I. K. Schuller, Phys. Rev. **B63**, 60403 (2001)
103. R. J. Celotta, R. Gupta, R. E. Scholten, and J. J. McClelland, J. Appl.Phys. **79**, 6079 (1996)
104. F. J. Himpsel, and J. E. Ortega, Phys. Rev. **B50**, 4992 (1994)
105. H. Roder, E. Hahn, H. Brune, J.-P. Bucher, and K. Kern, Nature **366**, 141 (1993)
106. D. D. Awschalom, and D. P. DiVincenzo, Phys. Today, April, 43 (1995)
107. J. C. Mallinson, The Foundations of Magnetic Recording (California: San Diego, 1987)
108. J. Tersoff, C. Teichert, and M. G. Lagally, Phys. Rev. Lett. **76**, 1675 (1996)
109. L.-S. Wang, H. Wu, and S. R. Desai, Phys. Rev. Lett. **76**, 4853 (1996)
110. M. Lederman, D. R. Fredkin, R. O' Barr, S. Schultz, and M. Ozaki, J. appl. Phys. **75**, 6217 (1994)
111. K. Runge, Y. Nozaki, Y. Otani, H. Miyajima, B. Pannetier, T. Matsuda, and A. Tonomura, J. appl. Phys. **79**, 5075 (1996)
112. T. A. Jung, R. R. Schlitttler, J. K. Gimzewski, H. Tang, and C. Joachim, Science **271**, 181 (1996)
113. D. B. McWhan, J. Synchrotron Radiat. **1**, 83 (1994)
114. J. A. Borchers, G. Nieuwenhuys, M. B. Salamon, C. P. Flynn, R. Du, R. W. Erwin, and J. J. Rhyne, J. Phys., Paris **49**, 1685 (1988)
115. J. F. MacKay, C. Teichert, D. E. Savage, and M. G. Lagally, Phys. Rev. Lett. **77**, 3925 (1996)
116. P. Grünberg, Top. appl. Phys. **66**, 303 (1989)
117. P. Krams, F. Lauks, R. L. Stamps, B. Hillebrands, and G. Güntherodt, Phys. Rev. Lett. **69**, 3674 (1992)
118. B. Aktaş, and M. Özdemir, Physica B **193**, 125 (1993)
119. C. Turtur, and G. Bayreuther, Phys. Rev. Lett. **72**, 1557 (1994)
120. M. Straub, R. Vollmer, and J. Kirschner, Phys. Rev. Lett. **77**, 743 (1996)
121. P. Rudolf, F. Sette, L. H. Tjeng, G. Meigs, and C. T. Chen, J. Magn. magn. Mater. **109**, 109 (1992)
122. G. Schutz, P. Fischer, E. Goering, K. Attenkofer, D. Alileos, and W. Roßl, Synch. Rad. News **10**, 13 (1997)
123. P. L. Trouilloud, B. Petek, and B. E. Argyle, IEEE Trans. Magn. **30**, 4494 (1994)
124. J. Levy, V. Nikitin, A. Cohen, N. Samarth, R. Garcia, and D. Awschalom, 1996,
125. E. Betzig, J. K. Trautman, R. Wolfe, E. M. Gyorgy, P. L. Finn, M. H. Kryder, and C.-H. Chang, Appl. Phys. Lett. **61**, 142 (1992)
126. S. Schultz, and T. J. Silva, SPIE Proceedings **1639**, 31 (1992)
127. M. Hehn, K. Ounadjela, J.-P. Bucher, F. Rosseaux, D. Decanini, B. Bartenlian, and C. Chappert, Science **272**, 1782 (1966)
128. Z. Zhang, P. C. Hammel, and P. E. Wigen, Appl. Phys. Lett. **68**, 2005 (1996)
129. M. B. Ketchen, and J. R. Kirtley, IEEE Trans. appl. Supercond. **5**, 2133 (1995)
130. D. T. Pierce, M. R. Scheinfein, J. Unguris, J. J. McClelland, and R. J. Celotta, Mater. Res. Soc. Symp. Proc. **151**, 49 (1989)
131. T. Duden, and E. Bauer, Phys. Rev. Lett. **77**, 2308 (1996)
132. J. Stohr, J. Electron Spectrosc. **75**, 253 (1995)

133. C. Beeli, B. Doudin, and P. Stadelmann, Phys. Rev. Lett. **75**, 4630 (1995)
134. S. F. Alvarado, Phys. Rev. Lett. **75**, 513 (1995)
135. H. N. Bertram, and J.-G. Zhu, Solid St. Phys. **46**, 271 (1992)
136. K. Aric, B. K. Menon, B.K. Gupta, Nanotechnology: A data storage perspective. Nanostructured Materials **11**, (1999)
137. D. N. Lambeth, E. M. T. Velu, G. H. Bellesis, L. L. Lee, and D. E. Laughlin, J. appl. Phys. **79**, 4496 (1996)
138. S. Y. Chou, P. R. Krauss, and L. Kong, J. appl. Phys. **79**, 6101 (1996)
139. M. H. Kryder, Mater. Res. Soc. Bull. **17**, 60 (1996)
140. S. S. P. Parkin, and T. Rabedau, Appl. Phys. Lett. **68**, 1162 (1996)
141. J. Z. Sun, L. Krusin-Elbaum, A. Gupta, G. Xiao, P. R. Duncombe, W. J. Gallagher, and S. S. P. Parkin, J. appl. Phys. **81**, 5343 (1997)
142. W. J. Gallagher, S. S. P. Parkin, Y. Lu, X. P. Bian, A. Marley, K. P. Roche, R. P. Altman, S. A. Rishton, C. Jahnes, T. M. Shaw, and G. Xiao, J. appl. Phys. **81**, 3741 (1997)
143. M. Johnson, J. appl. Phys. **75**, 6714 (1994)
144. A. Fert, and S.-F. Lee, Phys. Rev. **B53**, 6554 (1996)

2 AMR and GMR Layers and Multilayers for Magnetic Field Sensors

Anatoliy G. Ravlik

National Technical University – Kharkiv Polytechnical Institute,
21 Frunze st., Kharkiv, 61002, Ukraine

Abstract. The results of research on nanostructured film materials, both mono- and multilayerd showing anistropic or "giant" magnetoresistance (AMR and GMR respectively) are reviewed. The investigations carried out resulted in the development of some remarkable AMR elements such as "longitudinal" ones with increased sensitivity, multilayered film elements with crossed easy axes capable to detect magnetic field direction, etc. Some developed AMR sensor mockups are briefly described. In particular, it is reported about low magnetic field sensor with lowest level of about 0.002 A/cm. The dependences of "giant" magnetoresistive ratio and related characteristics on nonmagnetic interlayer thickness and bilayer number in multilayered nanostructures $(\mathrm{Co/Cu})_n$ are discussed. Some fields of GMR element applications are outlined.

2.1 Introduction

The detection and measurement of magnetic fields is problem in engineering and science research. The principles of operation of magnetosensitive elements (MSEs) are very diverse. For the last decades new magnetic sensors based on magnetoresistive effects that are observed in metallic ferromagnetic films and multilayers of submicrometer thickness have been intensively investigated and developed. These materials were designated as nanostructured films. Their electrical resistance varies depending on external magnetic field that affects distribution of magnetic moments in the MSE. Two types of magnetoresistive effects are mainly observed in the materials. These are "anisotropic magnetoresistance" (AMR) [1,2], which was described by W.Thomson in 1857 [2], and "giant magnetoresistance" (GMR) which has recently been discovered [3]. AMR depends on relative orientation of magnetization vector $\boldsymbol{I}_S$ and current density vector $\boldsymbol{j}$. As a rule the electrical resistance increases when $\boldsymbol{I}_S$ and $\boldsymbol{j}$ are parallel (or antiparallel) and decreases if $\boldsymbol{I}_S \perp \boldsymbol{j}$. The reverse magnetoresistive effect is only observed in some ferromagnets. The AMR in polycrystalline or amorphous magnets is evaluated by the magnetoresistance anisotropy coefficient $(\Delta\rho)_m$. The maximal resistivity variation is found under a change from the longitudinal to the transverse saturating magnetization relative to $\boldsymbol{j}$. The normalized $(\Delta\rho/\rho)_m$ values are equal to 1–6 % at room temperature for 3d-metals and alloys [1,2],[4]–[6]. The microscopic theory of the AMR effect has not been developed in detail [6]. The AMR effect is supposed to be caused by spin-dependent scattering of conducting electrons in the volume of ferromagnets.

The GMR effect is observed in multilayer systems "ferromagnetic film – nonmagnetic metal interlayer". The GMR is not anisotropic in the sense that it is independent of mutual $\boldsymbol{I}_S$ and $\boldsymbol{j}$ orientation in plane of a specimen. The GMR values reach some tens of percent mainly in monocrystalline multilayered systems. As a rule, these values are lower in polycrystalline systems. The GMR is considered to be caused by spin-dependent scattering of the electrons at interfaces between ferromagnetic films and nonmagnetic interlayers, moreover the value of the effect depends on $\boldsymbol{I}_S$ orientation in neighboring ferromagnetic layers, namely this value is the highest if $\boldsymbol{I}_S$ vectors are antiparallel. Thus, GMR is mainly a surface effect in comparison to AMR, the latter being a volume one.

The MSEs of the early 70-ies based on the AMR effect have been used for detection of motion of the bubble domains in CMD-devices. A little later they were used in magnetic recording technique for reading-out information recorded on tapes, disks etc. [1,4]; in automatics and magnetometry [4,5],[7]–[9], in operative memory devices based on bistable ferromagnetic films [10]. Some information is available on applying the AMR sensors for registration of stray fields caused by defects being present in materials and that should be revealed by nondestructive control.

The AMR elements have many essential advantages in comparison to MSEs of other types. They have high sensitivity, yielding only to the fluxgate the threshold sensitivity of which is approximately of 10^{-5} A/cm. The output signals of AMR element do not dependent on the measuring field frequency in comparison to the those of the induction MSEs. The active zone of AMR element is very small (order of 0.1–1 mm^2) and may be decreased in some cases (e.g. in CMD devices, reading-out magnetic heads) down to 10–100 μm^2. The advantage of miniature dimensions of AMR elements is combined with low energy consumption (1–50 mWt). The AMR elements function in a wide temperature range of (90–450 K). The high technological efficiency of production of AMR sensors which is being ensured by the thin film and semiconductor microelectronic technology is of doubtless advantage. In addition, materials of moderate purity (not higher than 99.95%) are suitable for fabrication of AMR elements. This is a great advantage of AMR elements in comparison to the semiconductor Hall and magnetoresistive MSEs. Finally, the ferromagnetic MSEs are radiation-resistent. The indicated advantages of AMR elements allowed for the wide application of AMR sensors especially in the field of automatics and magnetometry.

The GMR elements also have the advantages indicated above. In addition, their magnetoresistance values are higher than those of AMR elements. The GMR elements can be magnetized being subjected to only a very low magnetic flux. Their thermal stability may be a little bit higher than that of AMR elements. However, the preparation procedure of GMR elements is relatively complicated and at the moment it is not brought to the level the AMR elements possess. Particularly, the scatter of characteristics of GMR elements is considerable during the process of their manufacturing. Therefore, the GMR sensors are not as widely used as AMR ones. In spite of the fact that the AMR elements have found wide industrial application and GMR elements have some technological drawbacks the research of both types of elements still remains to be an actual problem and has both technological and physical aspects, which are closely interconnected. The selection of the alloys and the combinations of the film materials used for nanostructured systems, the optimization of various techniques used for the preparation of films and multilayer system and other similar purposes are referred to the first aspect. The second aspect is related to the problems that occupy the intermediate position between basic and applied tasks. Particularly, the various dimensional effects that develop with variation in thickness of ferromagnetic layers and/or nonmagnetic interlayers and in width of an element stripe and other parameters of nanostructured systems, attract attention of the researchers.

The results of the research of the considered problems are summarized in these notes. These investigations have mainly been carried out by the Department for Metal and Semiconductor Physics of NTU "KhPI" over the last years.

2.2 Experimental Procedure

The magnetosensitive materials used for the AMR elements were manufactured of polycrystalline films of 50–200 nm thick deposited onto the glass ceramic or oxidized silicon substrates using various thin film techniques, namely vacuum evaporation (when a charge is heated by high frequency inductor or electron beam), the triode ion-plasma sputtering and magnetron sputtering. The Ni-Fe alloys of various composition were used to ensure the films with acceptable magnetoresistance level – $(\Delta\rho/\rho)_m$ up to 3.5 % and low coercive force. The Ni-Fe-Co alloys were used in some modeling experiments in order to obtain films with increased values of the magnetic anisotropy field. Let's note that we also prepared Ni-Co films with $(\Delta\rho/\rho)_m$ up to 5.5–6 % but they had high coercivity. The Ta, Mo and SiO films were used as nonmagnetic interlayers in the layered systems intended for preparation of AMR elements. The MSEs were shaped using photolithography. Some specimens and the MSEs model were shaped using masks.

The magnetosensitive materials showing GMR effect were prepared employing a combination of two techniques: the triode ion-plasma sputtering and magnetron sputtering to obtain ferromagnetic layers and nonmagnetic interlayers, respectively. The films were deposited in argon atmosphere at a pressure of 0.3 to 0.4 Pa. Prior to filling with argon the minimum pressure in working chamber was of 1 to 2×10^{-4} Pa. The films were deposited on uncooled glass substrates. The multilayer films (Ni-Fe)/Cu, (Ni-Fe-Co)/Cu and (Co/Cu) were tested, but only the last one had suitable magnetoresistive properties. Therefore, our attention was concentrated on Co/Cu system. The magnetic and magnetoresistive characteristics were measured on samples that were produced by deposition through masks and had the form of stripes with dimensions of $(0.6\text{–}1)\times20\,\mathrm{mm}^2$.

The magnetic hysteresis loops of the materials used for the AMR and GMR films were measured by highly sensitive vibrating sample magnetometer and their magnetoresistive properties were tested on standard measuring device. More detailed information on experimental procedure will only be given in appropriate sections.

2.3 Sophisticated Elements Based on Anisotropic Magnetoresistance

It was noted above that AMR effect observed in soft magnetic films has been studied perfectly well and MSEs based on this effect are widely used in industry. Nevertheless, a number of problems remained unsolved at the time of the commencement of the work being reviewed in these notes. For example, dimensional effects of domain structure were investigated insufficiently. In addition, the traditional MSEs had some essential drawbacks. Therefore we made the attempts to eliminate them. The results obtained in this field are described in the following subsections.

2.3.1 Elements of Increased Sensitivity ("Longitudinal" Ones)

The principles of operation of the simplest AMR elements applied in routine practice are illustrated in Fig. 2.1a. As a rule, magnetically uniaxial film is used in case when easily magnetized axis, i.e. the easy axis ($\boldsymbol{EA}$), is parallel to the long side of a stripe $\boldsymbol{l}$. The sense current flowing in the same direction with density $\boldsymbol{j}$ passes through the film. The external field $\boldsymbol{H}_0$ to be measured is applied perpendicular to $\boldsymbol{l}$. Let such an element be referred to as "transversal" one. If the film is assumed to be single-domain then the magnetization vector $\boldsymbol{I}_S$ deviates from its initial position ($\boldsymbol{I}_S \perp \boldsymbol{EA}$) at some angle θ as the $\boldsymbol{H}_0$ is applied. As a result the voltage drop on the element becomes such as in [1,4,7]

$$\Delta U = j(\Delta\rho)_m l H_0^2/(H_K + H_D + H_B), \tag{2.1}$$

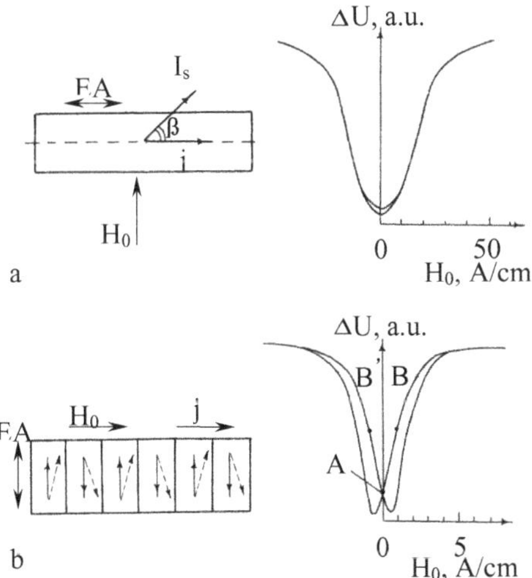

Fig. 2.1. Diagrams and the dependencies of the output signals ΔU (in arbitrary units) on the measured external field H_0 for (a) "transversal" and (b) "longitudinal" magnetoresistive elements. j-current density; EA – the easy axis

where l is the stripe length; $(\Delta\rho)_m$ – is the maximum change in electric resistivity measured when the stripe is magnetized in succession up to the saturation in direction parallel and perpendicular to $\boldsymbol{j}$, the anisotropy field is $H_K = 2K/I_S$; K is the anisotropy constant; the demagnetizing field is $H_D = NI_S$; N is the demagnetization factor; $N \approx h/b$, h and b being the film thickness and the stripe width, accordingly; H_B is the bias field strength. This approach is used in some cases to suppress hysteresis of the ΔU dependent on H_0. The bias field is usually $H_B > H_K$.

The sensitivity of the element is defined as $S_U = \Delta U/H_0$. In the linear section of the dependence Eq.(2.1) it can be estimated using the expression [7,8]:

$$S_U = j(\Delta\rho)_m l H_0/(H_K + H_D + H_B), \tag{2.2}$$

As it follows from Eq.(2.1) and Eq.(2.2) ΔU and S are in direct proportion to the stripe length l. Therefore, the elements used in the field of automatics and magnetometry are shaped as meanders. These are connected into 2- or 4-section differential circuit in order to compensate constant voltage component that is independent of the field being measured but is sensitive to temperature.

It is clear from (1) and (2) that the "transversal" element under consideration has a low sensitivity. This is due to the necessity to overcome a considerable demagnetizing field H_D when the stripe is magnetized along hard magnetization axis ("hard" one). It is obvious that the applied H_B also enhances "rigidity" of an element.

We have proposed [11] a "longitudinal" element where in contrast to the "transversal" one the $\boldsymbol{EA}$ is parallel to the short side of a stripe while the external field H_0 is oriented along its long side. The element has a domain structure. The "open" structure with parallel-sided domains is the most preferred one (Fig. 2.1a, 2.2a). This configuration is similar to one proposed by Kittel [12] for cross-section of the film with perpendicular anisotropy. In the shown $\boldsymbol{I}_S$ distribution the magnetic state of such an element will be altered by non-uniform rotation of the magnetization in the domains.

In order to change purposefully the domain structure and thus magnetoresistive characteristics we have studied the magnetization distribution for the film stripes of thickness $h = 50$–$200\,\text{nm}$ and width within $b = 5$–$200\,\mu\text{m}$. The anisotropy field was set choosing the appropriate alloy composition and the film preparation conditions. The H_K values being varied within 2 to 65 A/cm range. The domains were defined using the magnetic colloid technique. The following dimensional effects have been revealed in the domain structure when varying b.

In films with $H_K \approx 20\,\text{A/cm}$ it is just the "closed" rhombic-type domain structure which is the equilibrium one at $b = 5$ to $20\,\mu\text{m}$ (Fig. 2.2a–c) when the width exceeds the upper value the rhombic structure is transformed into a "closed" one that is similar to well known domain configuration proposed by Landau and Lifshits (Fig. 2.2d). The further increasing of b value is accompanied by a complication of the $\boldsymbol{I}_S$ distribution and the appearance of a configuration with incomplete magnetic flux closing (Fig. 2.2c). Such structure becomes more stable as HK increases. In smaller anisotropy films ($H_K \approx 2.5$ to $4\,\text{A/cm}$)Landau-Lifshits structure appears to be greater for b than $100\,\mu\text{m}$. Moreover at $b < 20\,\mu\text{m}$

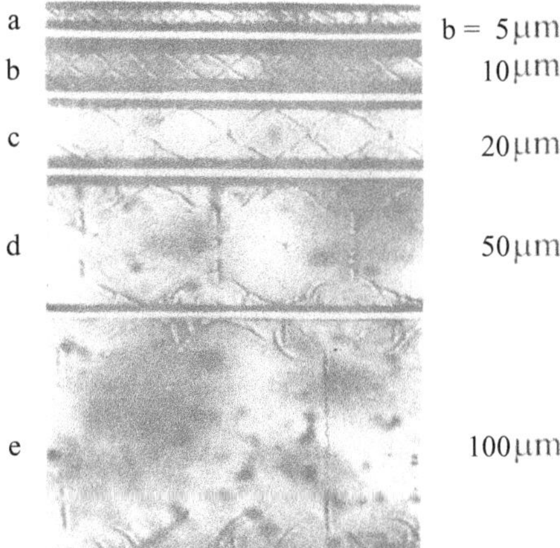

Fig. 2.2. Domain structure of the film stripes with thickness of 100 nm and different width b, μm: a–5, b–10, c–20, d–50, e–100. (as found in [13] and by V.N.Samophalov)

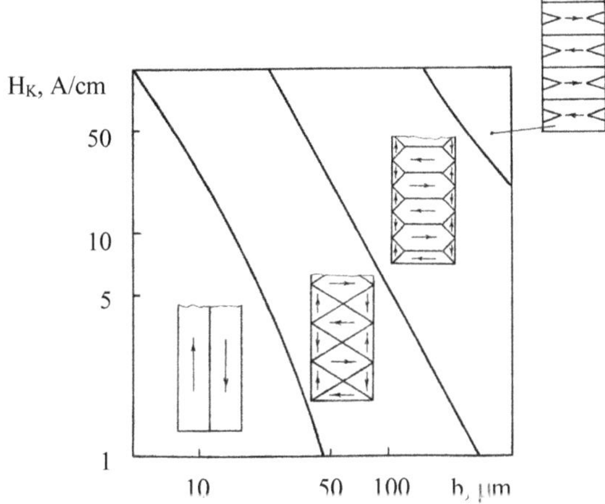

Fig. 2.3. Phase diagram showing existence of domain structures in the film stripes with different width b and anisotropy field H_k

the rhombic-type structure becomes unstable in such films and it is substituted by configuration with magnetization parallel to a long side of the stripe, moreover the structure tends to be a single-domain one if b decreases. It is known [1,4,7,8] that output signal of AMR elements depends on $\sin^2\theta$ (or $\cos^2\theta$) therefore such $\boldsymbol{I}_S$ distribution is unsuitable for the "longitudinal" elements.

The domain structures considered above are schematically shown in coordinates $H_K - b$ of the phase diagram (Fig. 2.3). This diagram is schematic. The position of the lines is approximate and can be varied with change in the film thickness. Nevertheless, it gives correct and clear notion of the domain structure for the film stripes. The "open" Kittel structure (Fig. 2.1a, 2.2a) is not shown in the diagram. It is quite possible that such configuration may be stable for high H_K or low magnetization value of the film. It is also not clear whether transition from rhombic-type structure to Landau-Lifshits one and after that to configuration with incomplete flux closing is abrupt or gradual. Additional experiments are necessary in order to clear up this problem.

The studied dimensional effects of the domain structure are observed due to the variations in the magnetic energy components, such as the anisotropy energy, domain wall and magnetostatic (demagnetizing field) energy depending on b. Those effects are two dimension analogs of the variations in the domain structure peculiar for the films of different thickness h with $\boldsymbol{EA}$ perpendicular to the film plane. These variations have been calculated by Kittel [12]. The effects studied by us are more diverse since they are affected in addition to h value by b parameter. Note, that some preliminary study results concerning those effects have been described earlier [13].

The experimental $\Delta U(H_0)$ dependencies for "longitudinal" elements confirm their considerably higher sensitivity as compared to "transversal" ones (cf. Figs. 2.1a, 2.1b). For MSEs with optimized parameters (h, b, H_K, I_S) the distinction is as high as two decimal orders. It should be noted that the dependence is of a rather linear character within a wide range (sections AB, AB′). This dependence differs obviously from the parabolic one obtained using the uniform rotation model (see Eq.(2.1)). Perhaps this linearization is due to the volume change of domains caused by the wall displacement in closing domains of the stripes having the Landau-Lifshits domain structure. A further study is necessary to elucidate this problem.

2.3.2 Elements with Decreased Hysteresis

As it is seen from Fig. 2.1b the "longitudinal" elements exhibit substantial hysteresis of AMR characteristics. It is caused by magnetic hysteresis under alternating magnetization reversal. This disadvantage can be eliminated by using the elements based on multilayer films consisting of ferromagnetic layers separated by nonmagnetic interlayers (dielectric or metal). The coercive force H_C of such films is considerably less than that of the monolayered ones [14,15]. This effect has been caused by the decrease in the domain wall energy of a multilayer system. The decrease of H_C should be correlated with the AMR hysteresis. In fact the magnetoresistance hysteresis is as low as 0.05 A/cm in the films containing ferromagnetic layers of 100–150 nm thick and nonmagntic interlayers of 50 nm thick. At the interlayer thickness of about 20 nm the single-domain state is realized in each layer due to the increased contribution of the magnetostatic interaction energy. In this case magnetization is oriented along $\boldsymbol{EA}$ and the $\boldsymbol{I}_S$ vectors are antiparallel in adjacent layers. Such multilayer films are very suitable for manufacturing of various AMR elements and sensors based on them.

2.3.3 Elements Determining the Magnetic Field Direction

It is known that AMR effect is an even one and cannot be used for immediate detection of the field sign, i.e. its direction. The known solutions of this problem are associated with considerable MSEs complications and measurement procedures. There are, for example, elements where $\boldsymbol{EA}$ is oriented at angle of 45° to the longitudinal axis of a stripe and H_0 is directed perpendicular to the $\boldsymbol{EA}$ [1,7,8]. In this case the angle between $\boldsymbol{I}_S$ and $\boldsymbol{j}$ changes as H_0 is reoriented in the opposite direction and the ΔU values are different. This is also possible for the elements where a highly conductive metal layer (most often Au) is deposited on ferromagnetic film and then using the photolithography technique a system of narrow stripes (so-called "barber-pole") is formed oblique to the magnetic stripe axis [1,5,8]. In the element the average current density is deflected from $\boldsymbol{I}_S$ to the long axis of a ferromagnetic stripe. These elements though being suitable for the field sign detection exhibit in addition to the above mentioned drawback a low sensitivity since they belong to "transversal" MSEs. It is worth to note that the longitudinal element can also be used to determine the field sign parallel to H_0. A higher sensitivity is attained in this case since H_0, H_B and H_K are commensurable. However, it should be kept in mind that in any case the emergence of the bias field means a complication of AMR elements since either special current stripes or miniature permanent magnet are to be used [1,4,5]. A solution that would provide the field sign detection taking into consideration design and technology should be as simple as possible is highly desirable. Such solution has been found by us [17] using the multilayer films with crossed $\boldsymbol{EA}$s (Fig. 2.4). The stripe-shaped elements are formed of these films using photolithography. The stripe longitudinal axis directions are selected so that they form with $\boldsymbol{EA}_1$ and $\boldsymbol{EA}_2$ some angles $\leq 90^o$. As seen from Fig. 2.1 the field to be measured is applied along the stripe long side thus the element is also a "longitudinal" one.

The domain structure and magnetoresistivity of the stripe elements based on the films with crossed $\boldsymbol{EA}$s have been studied in [17,18]. The single-domain state has been found to be realized in these elements within wide field range from the saturation field $+H_S$ (or $-H_S$) to the negative (resp. positive) threshold field $-H_t$ (or $+H_t$). The single-domain state stability is due to the partial closing of the magnetic flux on the short element side. In the absence of the field H_0 to be measured the magnetization vector $\boldsymbol{I}_{S1}$ and $\boldsymbol{I}_{S2}$ are almost parallel to

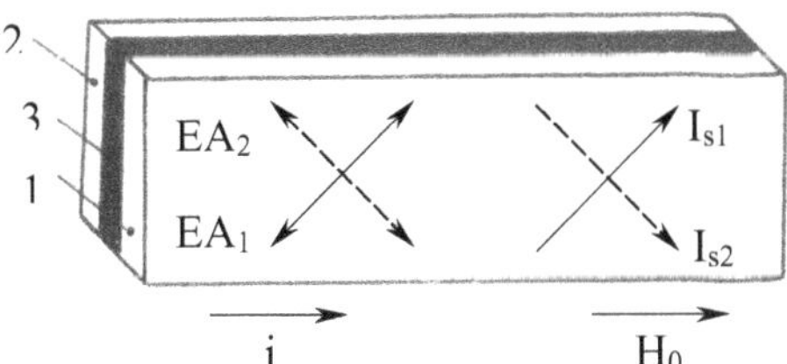

Fig. 2.4. Magnetoresistive element with crossed easy magnetization axes EA_1 and EA_2. I_{S1} and I_{S2} – magnetization vectors in the 1st and 2nd layer,respectively

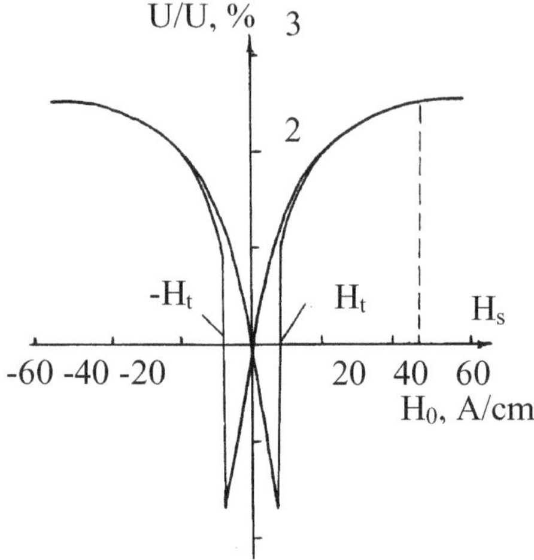

Fig. 2.5. Dependence of the output signal ΔU on the external field H_0 for the element with crossed easy magnetization axes

$\boldsymbol{EA}_1$ and $\boldsymbol{EA}_2$, respectively. As the H_0 increases from 0 to $+H_S$ (from left to right in Fig. 2.4) the angle between $\boldsymbol{I}_{S1}$ and $\boldsymbol{I}_{S2}$ vectors diminishes. As a result the output signal increases (Fig. 2.5). As the field drops from $+H_S$ to 0 the $\boldsymbol{I}_S$ vectors in each layer deviate towards $\boldsymbol{EA}$s. This is accompanied by reversible variation of the magnetoresistivity. The hysteresis does not exceed 1% of the signal level in weakly anisotropic films and 5% in strongly anisotropic ones. As the H_0 field direction is changed to the opposite one magnetoresistivity change remains reversible within the range of 0 to $-H_t$.

Thus, the reversible rotation of $\boldsymbol{I}_{S1}$ and $\boldsymbol{I}_{S2}$ takes place within the wide field range from the positive (or negative) saturation field to the negative (resp. positive) threshold field H_t. The output signal sign is positive or negative depending on the H_0 one. Just the ΔU sign is sensitive to the H_0 field direction that is principal advantage of the element with crossed magnetization axes proposed by us. Therefore, the use of that element allows us not only to measure the field strength but also to determine its sign without applying bias fields or resorting to any additional design and technological complications.

However, a sharp irreversible magnetoresistivity change occurs as the threshold value of the field is attained. For example, the $\Delta\rho/\rho$ increases from $\mp(0.9\text{–}1)\%$ to $\pm(0.8\text{–}0.9)\%$ in strongly anisotropic films within the field strength range of $\Delta H_0 \leq 0.01$ A/cm near H_t. This effect is caused by the single-domain structure failure. The study of that structure [18] has shown that the domains of reverse magnetization grow in a jump-like manner starting from the stripe ends when the field is close to H_t.

2.3.4 Elements with Increased Upper Level of Measuring Fields

It is seen from Eq.(2.1) when H_0 becomes close to $H_K + H_D + H_B$ the output signal ceases to change. For "longitudinal" element the H_K value determines the upper limit of measuring field. All described above elements are based on the films having planar anisotropy with H_K values from 2 to 65 A/cm. In order to measure stronger fields the "transverse" element with high bias fields can be, of course, used but this is hardly reasonable since it will mainly result in design complication and also in lowering of sensitivity.

To elevate the upper field measurability level we have proposed to make use of "overcritical" films [19,20]. These are specific in that a perpendicular anisotropy with $\boldsymbol{EA}$ in normal to the film plane is present therein. As a result in accordance with Kittel calculations [12] the $\boldsymbol{I}_S$ goes out of the film plane at thickness exceeding the critical one $h > h_C \approx 0.2$ to $0.5\,\mu$m. The stripe domain structure with Landau-Lifshits $\boldsymbol{I}_S$ distribution in the film cross-section appears under this condition. The magnetization in the main domains of this structure is parallel to the $\boldsymbol{EA}$. As H_0 is applied along current density direction the $\boldsymbol{I}_S$ vector deviates reversibly towards this direction, thus causing a change of U on the element. The saturation takes place in the field $H_0 = H_K$. Note, that if the H_0 is applied within the element plane and perpendicular to the current, the $\boldsymbol{I}_S$ deviation towards that plane does not result in any U change, since the angle θ remains unchanged in this case (see (1)). Thus, the element based on the overcritical film is a "longitudinal" one. Since the H_K value for such films is much higher than for those with planar $\boldsymbol{I}_S$ orientation and attains several hundreds A/cm, there is a considerably widened possibility to measure "moderate fields" using the MRE (Fig. 2.6).

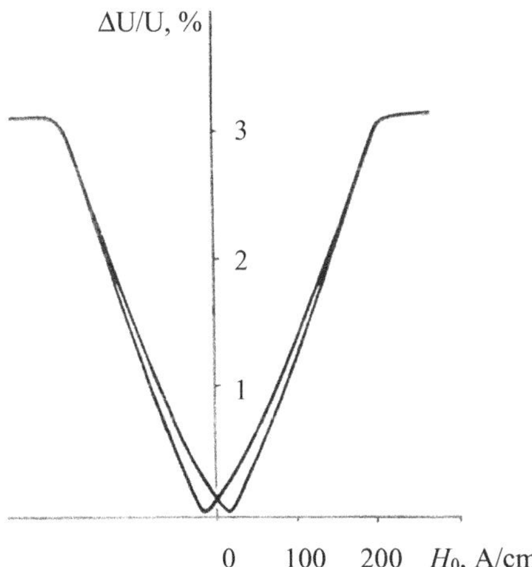

Fig. 2.6. Dependence of the output signal ΔU on the external field H_0 for the element based on the film with perpendicular anisotropy.

2.3.5 Developed AMR Sensor Mockups

Based on proposed and studied MSEs, the various mockups of magnetic sensors including both detectors and measuring transducers have been developed and tested.

- The r.p.m. sensor using multilayered film longitudinal element has been designed. The element consists of 4 meanders (0.5 mm size) connected on bridge circuit. There is miniature permanent magnet in the sensor. If any magnetic non-uniformity (e.g. a tooth of a gear) moves near the sensor, a field of the magnet is varied resulting in altering of the output signal. Also it depends on distance between rotating part and sensor. Thus, it can be used as analog transducer for the measuring of this clearance. The sensitivity of the sensor is $S = 40\,\text{mV}\cdot\text{cm/A}$.
- The low magnetic field sensor has also been designed. Their sensitive element is made of multilayer film with $(\Delta\rho/\rho)_m \approx 3$–4% and $H_C \approx 0.1\,\text{A/cm}$. The element is represented by 4-meander (3 mm size) bridge with bias stripe producing $H_B \gg 1\,\text{A/cm}$. The sensor allows to measure direct and alternating magnetic field with lowest level of about $0.002\,\text{A/cm}$ and has $S \approx 20\,\text{mV}\cdot\text{cm/A}$.
- The MSEs for the reading-out magnetic heads have been developed. The multilayer films with crossed $\boldsymbol{EA}$s have been used for these elements. It has been shown that their sensitivity is high enough to detect differently directed stray fields of magnetized spots on magnetic tape. A sign of output signal is dependent on field direction.

It should be noted that part of this research had been carried out in cooperation with Research Institute for Electrical Measuring Devices (St.-Petersburg, Russia). Using AMR film materials developed and prepared by NTU "KhPI" the following models of other sensors and devices based on available ones have been designed and tested: 1 – milliteslameter with longitudinal elements for measuring of induction up to 2 mT; 2 – non-contacting current gauge; 3 – linear displacement sensor; 4 – angular displacement sensor; 5 – object positioning sensor.

The testing results obtained for the described MSEs-based sensors confirmed the advantages of the elements developed by us.

2.4 Investigations of $(\text{Co/Cu})_n$ Multilayer Nanostructure Showing "Giant" Magnetoresistance (GMR)

It has been mentioned above that the problem related to the fabrication of the multilayered GMR elements ensuring good reproducibility of the characteristics is very difficult to solve. This is caused by simultaneous influence on the magnetoresistance of several parameters of multilayer system, namely, thickness of ferromagnetic layers and nonmagnetic interlayers, their atomic-crystalline structure, state of the interfaces etc. The GMR effect in multilayers is observed if $\boldsymbol{I}_S$

vectors in neighboring magnetic layers are reoriented from parallel to antiparallel arrangement during the magnetic field alteration. The periodical antiferromagnetic interaction arising between magnetic layers under variation of nonmagnetic interlayer thickness (the so-called oscillation of exchange [21]) promotes the antiparallel orientation of $\boldsymbol{I}_S$ vectors. Thus, we projected to study the dependencies of magnetoresistance value on Cu interlayer thickness. First of all these investigations are necessary because the peculiarities of a definite preparation procedure can affect the dependencies. Secondly, the magnetoresistance and related characteristics of the structures with various bilayer numbers (n) should be studied. The obtained results are considered in the following subsections.

2.4.1 Influence of Cu Interlayer Thickness on Magnetoresistive Ratio

The multilayered Co/Cu films have been studied. The thickness of the Co layers was equal to 1 nm; the thickness of the Cu layers varied from 0.7 to 10 nm. The X-ray diffraction analysis showed that the films had a layered polycrystalline structure; the X-ray reflectometry was used to estimate the surface roughness [22,23]. The properties of the multilayers were studied for the films both with and without a Fe buffer layer of 6 nm thick. The magnetoresistance was measured in the film plane at a room temperature in a maximum magnetic field of 12 kOe (960 kA/m).

It was shown that in order to obtain high magnetoresistance for multilayers deposited on different substrates a low surface roughness of the substrates is needed. The use of substrates with ad hoc prepared surface (glass coated with an amorphous film) allowed us to increase of the magnetoresistance up to 14%. It is pertinent to add that this value may be increased to 20% by optimum tempering [24]. An increase in the magnetoresistance was also observed for the multilayers with Cu layers of other thickness, irrespective of the presence or absence of the Fe buffer layer. No differences in the degree of the surface roughness σ were revealed by X-ray analysis of the substrate surfaces; for all the substrates, σ was equal to 2.2–2.3 nm. Since the used X-ray method characterizes only a roughness with a relatively long period (10–20 nm), we assume that roughness with a shorter period and a height of about 1 to 2 atomic sizes are most important for obtaining high giant magnetoresistance.

Fig. 2.7 shows the magnetoresistance of $(\mathrm{Co/Cu})_{30}$ films as a function of the thickness (t) of Cu layers. The films were deposited on a smoother substrate (glass covered by an amorphous film) with (1) and without (2) Fe buffer layer. The comparison of the curves with $\Delta R/R_s(t)$ dependencies available in the references for multilayers with the preferred crystallographic orientation shows that no pronounced oscillations with a period of ~1 nm, typical for the epitaxial multilayers (Co/Cu) [25]–[27], are present. A pronounced maximum at $t = 2$ nm ($\Delta R/R_s(t) = 13$–14%) is observed for the multilayers with Fe buffer layer (1). A weak maximum (to 6%) at t=3.5–4 nm and a subsequent progressive decrease in $\Delta R/R_s(t)$ to 2% at t=10 nm occur with increase in Cu-layer thickness. No

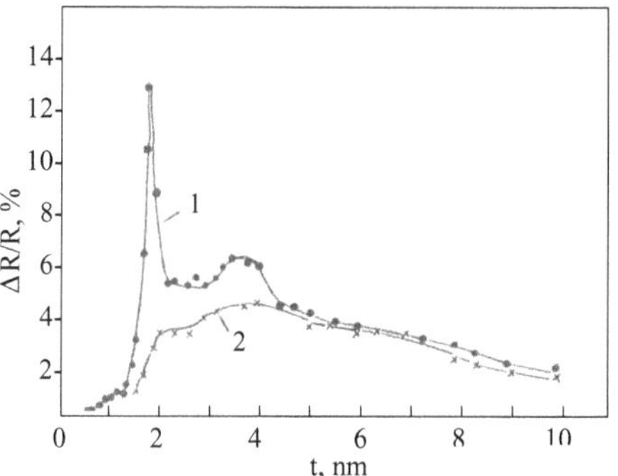

Fig. 2.7. Magnetoresistance ratio $\Delta R/R_s$ as a function of the thickness of Cu layers t for $[Co(1\,nm)/Cu(t)]_{30}$ multilayers deposited on a substrate (1) with the Fe buffer layer (1) and (2) without it.

maximum in the $\Delta R/R_s(t)$ dependence (2) at t=2 nm is observed for the multilayers without Fe buffer layer. At t=4 nm, a weak diffusive maximum ($\Delta R/R_s(t)$ = 4.5 %) is observed, and at $t > 5$ nm the curve 1 virtually coincides with curve 2. Thus, the dependence may be divided into two sections: at $t < 5$ nm, the Fe buffer layer substantially affects the magnetoresistance, and at $t > 5$ nm, its effect is reduced.

The behavior of the observed dependencies can be accounted for the following reasons. The antiferromagnetic coupling and the different coercive forces of the layers contribute to the antiparallel arrangement of magnetization vectors in polycrystalline multilayers. The different values of the coercive force of the layers are indirectly confirmed by the fact that the coercive force of multilayers changes substantially as the number of layers increases. This effect will be described in detail below (in subsec. 2.4.2). The different coercive forces of the layers may substantially affect the $\Delta R/R_s(t)$ dependence. If, with increasing of t, the antiferromagnetic coupling alternates with ferromagnetic one that leads to maxima and minima of the magnetoresistance, the state with antiparallel arrangement of $\boldsymbol{I}_S$ at different coercive forces of the layers can be observed at any thickness of Cu layers t.

This leads both to the absence of clearly defined minima in the $\Delta R/R_s(t)$ curve and to the smoothing of the curve at large thicknesses t, at which interlayer exchange coupling is low and GMR is due to uncoordinated layer-by-layer magnetization reversal, causing the antiparallel alignment of $\boldsymbol{I}_S$ in some regions of adjacent layers. The different effect of the Fe buffer layer on the magnetoresistance of thin and thick Cu layers confirms the conclusions of some authors (see, for instance, [28]) about the effect of the buffer layer on antiferromagnetic coupling. At high t, virtually no effect of the buffer layer on the magnetoresistance is observed because of the weaker exchange coupling due to large distances between the magnetic layers. At low t, the effect is greater and perhaps is realized through the formation of a certain structure of thin pseudomorphic Co and Cu layers.

Thus, well-reproducible explicit maximum is observed in multilayers with the Fe buffer layer at t=2 nm. In the following experiments the Cu interlayers of 2 nm thick have been used.

2.4.2 Nanostructures with a Variable Number of Bilayers

Since the GMR is mainly caused by scattering of the electrons on the interfaces, it is expected that magnetoresistance ratio will increase with increase in the quantity of bilayers. In addition, the application of GMR materials is based on the shape and the characteristics of the field dependence of the magnetoresistance, which are closely related to the processes of magnetization reversal. For multilayered structures containing two ferromagnetic layers successful attempts were made to theoretically explain the type of the resistance magnetic fild relationship using calculated hysteresis loops (see e.g., [29]). For multilayered systems with a large number of layers such an approach is an inefficient so far. In this research, the changes in the field dependence of the electrical resistance and magnetic hysteresis loops that are observed with increase in the number of the system layers were investigated experimentally in the multilayered Fe/(Co/Cu)$_n$ films, and a model of the magnetization reversal accounting for the features of magnetoresistive loops was suggested.

The multilayered structures with variable number of Co(1 nm)/Cu(2 nm) and Fe buffer layers (6 nm thick) were prepared. The n-value varied within the system from 2 to 30. When measuring the magnetoresistive properties the current passed along the stripe, and the magnetic field was oriented perpendicularly to the current (component) in the film plane. The maximum magnetoresistive ratio $\Delta R/R_{max}$ was determined under the quasistatic magnetization reversal beginning from fields exceeding the saturation field for electrical resistance. The $\Delta R/R_{max}$ values were calculated using the formula $\Delta R/R_{max} = (R_{max} - R_S)/R_S$, where R_{max} is the maximum value of the resistance in the process of magnetization reversal. In addition, the magnetic hysteresis loops were obtained using a vibrating-sample magnetometer and the $\Delta R/R = f(H)$ dependencies were compared for the same sample. Moreover, ΔR is $R_H - R$, R_H is the electrical resistance in the field H, and R is a minimum value of the resistance. The investigations of the structure of the samples which was carried out by X-ray diffractometry techniques and the electron transmission microscopy using oblique cross-section show that the samples were polycrystalline and exhibit a fairly well-defined laminated structure. The layer surface was wavy with a period of 10–20 nm along the surface.

The changes in the magnetoresistive properties of a multilayered structure with increase in the bilayer numbers were noted earlier and both an increase in the magnetoresistive ratio and its decrease were observed in different systems [30]–[32]. Note that this phenomenon was not studied in detail and its origin remains to be clarified. We performed systematic measurements of the maximum magnetoresistive ratio as well as comparative investigations of magnetic hysteresis loops and $\Delta R/R = F(H)$ behavior. The choice of (Co/Cu)$_n$ system was accounted for the fact that this system is characterized by a sufficiently

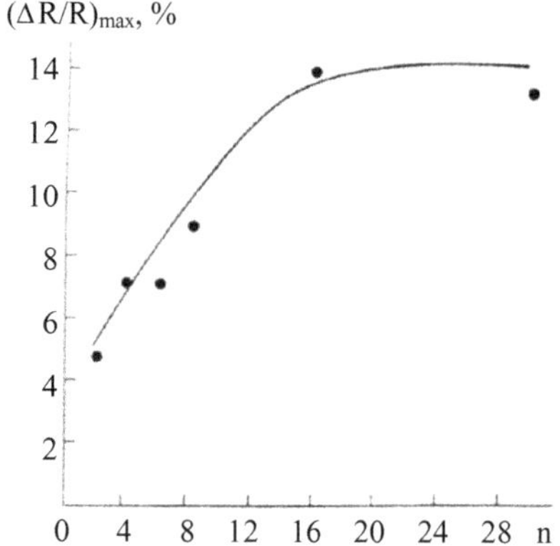

Fig. 2.8. Dependence of the magnetoresistive ratio $(\Delta R/R)_{max}$ on the number n of Co/Cu bilayers in the Fe(6 nm)/[Co(1 nm)/Cu(2 nm)]$_n$ multilayer structure.

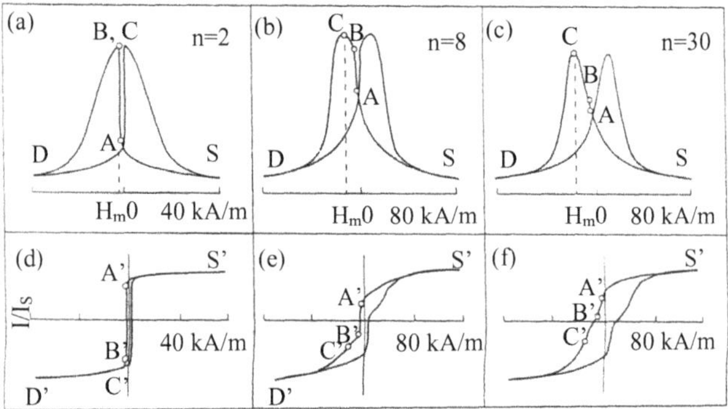

Fig. 2.9. (a–c) – field dependencies of $\Delta R/R = f(H)$; and (d–f) – magnetic hysteresis loops for Fe(6 nm)/[Co(1 nm)/Cu(2 nm)]$_n$ multilayer structure with a variable number of bilayers n.

high GMR effect and comparatively small fields in which the resistance jump is observed and it exhibits good reproducibility in fabrication.

Fig. 2.8 shows the $\Delta R/R_{max}(n)$ dependence for the series of samples. Thus, $\Delta R/R_{max}$ reaches 14% for $n = 16$, the latter value being retained up to $n = 30$. With the increase in the number of bilayers, there also occurs a change in the character of the field dependencies $\Delta R/R(H)$, in which hysteresis and maximum in the reverse magnetic field H_m were revealed. These curves for the number of bilayers equal to 2; 8 and 30 are shown in Fig. 2.9a–2.9c. In all curves one can separate several characteristic segments with different slopes: (1) segment SA with a gradual increase in the resistance; (2) segment AB with a jump-like increase; (3) segment BC with a more gradual increase in the resistance and attainment of the maximum value $R_H = R_{max}$ (for n=2, points B and C coincide, i.e. this segment is absent); and (4) segment CD with a gradual decrease in the resistance, tending to R_S in high field. The point A lies in the

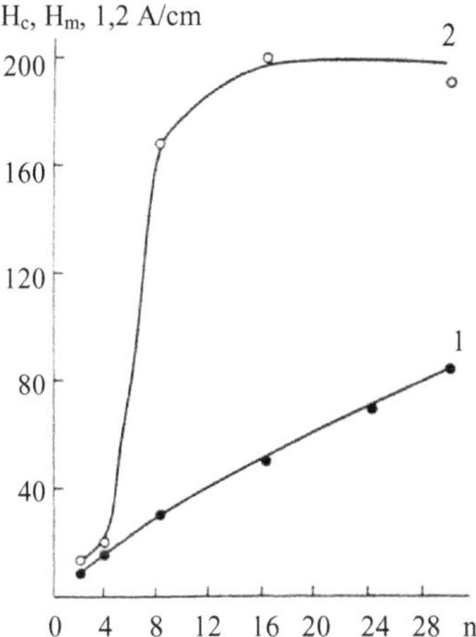

Fig. 2.10. Dependence of (1) the coercive force H_c and (2) the maximum magnetoresistance field H_m on the number of bilayers n for the Fe(6 nm)/ [Co(1 nm)/ Cu(2 nm)]$_n$ multilayer structure.

region of small negative fields, and the point C corresponds to the field of the maximum resistance H_m. The resistance change in the above segments and their slopes depend on the number of layers. The increase in the number of layers brings about an extension of the hysteresis regions and a shift of the segment boundaries (points A, B, and C) toward higher fields. The maximum field H_m reaches 180–200 Oe for n=16 and 30 (Fig. 2.10). The H_C value of the system grows too (Figs. 2.9, 2.10).

The comparison of the magnetic hysteresis loops (Figs. 2.9d–2.9f) with the magnetoresistive hysteresis loops (Figs. 2.9a–2.9c) shows that certain correlation between the position of the boundaries of some separate regions in magnetic hystersis loop (points A′, B′, and C′) and the specific points of the $\Delta R/R(H)$ dependencies (points A, B, and C) is observed. An abrupt growth of electrical resistance (section AB) corresponds to a sudden change of magnetization ΔI in section A′B′ of the magnetic hysteresis loop. Here the magnetization reversal of the buffer layer and adjacent Co layer occurs apparently through displacement of the domain walls. This assumption is also corroborated by the decrease of the ΔI contribution to the total magnetization in this section with an increase in number of bilayers in the system and by the total absence of a similar section in the case of magnetization reversal of multilayer structures deposited directly on glass without a buffer layer. Note that the highest sensitivity of resistance to the change in the magnetic field is observed exactly in this section. The growth of H_C in the multilayered structures with increase in the number of layers can be interpreted as the appearance of layers in the system that are switched over at higher threshold fields in comparison to those that were formed previously. This assumption can be partly substantiated by the fact that with increase of a distance to the substrate, the surface waviness (roughness) that retards the motion of the domain walls in the layers is enhanced [33].

Taking into account the fact that the coercive force increases from lower to upper layers and the magnetization reversal of each layer occurs through displacement of the domain walls, a simplified diagram of magnetization reversal of multilayer structure can be suggested as follows. If, beginning with the state of the remanent magnetization obtained after the value of magnetic field is decrease from that which is close to the saturation field, the field is applied in negative direction, this triggers a reorientation of magnetization vectors in those regions of the sample that are in the most unstable state. First, this will be the buffer layer with the adjacent Co layer. (It is necessary to add that the coercive force was 15 Oe for the Fe film, 11 Oe for the Co film and decreases to 5 Oe for resulting system Fe + Co). After this first reorientation of $\boldsymbol{I}_S$, a pair of layers will appear with antiparallel alignment of the magnetization: the Fe + Co bilayer and the second Co layer (Fig. 2.11a). At this instant, a sudden change in the electrical resistance is observed. Since the mutual orientation of the $\boldsymbol{I}_S$ vectors in this pair is relatively stable because of the weak antiferromagnetic interaction, the reorientation of $\boldsymbol{I}_S$ with a further increase in the reversing field is likely to occur in the third Co layer rather than in the second Co layer, not to break the pair. Thus, two more interfaces will appear along which the $\boldsymbol{I}_S$ vectors are antiparallel (Fig. 2.11b). In this case, the relative volume of the film with antiparallel alignment of magnetization in neighboring layers increases and, correspondingly, the volume with a parallel alignment of magnetization decreases. For this reason, the electrical resistance of the whole multilayered structure will grow. The above-described process, i.e., the reversal of $\boldsymbol{I}_S$ in the odd-numbered Co layers, will continue until the reorientation of $\boldsymbol{I}_S$ occurs in the last odd-numbered Co layer. At this instant, the regions with antiparallel alignment of magnetization in the nearest layers occupy the greatest possible volume (in the ideal case) and the electrical resistance should be a maximum one. Note that we consider an idealized region of multilayered structure. In real polycrystalline films, the pro-

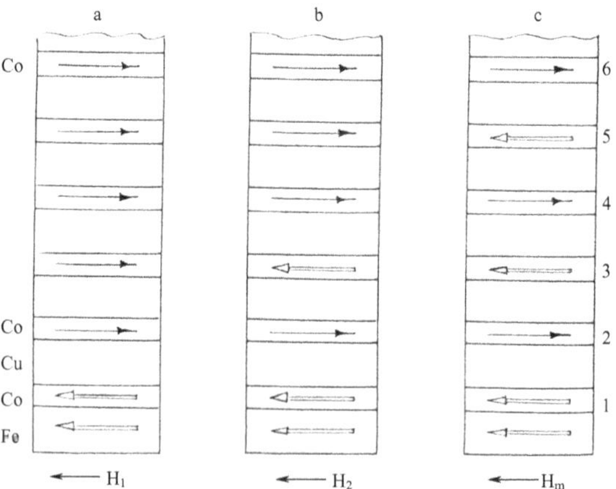

Fig. 2.11. Magnetization reversal diagram of multilayered structure in the region of reverse fields $H_1 < H_2 < H_m$.

cess of magnetization reversal can differ somewhat from one region to another, and in fact, we observe a certain averaged picture.

The suggested model properly describes the behavior of the magnetic and magnetoresistive characteristics in A′-B′-C′ and A-B-C sections. In other regions it is necessary to take into account the magnetization reversal processes caused by domain-wall motion processes. Thus, when the field is decreased from the saturation state to zero (section S′A′), the magnetization vectors supposedly rotate toward random local axes and, as a consequence, areas with a noncollinear arrangement of $\boldsymbol{I}_S$ appear in neighboring layers. In this case the electrical resistance of the structure somewhat increases (section SA). Then up to the point C the magnetization reversal occurs with the prevalence displacement processes as discussed above. The magnetic behavior in the section C′D′ is interpreted more difficult because of the lack of any distinguishing features in the field dependencies of I/I_S and $\Delta R/R$. The magnetization reversal in this case is most likely to occur first via irreversible rotation of $\boldsymbol{I}_S$ in the even-numbered layers and then via reversible rotation toward the field direction in all the layers. This gives a smooth decrease of the electrical resistance.

Thus, mechanism of the magnetization reversal responsible for the specific shape of the field dependence of the magnetoresistive ratio $\Delta R/R(H)$ is suggested. This mechanism implies that the reorientation of the magnetization vectors towards the field direction occurs in alternating magnetic layers and this gives rise to antiparallel arrangement of $\boldsymbol{I}_S$ in the film volume in a certain reversing field (the field of maximum magnetoresistance). The mechanism described above can also be realized in other systems with weak exchange interaction and variable coercive force specified intentionally. This makes it possible to control the magnitude of a field of maximum magnetoresistance.

2.4.3 Possible Fields of Application

It has been shown above, that the satisfactory GMR characteristics can be achieved in investigated $(Co/Cu)_n$ polycristalline multilayers. The magnetoresistive ratio is reached up to 14 % for $n \geq 15$. Moreover the $(\Delta R/R)_m$ value increases up to 20 % after tempering. The procedure of obtaining such multilayers is reproducible. The shown characteristics are comparable with those obtained by other investigators. It is important that the shape of $(\Delta R/R)$ dependencies on H can be altered by varying n-value.

Although application of GMR multilayers presents certain problems and at the present time they cannot find such a wide application as AMR films the research and development of the GMR elements are expedient without doubt. On the basis of the results obtained from the work done the fields of application of GMR elements can be outlined.

- The film structure at n=2 can be used for threshold elements. Indeed, jump-like resistivity change is observed for low fields. Despite the fact that $(\Delta R/R)_m$ value of this film is less then 2.5 %, its sensitivity $S = (\Delta R/R)/H$

reaches up to $2\times10^{-2}\,(\mathrm{A/cm})^{-1}$. This value is comparable with characteristics of the best specimens described in the published papers. It is worth to note that the behavior of the films described above is similar to that of the so-called "spin-valves" multilayers [34].

- The use of GMR multilayers for precise measurement of the magnetic fields is more complicated problem because the element can possess single-valued dependences ($\Delta R/R$) on H. However, the investigated multilayers have an explicit hysteresis both for the complete and for the partial cycles of magnetization reversal. In order to overcome this disadvantages, we tried to use the bridge circuit with bias field applied to the stripe elements. These attempt was found successful for some mockups that have linear and most non-hysteresis dependence of output signal on measuring field in the range of $\pm270\,\mathrm{A/cm}$.

2.5 Conclusions

The film mono- and multilayers showing AMR effect are widely applied for fabrication of sensitive elements used in magnetic sensors. The drawbacks of the traditional elements have been analyzed. Basing on the results of the carried out investigations the new elements free of these drawbacks have been developed as it is shown in the reviewed materials: these are the "longitudinal" element having high sensitivity and the element with crossed easy axes by which the direction of magnetic field can be determined. Some other improvements of elements have also been introduced.

The multilayer films possessing GMR effect are rather promising as nanostructured materials that find new fields of application. However, these materials develop numerous peculiarities of magnetoresistance that are insufficiently studied. Some peculiarities of magnetoresistance in $(\mathrm{Co/Cu})_n$ multilayers depending on the copper film thickness and number of bilayers have been investigated

Various dimensional effects observed in nanostructures have essential meaning for regulation of magnetoresistance and magnetic characteristics. Some of them were studied in the reviewed materials. These are the dependencies of magnetoresistive ratio on interlayer thickness and number of bilayers and also the change of domain structure types depending on stripe width resulting in alteration of magnetoresistance values in "longitudinal" elements and in those with crossed easy axes etc.

Acknowledgments

The author is grateful to his collaborators, namely S.T. Roschenko, I.G. Shipkova, V.N. Samophalov, L.I. Lukashenko and others for their important contribution to the research described in these notes. He would also like to express his gratitude to L.N. Gerasimuk and V.A. Sagaydak for testing the magnetic head mockup. The author also acknowledges the support of this work and the notes by the Foundation for Basic Research of Ukraine (Project N 4/157 and

4.1/133), Science and Technology Center in Ukraine (Project N 2266) and Organizing Committee of the International Workshop on "Nanostructured Magnetic Materials and Their Application".

References

1. S.Kh. Karpenkov: *Thin Film Magnetic Transducers* (Radio y svyaz', Moskow 1985)- in Russian
2. R. Bozorth: *Ferromagnetism* (D.Van Nostrand Co. Inc., Toronto-N.-Y.- London 1951)
3. S.S. Parkin, Z.G.Li, D.J.Smith: Appl.Phys.Lett. **58**, 2710 (1991)
4. S.Kh.Karpenkov: *Thin Film Magnetic Storage* (Radio y svyaz', Moskow 1993)- in Russian
5. H. Schaumburg: *Sensoren* (B.G.Teubner, Stuttgart 1992)- in German
6. S.V.Vonsovskiy: *Magnetism* (Nauka, Moskow 1971)- in Russian
7. W.Kwiatkowski, S, Tumanski: J.Phys.: E.Sci.Instrum. **19**, 502 (1986)
8. R.S.Hebbert, I. Svhwee: Rev.Sci.Instrum. **37**, 1321 (1966)
9. *Sensoranwendungen.* Ed, by H. Schaumburg (B.G.Teubner, Stuttgart 1995)- in German
10. J.M.Daughton and A.V. Pohm: Patent EP 0248355, Int.Cl.4G11C11/14, 1987
11. A.G.Ravlik, S.T.Roschenko, V.N.Samophalov et al.: Author. Certificate of the USSR N1774296, Int.$Cl.^5$ G01R33/06, 1992
12. Ch.Kittel: Phys.Rev.**70**, 965 (1946)
13. L.I.Lukashenko, S.V.Potapov, A.G.Ravlik et al.: J.Magn. and Magn.Mater.**116**, 70 (1992)
14. H.Clow: Nature, **94**, 1035 (1962)
15. L.S.Palatnik, Yu.V. Zolotnitskiy, A.G.Ravlik: Fizika Tverdogo Tela (FTT), **8**, 3714 (1965)- in Russian. English version of the FTT is entitled "Solid State Physics"
16. M.S.Cohen: "Ferromagnetic Properties of Film". In: *Handbook of Thin Film Technology* v.2 Ch.17. ed. By L.Maissel and Glang (McGraw Hill Hook Co., N.-Y. 1970)
17. A.G.Ravlik, S.T.Roschenko, V.N.Samophalov et al.: Patent of Russia N1764423 A1 Int.Cl.-^{5}5G01R33/05, 1992
18. V.N.Samophalov, L.I. Lukashenko: J.Magn. and Magn.Mater. **128**, 354 (1993)
19. L.S.Palatnik, M.Ya.Fuks, L.I.Lukashenko, A.G.Ravlik, A.A. Kozma: Phys.Stat.Sol. **17**, 543 (1966)- in German
20. A.G.Ravlik, B.A.Avramenko: Fizika Metallov y Metallovedenie (FMM), **85**, 62 (1998)- in Russian. English version of the FMM is entitled "The Physics of Metals and Metallography"
21. D.M.Edwards, J.Matton: J.Magn. and Magn.Mater. **93**, 85 (1991)
22. S.T.Rocshenko, I.D.Shipkova, A.A.Kozma, V.I.Pinegin: FMM, **85** , 65 (1998)
23. V.A.Vinogradov, I.A.Britov, A.Ya.Grudskiy: *Mirror X-Ray Optics* (Mashinostroenie, Leningrad 1989)- in Russian
24. S.T.Roschenko, A.G.Ravlik, I.G.Shipkova: "Annealing influence on magnetoresistance properties of nanostructured $Fe/(Co/Cu)_n$", In.: Thin Films in Electronics (Proc. Of the 12th International Colloq. at Kharkiv, Ukraine (NSC "KhPTI", Kharkiv, April 23-27, 2001), pp.60-63.
25. S.S.P.Parkin, R.Bhadra, K.P.Roche: Phys.Rev.Lett. **66**, 2152 (1991)
26. W.F.Jr.Egelgoff, M.T.Kief: Phys.Rev.B.: Condens.Matter, **45**, 7795 (1992)

27. S.K.J.Lenczowski, M.A.M. Gijs, R.J.M. Van der Veerdonk et al.: "Giant Magnetoresistance in Co/Cu (100) Multilayers". In.: *Magnetic Films and Surfaces, Digest of the* 14th *Internat. Colloq. at Dusseldorf, Germany (H.Heine Univ. Dusseldorf, 29August-2 September 1994)*, A12, pp.21-22.
28. M.Jimbo, T.Kanda, T.Goto et al.: J.Magn. and Magn.Mater. **126**, 422 (1993)
29. D.Wei: IEEE Trans. On Magn. **38**, Pt2. N2, 552 (1998)
30. A.Dinia, S.Zoll, K.Ounadjeva: "Temperature and Number of Periods Dependence on the Magnetoresistance in Co/Ru Superlattices". In: Digest [27] pp.69-70
31. M.Mulloy, E.Kolb, C.Dupas et al.: "Magnetoresistance of Gold-Cobalt Multilayers as a Function of the Number of Cobalt Layers": ibid. pp.96-97
32. T.Valet, P.Galtier, J.C. Jacquet et al.: J.Magn. and Magn. Mater. **121**, 402 (1993)
33. A.R.Modak, S.S. Parkin, D.J.Smith: J.Magn. and Magn.Mater. **129**, 415 (1994)
34. B.Dieny: J.Magn. and Magn.Mater. **136**, 335 (1994)

3 Model Calculation of the Giant Magnetoresistance in Multilayers with an Arbitrary Number of Layers

M.Ye. Zhuravlev[1,2], W. Schepper[1], S. Heitmann[1], H.O. Lutz[1], A.V. Vedyayev[3,4], G. Reiss[1], and A. Hütten[1]

[1] Fakultät für Physik, Universität Bielefeld, Universität str. 25, 33501 Bielefeld, Germany
[2] Institute of General and Inorganic Chemistry, RAS, Leninskii prosp., 31, Moscow 117907, Russia
[3] CEA/Département de Recherche Fondamentale sur la Matiére Condensée, SP2M/NM, 38054 Grenoble, France
[4] Physical Department of Moscow State University, Moscow 119899, Russia

Abstract. We present a quantum statistical treatment of the Giant Magnetoresistance (GMR) in multilayers with an arbitrary number of layers and arbitrary angles between the magnetization vectors in the magnetic layers. The intrinsic potential and the exchange splitting of the conducting electron band is taken into account. The developed approach was successfully applied to the description of the GMR in combination multilayers.

3.1 Introduction

Giant magnetoresistance (GMR) is one of the transport phenomena in solid state physics which has stimulated extensive experimental as well as theoretical studies over the past decade due to its fundamental interest as well as its application potential [e.g. see 1]. In ferromagnetic/paramagnetic-sandwich layers the thicknesses of the paramagnetic layers determine the strength of the antiferromagnetic coupling, and hence the mutual orientation of the magnetization vectors of the magnetic layers in an external magnetic field. Therefore the specific form of the dependence of the GMR on an applied magnetic field depends on the thicknesses and the order of the layers. Systems with different thicknesses of the paramagnetic layers (combinational multilayers, CML) were investigated [2–4] and new shapes of GMR curves were found.

Besides *ab initio* calculations of the GMR on the basis of realistic band structures (c.f. e.g. [5]), model calculations are extensively used [22,7,8] which are able to give a transparent description of the physical phenomena through several parameters such as spin-dependent mean-free path, Fermi momentum, etc. In quantum statistical theories the transport characteristics are calculated within the Kubo linear response formalism.

It is widely adopted that the underlying mechanism of GMR is a coherent interplay of spin-dependent scattering of conducting electrons occurring in the bulk of the succesive magnetic layers or at the interfaces, and electron reflection from spin-dependent potential barriers between layers. The height of the barriers as well as their spin-dependence are different for s and d electrons.

In the present work, following the treatment presented in [8], the free-electron model is applied to the conducting s-electrons whose mass is supposed to be much smaller than the mass of the almost localized d-electrons. Therefore the s electrons carry the main part of the current . The mean-free path of the conducting s-electrons depends on their spin due to $s-d$ hybridization and the different density of d-states at the Fermi level as a consequence of the exchange splitting of the d-band. It was shown [9] also that conducting free-like electrons tunneling between ferromagnetic (Ni, Fe, Co) displays significant spin splitting. Therefore, transition metal based multilayers can display a spin-dependent modulation of the lattice potential as well as a spin-dependent scattering cross-section for the conduction electrons.

Our quantum statistical approach requires to construct the Green function (GF) of the system under consideration. As a serious part of the problem the matching of the GF at the interfaces arises. If such a multilayer contains only a few layers the GF can be constructed directly. For a large number of layers a more cumbersome approach has to be applied. A review of some matching methods for the GFs was presented in [10].

We consider the construction of an exact one-particle GF in the framework of quantum statistical theory [8] proposed for the description of the GMR in a three-layered system. We approach the problem in two stages. We solve the problem of GF matching at the interfaces for multilayers with arbitrary magnetization directions of the ferromagnetic layers, exchange splitting of the conducting electron band and intrinsic potential. The angles between the magnetization vectors of the ferromagnetic layers in ian external magnetic field were determined numerically on the basis of the extended Stoner-Wohlfarth model.

We apply the developed treatment to the description of the GMR in the CIP (current-in-plane) geometry in magnetron-sputtered "combination multilayers" (CML) of type $\mathrm{Py}_{1.8\mathrm{nm}}//\{[\mathrm{Cu}_{1.8\mathrm{nm}}/\mathrm{Py}_{1.6\mathrm{nm}}]_M/[\mathrm{Cu}_{0.9\mathrm{nm}}/\mathrm{Py}_{1.6\mathrm{nm}}]_M\}_Y$ (with $\mathrm{Py}{=}\mathrm{Ni}_{81}\mathrm{Fe}_{19}$).

In disordered binary alloys, like Py, the mean free path of the conducting electrons can be calculated in the coherent potential approximation on the basis of the $s-d$ model considered in [11]. The Hamiltonian of the model is the following:

$$\begin{aligned} H = &\sum_{k\sigma} E_s(k) a^{s\dagger}_{k\sigma} a^s_{k\sigma} \sum_{k\sigma} E_d(k) a^{d\dagger}_{k\sigma} a^d_{k\sigma} + \\ &\sum_{n\sigma} \epsilon^d_{n,\sigma} a^{d\dagger}_{n\sigma} a^d_{n\sigma} + \sum_i n\sigma \epsilon^{n,\sigma}_s a^{s\dagger}_{n\sigma} a^s_{n\sigma} + \sum_{n\sigma} \gamma^{sd}_n (a^{s\dagger}_{n\sigma} a^d_{n\sigma} + a^{d\dagger}_{n\sigma} a^s_{n\sigma}), \end{aligned} \tag{3.1}$$

where $a^{s\dagger}_{k\sigma}, a^s_{k\sigma}, a^{d\dagger}_{k\sigma}, a^d_{k\sigma}$ are the creation-annihilation operators for Bloch states, $a^{s\dagger}_{n\sigma}, a^s_{n\sigma}, a^{d\dagger}_{n\sigma}, a^d_{n\sigma}$ are the creation-annihilation operators for Wannier states, $E_s(k)$ and $E_d(k)$ are the kinetic energy of $s-$ and $d-$ electrons respectively, $\epsilon^s_{n,\sigma}$, $\epsilon^{n,\sigma}_d$ are the positions of s and d levels in site n, γ^{sd}_n is the hybridization constant; $\epsilon^s_{n,\sigma}$, $\epsilon^{n,\sigma}_d$ and γ^{sd}_n are the random potentials,

$$\gamma_n^{sd} = \begin{cases} \gamma^A & \text{with probability} \quad x \\ \gamma^B & \text{with probability} \quad 1-x \end{cases}$$

$$\epsilon_n^{s(d)} = \begin{cases} \epsilon^{A,s(d)} & \text{with probability} \quad x \\ \epsilon^{B,s(d)} & \text{with probability} \quad 1-x. \end{cases}$$

These potentials can take one of the two values with probability x or $1-x$, determined by the material. In the case of permalloy the components of the alloy are Ni and Fe; in the case of a pure transition metal, the second component represents the impurities which provoke the scattering of the electrons. In Born approximation the s-electron GF obeys the following equation:

$$G^\sigma_{ss}(k) = G^{(0)\,\sigma}_{ss}(k) + G^{(0)\,\sigma}_{ss}(k)\Sigma G^\sigma_{ss}(k) \tag{3.2}$$

where the s-electron self-energy is given by

$$\Sigma = x(1-x)(\epsilon^{A,s} - \epsilon^{B,s})^2 \sum_k G_0^{ss}(k) + x(1-x)(\gamma^A - \gamma^B)^2 \sum_k G_0^{dd}(k). \tag{3.3}$$

Therefore, the imaginary part of the self energy which determines the mean free path can be expressed through the density of states of s and d electrons as follows:

$$\operatorname{Im} \Sigma = x(1-x)(\epsilon^{A,s} - \epsilon^{B,s})^2 \rho^s_{E_F} + x(1-x)(\gamma^A - \gamma^B)^2 \rho^d_{E_F} \tag{3.4}$$

The real part of the self energy Σ determines the exchange splitting of the s-band induced by $s-d$ hybridization. In ferromagnetic transition metals the density of d states for spin-up and spin-down electrons can be rather large near the Fermi energy. The above consideration shows how the density of states of d-electrons influences the s electron mean free paths.

Now we pass on to the construction of conducting electrons' GF for the entire multilayer taking into account the exchange splitting of the conducting band and the spin-dependent character of the mean free path of the conducting electrons in the ferromagnetic layers.

3.2 Construction of the Green Function

Generalizing the treatment of a trilayer [12,13], we consider a system consisting of several alternating ferromagnetic and paramagnetic layers. The magnetization of the n-th ferromagnetic layer is allowed to subtend an angle γ_n with the quantization axis, and the GF of the conducting electrons is a 2×2 matrix with non-zero off-diagonal elements; this is a consequence of spinor nature of the GF (see H. E. Camblong, et al. [22]). The spin $\uparrow$ and spin $\downarrow$ electrons are supposed to form a free electron gas within each of the ferromagnetic materials. In the mixed κ, z representation [8] the GF G_n in the n-th layer obeys the following equation [12]:

$$\left[\left(\frac{\partial^2}{\partial z^2} + (k^0_{nF})^2 - \kappa^2 - E^{(0)}_n\right)\begin{pmatrix} 1 & 0 \\ 0 & 1 \end{pmatrix} - E^{(1)}_n \begin{pmatrix} \cos\gamma_n & \sin\gamma_n \\ \sin\gamma_n & -\cos\gamma_n \end{pmatrix}\right]$$

$$\times \begin{bmatrix} G^{\uparrow\uparrow}_n(z,z') & G^{\uparrow\downarrow}_n(z,z') \\ G^{\downarrow\uparrow}_n(z,z') & G^{\downarrow\downarrow}_n(z,z') \end{bmatrix} = \frac{2Ma_0}{\hbar^2}\delta(z-z')\begin{pmatrix} 1 & 0 \\ 0 & 1 \end{pmatrix}, \qquad (3.5)$$

where a_0 is the lattice constant, M is the electron mass, $(k^0_{nF})^2 = 2ME_{nF}/\hbar^2$, E_{nF} is the Fermi energy in the n-th layer, and κ is the in-plane momentum;

$$E^{(0)}_n = \frac{1}{2}\left(\Sigma^{\uparrow} + \Sigma^{\downarrow}\right), \quad E^{(1)}_n = \frac{1}{2}\left(\Sigma^{\uparrow} - \Sigma^{\downarrow}\right) \qquad (3.6)$$

and the real parts of the electron self-energy Σ^{σ} determine the exchange splitting of the s-electrons with spin σ, whereas the imaginary parts are proportional to the inverse mean free paths $l^{\uparrow,\downarrow,(para)}$ of the electrons. We use variation-of-constants method [14] to obtain the GF. The boundary conditions which should be imposed on the GF depend on the GMR geometry. For CIP we impose zero boundary conditions at the outer interfaces.

The first and the second columns of the GF are independent and we consider a solution for the first column of the matrix equation (3.5). We assume that the first layer is ferromagnetic and the last one is paramagnetic. The initial differential equation for the first column of (3.5) can be written as a system of differential equations of first order:

$$\left(\hat{I} \times \frac{\partial}{\partial z} + \hat{L}_n\right) G_n(z,z') = (2Ma_0/\hbar^2)\begin{pmatrix} 0 \\ \delta(z-z') \\ 0 \\ 0 \end{pmatrix}, \qquad (3.7)$$

where matrix $\hat{L}_n$ is as follows:

$$\begin{pmatrix} 0 & -1 & 0 & 0 \\ ((k^0_{nF})^2 - \kappa^2 - E^{(0)}_n - E^{(1)}_n\cos\gamma_n) & 0 & -E^{(1)}_n\sin\gamma_n & 0 \\ 0 & 0 & 0 & -1 \\ -E^{(1)}_n\sin\gamma_n & 0 & ((k^0_{nF})^2 - \kappa^2 - E^{(0)}_n + E^{(1)}_n\cos\gamma_n) & 0 \end{pmatrix}, \qquad (3.8)$$

$$G_n(z,z') = \begin{pmatrix} G^{\uparrow\uparrow}_n(z,z') \\ W^{\uparrow\uparrow}_n(z,z') \\ G^{\downarrow\uparrow}_n(z,z') \\ W^{\downarrow\uparrow}_n(z,z') \end{pmatrix} \quad \text{and} \quad \begin{matrix} W^{\uparrow\uparrow}_n(z,z') = \frac{\partial}{\partial z}G^{\uparrow\uparrow}_n(z,z') \\ W^{\downarrow\uparrow}_n(z,z') = \frac{\partial}{\partial z}G^{\downarrow\uparrow}_n(z,z'), \end{matrix} \qquad (3.9)$$

In the n-th ferromagnetic layer the corresponding homogeneous equation has four linear independent solutions which can be combined into a 4×4 matrix:

$$\Phi^{(0)}_n(z) =$$

$$\begin{pmatrix} C_{\gamma n}e^{ik_{n1}z} & C_{\gamma n}e^{-ik_{n1}z} & -S_{\gamma n}e^{ik_{n2}z} & -S_{\gamma n},e^{-ik_{n2}z} \\ ik_{n1}C_{\gamma n}e^{ik_{n1}z} & -ik_{n1}C_{\gamma n}e^{-ik_{n1}z} & -ik_{n2}S_{\gamma n}e^{ik_{n2}z} & ik_{n2}S_{\gamma n}e^{-ik_{n2}z} \\ S_{\gamma n}e^{ik_{n1}z} & S_{\gamma n}e^{-ik_{n1}z} & C_{\gamma n}e^{ik_{n2}z} & C_{\gamma n}e^{-ik_{n2}z} \\ ik_{n1}S_{\gamma n}e^{ik_{n1}z} & -ik_{n1}S_{\gamma n}e^{-ik_{n1}z} & ik_{n2}C_{\gamma n}e^{ik_{n2}z} & -ik_{n2}C_{\gamma n}e^{-ik_{n2}z} \end{pmatrix}, \quad (3.10)$$

where $\quad C_{\gamma n} = 1 + \cos\gamma_n, \quad S_{\gamma n} = \sin\gamma_n,$

$$\begin{aligned} k_{n1} &= \sqrt{(k_{nF}^{(0)})^2 - \kappa^2 - \Sigma_n^{\uparrow}} = \sqrt{(k_{nF}^{(\uparrow)})^2 - \kappa^2 + (2ik_{nF}^{(\uparrow)}/l_n^{(\uparrow)})}, \\ k_{n2} &= \sqrt{(k_{nF}^{(0)})^2 - \kappa^2 - \Sigma_n^{\downarrow}} = \sqrt{(k_{nF}^{(\downarrow)})^2 - \kappa^2 + (2ik_{nF}^{(\downarrow)}/l_n^{(\downarrow)})}. \end{aligned} \quad (3.11)$$

Similarly, the basis of the solution of the differential equation for a paramagnetic layer ($E_n^{(1)} = 0$) can be chosen as

$$\Psi_n^{(0)}(z) = \begin{pmatrix} e^{ik_n z} & e^{-ik_n z} & 0 & 0 \\ ike^{ik_n z} & -ik_n e^{-ik_n z} & 0 & 0 \\ 0 & 0 & e^{ik_n z} & e^{-ik_n z} \\ 0 & 0 & ike^{ik_n z} & -ik_n e^{-ik_n z} \end{pmatrix} \quad (3.12)$$

and

$$k_n = \sqrt{(k_{nF}^{(para)})^2 - \kappa^2 + (2ik_{nF}^{(para)}/l_n^{(para)})}. \quad (3.13)$$

Using Eqs.(3.10), (3.12) we construct four linear independent solutions continuous in the interval $0 \le z \le c_{N+1}$. For our purpose it is convenient to choose solutions of the homogeneous system so that two of them obey the boundary conditions on the left ($F^{(l)}(z)$) and two on the right ($F^{(r)}(z)$) interfaces:

$$\begin{aligned} F^{(l)}(z) &= \begin{bmatrix} c_n \le z \le b_n, \Phi_n^{(0)}(z) \cdot A_n^{(l)} \\ b_n \le z \le c_{n+1}, \Psi_n^{(0)}(z) \cdot R_n^{(l)} \end{bmatrix}, \\ F^{(r)}(z) &= \begin{bmatrix} c_n \le z \le b_n, \Phi_n^{(0)}(z) \cdot A_n^{(r)} \\ b_n \le z \le c_{n+1}, \Psi_n^{(0)}(z) \cdot R_n^{(r)} \end{bmatrix} \end{aligned} \quad (3.14)$$

where $A_n^{(l,r)}, R_n^{(l,r)}$ are the columns of numbers which provide the corresponding boundary conditions (left or right) and continuity of the functions $F^{(l)}(z)$ and $F^{(r)}(z)$. The thicknesses of the ferromagnetic ($c_n \le z \le b_n$) and paramagnetic ($b_n \le z \le c_{n+1}$) layers can be arbitrary (Fig. 3.1).

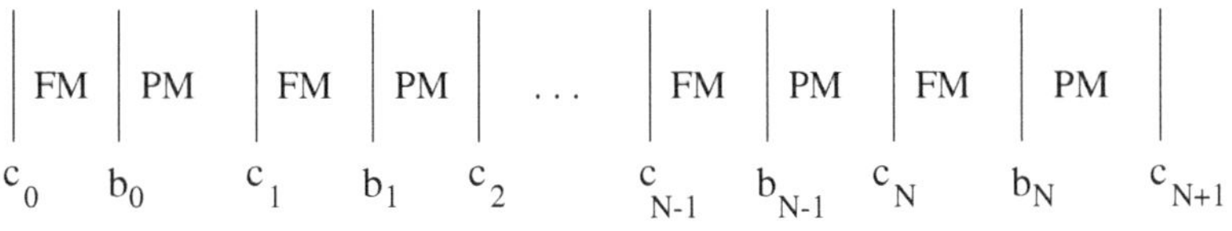

Fig. 3.1. b_i, c_i are the boundaries of the layers

We impose zero boundary conditions on the "right" and "left" solutions at the corresponding end points:

$$\Phi_0^{(0)}(z=0)\cdot A_0^{(l)} = \begin{pmatrix} 0 \\ \dots \\ 0 \\ \dots \end{pmatrix} \quad \text{and} \quad \Psi^{(0)}(z=c_{N+1})\cdot R_N^{(r)} = \begin{pmatrix} 0 \\ \dots \\ 0 \\ \dots \end{pmatrix} \tag{3.15}$$

where the second and the fourth elements of the columns can take arbitrary values since the boundary conditions were imposed only on the functions $G^{\uparrow\uparrow}, G^{\downarrow\uparrow}$, not on its derivatives. The conditions of continuity at the interfaces require

$$\begin{aligned}
&\Phi_n^{(0)}(b_n)A_n^{(l)} = \Psi_n^{(0)}(b_n)R_n^{(l)}; \qquad \Phi_{n-1}^{(0)}(b_{n-1})A_{n-1}^{(l)} = \Psi_{n-1}^{(0)}(b_{n-1})R_{n-1}^{(l)};\\
&\Phi_{n-2}^{(0)}(b_{n-2})A_{n-2}^{(l)} = \Psi_{n-2}^{(0)}(b_{n-2})R_{n-2}^{(l)}.\\[1em]
&\Psi_n^{(0)}(c_{n+1})R_n^{(l)} = \Phi_{n+1}^{(0)}(c_{n+1})A_{n+1}^{(l)}; \quad \Psi_{n-1}^{(0)}(c_n)R_{n-1}^{(l)} = \Phi_n^{(0)}(c_n)A_n^{(l)};\\
&\Psi_{n-2}^{(0)}(c_{n-1})R_{n-2}^{(l)} = \Phi_{n-1}^{(0)}(c_{n-1})A_{n-1}^{(l)}.
\end{aligned} \tag{3.16}$$

Therefore we get $A_n^{(l)}$ and $R_n^{(l)}$, for instance,

$$\begin{aligned}
&A_n^{(l)} = \Phi_n^{(0)^{-1}}(c_n)\Psi_{n-1}^{(0)}(c_n)R_{n-1}^{(l)} =\\
&\Phi_n^{(0)^{-1}}(c_n)\Psi_{n-1}^{(0)}(c_n)\Psi_{n-1}^{(0)^{-1}}(b_{n-1})\Phi_{n-1}^{(0)}(b_{n-1})A_{n-1}^{(l)} =\\
&\Phi_n^{(0)^{-1}}(c_n)\Psi_{n-1}^{(0)}(c_n)\Psi_{n-1}^{(0)^{-1}}(b_{n-1})\Phi_{n-1}^{(0)}(b_{n-1})\times\\
&\Phi_{n-1}^{(0)^{-1}}(c_{n-1})\Psi_{n-2}^{(0)}(c_{n-1})\Psi_{n-2}^{(0)^{-1}}(b_{n-2})\Phi_{n-2}^{(0)}(b_{n-2})\ldots\times\\
&\Phi_1^{(0)^{-1}}(c_1)\Psi_0^{(0)}(c_1)\Psi_0^{(0)^{-1}}(b_0)\Phi_0^{(0)}(b_0)A_0^{(l)}.
\end{aligned} \tag{3.17}$$

Similarly the $A_n^{(r)}$ and $R_n^{(r)}$ can be determined.

The left and right boundary conditions can be provided by a corresponding choice of the columns $A_0^{(l)}$ and $R_N^{(r)}$. We define four linear independent solutions $F^{(l,1)}$, $F^{(l,2)}$, $F^{(r,1)}$, $F^{(r,2)}$ of the homogeneous equation in the following way: the solutions $F^{(l,1)}(z)$ and $F^{(l,2)}(z)$ are determined by the two columns

$$A_0^{(l,1)} = \begin{pmatrix} 1 \\ -1 \\ 0 \\ 0 \end{pmatrix}, \quad A_0^{(l,2)} = \begin{pmatrix} 0 \\ 0 \\ 1 \\ -1 \end{pmatrix}. \tag{3.18}$$

Similarly, $F^{(r,1)}(z)$ and $F^{(r,2)}(z)$ are determined by

$$R_N^{(r,1)} = \begin{pmatrix} e^{-ikc_{N+1}} \\ -e^{ikc_{N+1}} \\ 0 \\ 0 \end{pmatrix}, \quad R_N^{(r,2)} = \begin{pmatrix} 0 \\ 0 \\ e^{-ikc_{N+1}} \\ -e^{ikc_{N+1}} \end{pmatrix}; \tag{3.19}$$

(note that $F^{(l,1,2)}(0) = F^{(r,1,2)}(c_{N+1}) = 0$). Then we construct the 4×4 fundamental matrix F, whose columns are the $F^{(l,1)}$, $F^{(l,2)}$, $F^{(r,1)}$, $F^{(r,2)}$. The elements of the matrix F are the continuous functions of the variable for $0 \le z \le c_{N+1}$. The matrix F is non-singular, so the inverse matrix F^{-1} is defined for all $0 \le z \le c_{N+1}$. We search for the GF, i.e. a solution of Eq.(3.7) which obeys the zero boundary conditions in the form:

$$G(z, z') = F(z) \int_0^z F^{-1}(s) f(s, z')\, ds + F(z) h(z') \tag{3.20}$$

where column $f(s, z')$ is a right-hand side of Eq.(3.7), and $h(z')$ should be chosen such as to provide the boundary conditions for the GF $G(z, z')$. From Eq.(3.20) one can easily get

$$h(z') = -(2Ma_0/\hbar^2) \begin{pmatrix} F_{12}^{-1}(z') \\ F_{22}^{-1}(z') \\ 0 \\ 0 \end{pmatrix} \tag{3.21}$$

Then the elements of the GF $G^{\uparrow\uparrow}, G^{\downarrow\uparrow}$ are the first and the third elements of the column:

$$G(z > z') = (2Ma_0/\hbar^2) F(z) \begin{pmatrix} 0 \\ 0 \\ F^{-1}(z')_{32} \\ F^{-1}(z')_{42} \end{pmatrix}, \tag{3.22}$$

$$G(z < z') = -(2Ma_0/\hbar^2) F(z) \begin{pmatrix} F^{-1}(z')_{12} \\ F^{-1}(z')_{22} \\ 0 \\ 0 \end{pmatrix}. \tag{3.23}$$

If the right-hand side of Eq.(3.7) equals

$$(2Ma_0/\hbar^2) \begin{pmatrix} 0 \\ 0 \\ 0 \\ \delta(z - z') \end{pmatrix}, \tag{3.24}$$

we get the the $G^{\uparrow\downarrow}, G^{\downarrow\downarrow}$ as the first and the third elements of the column:

$$G(z > z') = (2Ma_0/\hbar^2) F(z) \begin{pmatrix} 0 \\ 0 \\ F^{-1}(z')_{34} \\ F^{-1}(z')_{44} \end{pmatrix}, \tag{3.25}$$

$$G(z < z') = -(2Ma_0/\hbar^2) F(z) \begin{pmatrix} F^{-1}(z')_{14} \\ F^{-1}(z')_{24} \\ 0 \\ 0 \end{pmatrix} \tag{3.26}$$

Thus, we found the solution of the Eq.(3.5). Then the conductivity for CIP geometry can be found as

$$\sigma(z,z') = \frac{\hbar e^2}{4\pi N_s a_0^2} \sum_{\kappa} \kappa^2 \mathrm{Tr}[G_\kappa(z,z') - G_\kappa^\dagger(z',z)][G_\kappa^\dagger(z,z') - G_\kappa(z',z)], \quad (3.27)$$

where N_s is the number of sites in the (x,y)-plane. To calculate an average current J the integration over z, z' should be performed. Then the resistivity for given angles is proportional to $1/J$.

3.3 Calculation of the Spin Angles

To apply the preceding consideration to a specific multilayer we need to calculate the magnetic configuration of a multilayer in an external magnetic field. In the phenomenological simulation model the external field H_e and the easy axis of the magnetization are supposed to be in the layer plane. The model is basically an extension of the Stoner-Wohlfarth's model [15] in the single-domain limit including biquadratic coupling and allowing both different values of the directions γ_i of the spins in each layer as well as of the coupling constants. This model is very realistic and shows e.g. different angles for the top and down layer, because these spins are more weaker coupled into the whole stack [21].

The energy of a multilayer structure, consisting of N magnetic layers, is given by the the anisotropy, the Zeeman and the linear and quadratic exchange coupling terms:

$$\begin{aligned} E_j = & \sum_{i=1}^{N} t_i [k_{ui} \sin^2(\gamma_i - \xi_{ij}) - H_e M_i \cos\gamma_i] \\ & + \sum_{i=1}^{N-1} [J_{li} \cos(\gamma_i - \gamma_{i+1}) + J_{qi} \cos^2(\gamma_i - \gamma_{i+1})], \end{aligned} \quad (3.28)$$

where M_i is the magnetization of the ith layer, j is the index of a subsequent loop, where the easy directions are averaged. J_{li} is the linear exchange constant and J_{qi} the quadratic term, k_{ui} is the uniaxial anisotropy constant, ξ_{ij} the easy direction, t_i the thickness of layer i. Because no hysteresis has been observed in (Cu/NiFe)-multilayers, the anisotropy term in the energy equation has been neglected. A system of equations results from this energy by derivation with respect to the unknown angles:

$$\begin{aligned} \frac{\partial E}{\partial \gamma_i} &= 0 = f_i = H_e t_i M_i \sin\gamma_i + A_i|_{i<N} - A_{i-1}|_{i>1} \\ A_i &= J_{li} \sin(\gamma_{i+1} - \gamma_i)[1 + 2J_{qi} \cos(\gamma_{i+1} - \gamma_i)] \end{aligned} \quad (3.29)$$

The i-th equation contains two sin-terms except for the two edge equations (i=1,N), since every layer i is coupled to the preceding $(i-1)$ and the following one $(i+1)$. The equations are strongly coupled since a positive term in equation i is followed by a negative contribution in the equation $i+1$. The nonlinear system of equations $\mathbf{f}(\boldsymbol{\gamma}) = 0$ was solved according to an iterative procedure.

$$\mathbf{f}' \Delta\boldsymbol{\gamma} = -\mathbf{f} \tag{3.30}$$

$$\boldsymbol{\gamma}_{j+1} = \boldsymbol{\gamma}_j + \omega\Delta\boldsymbol{\gamma};\ \epsilon = \sum_{i=1}^{N} |\Delta\gamma_i| > 10^{-7} \tag{3.31}$$

$$\boldsymbol{\gamma} = \begin{pmatrix} \gamma_1 \\ \gamma_2 \\ \vdots \\ \gamma_N \end{pmatrix};\ \mathbf{f} = \begin{pmatrix} f_1 \\ f_2 \\ \vdots \\ f_N \end{pmatrix};\ \mathbf{f}' = \begin{pmatrix} \frac{\partial f_1}{\partial\gamma_1} & \frac{\partial f_1}{\partial\gamma_2} & \frac{\partial f_1}{\partial\gamma_3} & \cdots & \frac{\partial f_1}{\partial\gamma_N} \\ \frac{\partial f_2}{\partial\gamma_1} & \frac{\partial f_2}{\partial\gamma_2} & \frac{\partial f_2}{\partial\gamma_3} & \cdots & \frac{\partial f_2}{\partial\gamma_N} \\ \vdots & \vdots & \vdots & \ddots & \vdots \\ \frac{\partial f_N}{\partial\gamma_1} & \frac{\partial f_N}{\partial\gamma_2} & \frac{\partial f_N}{\partial\gamma_3} & \cdots & \frac{\partial f_N}{\partial\gamma_N} \end{pmatrix}.$$

The Jacobian-matrix $\mathbf{f}'$ consists of N rows, each row at most of 3 elements. It has the special tridiagonal form with nonzero elements only on the diagonal plus or minus column. The two edge equations (i=1,N) have only two elements. The equations are solved with regard to the special tridiagonal form, using the Crout-algorithm. The big advantage compared to the Gauss-Jordan procedure earlier used is the fact, that the computer time is much smaller because it increases only linearly with the number of equations.

For every iteration step the system of Eqs.(5.5) for the $\Delta\gamma_i$ values is solved: 50 equations for 50 layers. After this the angles must be updated according to the Eqs.(5.6) with $\omega = 1$. In the first three steps the ω-value was reduced to 0.1 to avoid slipping in secondary minima. The iterative loop is continued, until the sum ϵ over the changes $|\Delta\gamma_i|$ of the absolute angle drops below a default limit in the corresponding iteration step.

A crucial point in the minimization of the energy according to Eq.(5.5) is the choice of the start values for $\gamma_i|_{i=1\ldots N}$. It turned out, that a simple analytic approach can provide start values for the angles which lead to a fast and reliable detection of the energy minimum. For every new field H_e the previous distribution for the spins was used as a new initial distribution $\boldsymbol{\gamma}_j$, which then was corrected by the method of largest descent (Eq.(5.8)).

$$\boldsymbol{\gamma}_{j+1} = \boldsymbol{\gamma}_j - \gamma\nabla E_j(\gamma);\ \nabla E_j = \mathbf{f}_j(\gamma) \tag{3.32}$$

This method is an iterative procedure too. It was stopped if the energy from one iteration step to the next declines no more but increases again. The value of γ was chosen in such a way that the program spent about 10 steps in this loop. The introduction of this step has considerably reduced the sensitivity of the program to the initial values.

The scheme which we use for calculation of the angles contains a set of parameters. To determine all the parameters correctly we need to compare phenomenologically calculated GMR curves with the experimental field dependence of GMR [21]. In the present work we use this numerical approach only to extract the field dependence of the angles γ_i.

The Downhill-Simplex method was used as an alternative method to minimize the energy (Eq.(5.1)) if the anisotropy terms become dominant. In this method the external field H_e is increased stepwise, starting in the saturation region or with the virgin curve at $H_e = 0$. At every new field H_e, the previously calculated angles γ_i were used as a new initial distribution, also being corrected by the method of largest descent (Eq.(5.8)). The GMR contributions for only one easy axis distribution ξ_{ij} are averaged in a subsequent step within a predefined region. It shall be noted here that this procedure gives only an estimate since spatial fluctuations of the anisotropy parameters do not enter into the minimization process. This simulation describes a physical situation in which the multilayer consists of independent grains without any inter-grain coupling. A simulation by lattice methods [3] is more flexible in adapting to the real experimental situation and may be necessary in the future.

In a previous paper [10] the method of largest descent (Eq.(5.8)) was used as the only algorithm to find the minimum, however, according to our experience, this method is much too simple.

3.4 Calculation of the GMR and Application to Some Measured Characteristics

We applied our theory to describe the GMR in magnetic multilayers of different structure, consisting of Py/Cu-sandwiches. The multilayers were prepared by dc magnetron sputtering in high vacuum [3]. The electrical resistance has been measured at 295 K using a four-probe geometry in a magnetic field.

Numerical calculations have been performed to explore the magnetic reversal process of each individual magnetic layer of these multilayers on the basis of an extended Stoner-Wohlfarth model in the single domain limit according to the consideration presented in Section 3.3. Thus, the field dependence of the angles between adjacent magnetization vectors is known and can directly be used as input for the calculation of the GMR versus field characteristics following the scheme presented in Section 3.2. First of all we reproduced the field-dependence of GMR in the "base" system, i.e. $\{Cu_{0.9nm}/Py_{1.6nm}\}_M$ at the first and $\{Cu_{1.8nm}/Py_{1.6nm}\}_M$ at the second antiferromagnetically coupling maximum. The comparison of the normalized experimental and theoretical GMR as function of magnetic field is presented at Fig. 3.2 (first maximum, M=4) and Fig. 3.3 (second maximum, M=4). In comparison to the experimental data of CML with M=1 the corresponding quantum statistical calculation is displayed in Fig. 3.4. The difference between the experimental and the calculated GMR-amplitude is only about 1%, and the field dependencies are in excellent agreement. The fact, that a minimum layer sequence of $[Py_{1.6nm}/Cu_{1.8nm}]_1$ $[Py_{1.6nm}/Cu_{0.9nm}]_1$ $Py_{1.6nm}$ is sufficient in these calculations to match the GMR characteristics of the whole $Py_{1.8nm}//\{[Cu_{1.8nm}/Py_{1.6nm}]_1/[Cu_{0.9nm}/Py_{1.6nm}]_1\}_{20}$ multilayer stack can be taken as an indication of the dominat-

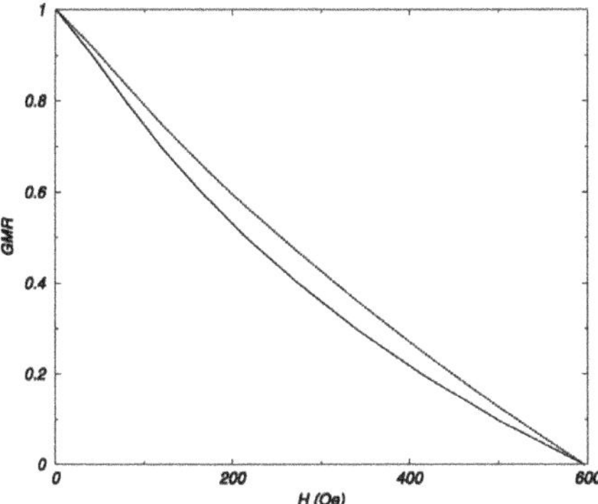

Fig. 3.2. Measured room temperature GMR vs field dependence of a $Py_{1.8nm}$//$\{Cu_{0.9nm}/Py_{1.6nm}\}_{40}$ multilayer at the first antiferromagnetic coupling maximum (solid line) and the corresponding quantum statistical calculation for $[Py_{1.6nm}/Cu_{0.9nm}]_4$ (dotted line). The transport parameters used are: $l^{\uparrow}_{Py}$=6.1 nm, $l^{\downarrow}_{Py}$=0.9 nm, $l_{Cu} = l^{\uparrow}_{Cu} = l^{\downarrow}_{Cu}$=33 nm and $k^{\uparrow}_{FPy} = k^{\downarrow}_{FPy} = k_{FCu}$ =0.1 nm^{-1}.

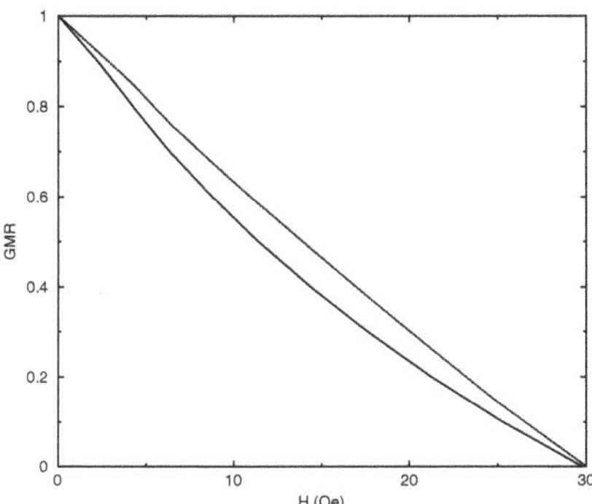

Fig. 3.3. Measured room temperature measured GMR vs field dependence of a $Py_{1.8nm}$//$\{Cu_{1.8nm}/Py_{1.6nm}\}_{40}$ multilayer at the second antiferromagnetic coupling maximum (solid line) and the corresponding quantum statistical calculation for $[Py_{1.6nm}/Cu_{1.8nm}]_4$ (dotted line). The transport parameters are the same as in Fig. 3.2.

ing nearest-neighbor coupling. At last, in Fig. 3.5 the measured and calculated field dependencies of the GMR for CML with M=2 are shown. We note that this good match between measured and calculated GMR characteristics was achieved using the same spin-dependent mean-free paths for the spin-up and spin-down electrons in Py in all systems.

So far, our treatment has been applied to the CIP geometry; the amplitude of the GMR and the shape of the curves depend on the layer thicknesses and their order in the multilayers. In the CPP geometry the order of the layers influences the GMR in particular if the mean free path is comparable to or exceeding the layer thicknesses [2,18], resulting in a break-down of the so-called resistor model [19]. We can try to estimate to what degree the contributions of each magnetic

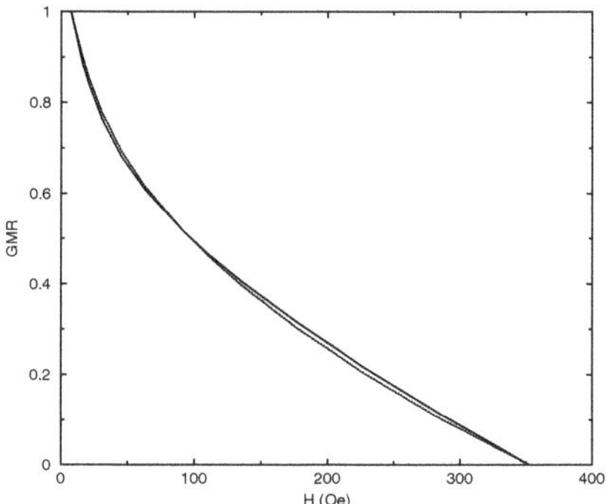

Fig. 3.4. Measured room temperature GMR vs field dependence of $Py_{1.8nm}//\{[Cu_{1.8nm}/Py_{1.6nm}]_1/[Cu_{0.9nm}/Py_{1.6nm}]_1\}_{20}$ (solid line) and the corresponding quantum statistical calculation for $[Py_{1.6nm}/Cu_{1.8nm}]_1[Py_{1.6nm}/Cu_{0.9nm}]_1$ $Py_{1.6nm}$ (dotted line). The transport parameters are the same as in Fig. 3.2.

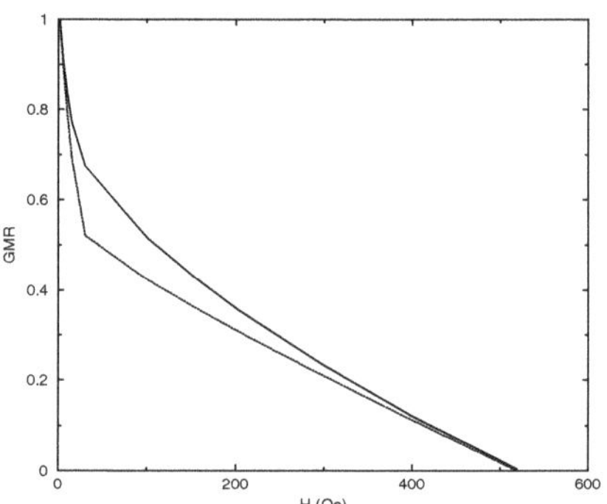

Fig. 3.5. Measured room temperature GMR characteristic of $Py_{1.8nm}//$ $\{[Cu_{1.8nm}/Py_{1.6nm}]_2/[Cu_{0.9nm}/Py_{1.6nm}]_2\}_{10}$ (solid line) and the corresponding quantum statistical calculation for $[Py_{1.6nm}/Cu_{1.8nm}]_2[Py_{1.6nm}/Cu_{0.9nm}]_2$ $Py_{1.6nm}$ (dotted line). The transport parameters are the same as in Fig. 3.2.

layer to the conductivity are independent of one another and if the conductivity (3.27) can be represented as a linear combination of separate contributions. Consider, for example, a multilayer containing five magnetic layers. We can guess the angles of the magnetization of the ferromagnetic layers from the magnetization vector of the first ferromagnetic layer. Assume that the conductivity of the multilayer can be represented as

$$\sigma(H) = \sigma_0 + \sigma_2(H) + \sigma_3(H) + \sigma_4(H) + \sigma_5(H) \tag{3.33}$$

where $\sigma_j(H)$, $j = 2, 3, 4, 5$, is the contribution from the $j - th$ magnetic layer, for a given magnetization direction of this layer with the directions of the magnetization of the other layers kept fixed. σ_0 is the field-independent contribution to the GMR (for instance, from the paramagnetic layers and the first magnetic

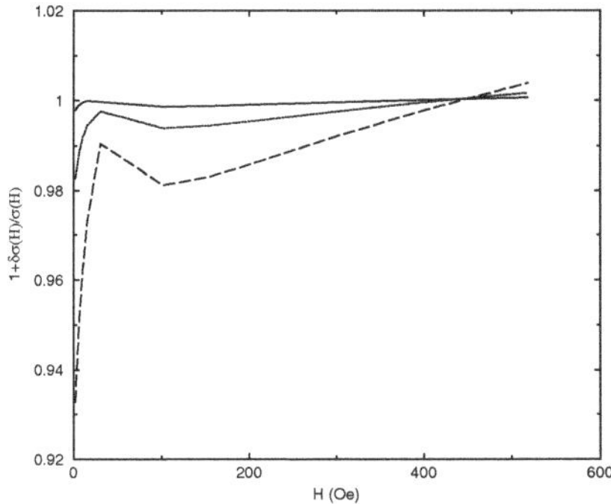

Fig. 3.6. $1 + \delta\sigma(H)/\sigma(H)$ for $[\mathrm{Py}_{1.6nm}/\mathrm{Cu}_{0.9nm}]_5$. The transport parameters are the following. $k_F = 1.0$ Å$^{(-1)}$ for all curves, $l^{\uparrow} = 7.0$ Å, $l^{\downarrow} = 40.0$ Å, $l^{para} = 80.0$ Å, solid line; $l^{\uparrow} = 15.0$ Å, $l^{\downarrow} = 80.0$ Å, $l^{para} = 160.0$ Å, dotted line; $l^{\uparrow} = 18$ Å, $l^{\downarrow} = 130.0$ Å, $l^{para} = 370.0$ Å, broken line.

layer). Therefore, we define the following functions:

$$\begin{aligned}
\sigma^{(2)}(H) &= \sigma_0 + \sigma_2(H) + \sigma_3(H_0) + \sigma_4(H_0) + \sigma_5(H_0) \\
\sigma^{(3)}(H) &= \sigma_0 + \sigma_2(H_0) + \sigma_3(H) + \sigma_4(H_0) + \sigma_5(H_0) \\
\sigma^{(4)}(H) &= \sigma_0 + \sigma_2(H_0) + \sigma_3(H_0) + \sigma_4(H) + \sigma_5(H_0) \\
\sigma^{(5)}(H) &= \sigma_0 + \sigma_2(H_0) + \sigma_3(H_0) + \sigma_4(H_0) + \sigma_5(H)
\end{aligned} \tag{3.34}$$

Here H_0 is some fixed value of the external magnetic field. The $\sigma_j(H)$ depend on the external magnetic field H only through the respective angle. They are equal to $\sigma(H)$ if $H = H_0$ and all angles except that for the j-th layer are fixed. If the linear representation (3.33) really holds, then

$$\sigma(H) = \sigma^{(2)}(H) + \sigma^{(3)}(H) + \sigma^{(4)}(H) + \sigma^{(5)}(H) - 3\sigma^{(j)}(H_0) \tag{3.35}$$

where j may be 2 or 3 or 4 or 5 since $\sigma^{(2)}(H_0) = \sigma^{(3)}(H_0) = \sigma^{(4)}(H_0) = \sigma^{(5)}(H_0)$. Therefore the value

$$\begin{aligned}
&\delta\sigma(H)/\sigma(H) = \\
&\Big(\sigma(H) - \sigma^{(2)}(H) - \sigma^{(3)}(H) - \sigma^{(4)}(H) - \sigma^{(5)}(H) + 3\sigma^{(j)}(H_0)\Big)/\sigma(H)
\end{aligned} \tag{3.36}$$

can be taken as a measure of the influence of the interfaces. We calculated the function $1 + \delta\sigma(H)/\sigma(H)$ for various values of mean free paths (Fig.3.6), and found visible deviations from linearity for long mean free paths and magnetization directions far from parallel.

3.5 Conclusion

The GMR can be observed in various systems. It is not surprising that systems with more complicated geometrical structure display a more interesting behavior [20,21]. We presented a quantum statistical treatment of the GMR in magnetic multilayers with arbitrary thicknesses of the layers and arbitrary angles between the magnetization vectors of the ferromagnetic layers. Combined with a numerical calculation of the respective angles, our approach demonstrates a good agreement with experimental GMR data even for complex multilayered structures. We also investigated for which mean free paths and angles between the magnetization vectors the contribution of the separate layers in the CIP GMR is not linear. This violation of linearity is analogous to a break-down of the resistor model in CPP geometry of GMR.

The following features can be incorporated into the proposed scheme. First of all the spin-dependent scattering potential on the interfaces can be taken into account. In that case the condition of the continuity (3.16) should be replaced by the corresponding jump of the functions $F^{(l)}, F^{(r)}$ on the interfaces. The constructed GF can be used also for consideration of the GMR in the CPP geometry. The well-known problem of vertex correction in the CPP geometry of the GMR can be treated with the help of effective fields ([7,13]). Also the tunneling magnetoresistance can be considered in the same manner with necessary changes in parameters of Eq.(3.5).

Acknowledgments

The work has been supported by the Deutsche Forschungsgemeinschaft in the "Forschergruppe Nanometer-Schichtsysteme". A. V. Vedyayev acknowledges the CENG DRMC SP2M and Bielefeld University for hospitality and the Russian Foundation of Fundamental Research for financial support. M. Ye. Zhuravlev is grateful to Bielefeld University for hospitality.

References

1. Special issue of J. of Magn. Magn. Mater. **200**, No. 1-3 (1999)
2. W.-C. Chiang, Q. Yang, W. P. Pratt Jr., R. Loloee, J. Bass, J. Appl. Phys. **81**, 4570 (1997)
3. S. Heitmann, A. Hütten, T. Hempel, W. Schepper, G. Reiss, J. Appl. Phys. **87**, 4849 (2000)
4. S. Heitmann, A. Hütten, T. Hempel, W. Schepper, G. Reiss, J. Magn. Magn. Mater. **226-230**, 1752 (2001)
5. B. Yu. Yavorsky, I. Mertig, A. Ya. Perlov, A. N. Yaresko, V. N. Antonov, Phys. Rev. B **62**, 9586 (2000)
6. P. M. Levy, S. Zhang and A. Fert, Phys. Rev. Lett. **65**, 1643 (1990); H. E. Camblong, P. M. Levy and S. Zhang, Phys. Rev. B **51**, 16052 (1995);
7. P. M. Levy, H. E. Camblong, and S. Zhang, J. Appl. Phys., **75**, 7076 (1994)
8. A. Vedyayev, B. Dieny, N. Ryzhanova, Europhys. Lett. **19**, 329 (1992); A. Vedyayev, C. Cowache, N. Ryzhanova and B. Dieny, J. Phys.: Condens. Matter **5**, 8289 (1993)
9. M. B. Stearns, J. of Magn. Magn. Mater. **5**, 167 (1977)
10. R. Perez-Alvarez, F. Garcia-Moliner, V. R. Velasco, J. Phys.: Condens. Matter **7**, 2037 (1995)

11. F. Brouers, A. Vedyayev, M. Giorgino, Phys. Pev. B **7**, 380 (1973)
12. A. Vedyayev, B. Dieny, N. Ryzhanova, J. B. Genin, C. Cowache, Europhys. Lett. **25**, 465 (1994);
13. A. Vedyayev, N. Ryzhanova, B. Dieny, P. Dauguet, P. Gandit, J. Chaussy, Phys. Pev. B **55**, 3728 (1997)
14. E.A. Coddington, N. Levinson: *Theory of ordinary differential equations*, (McGraw-Hill Book Company, New York-London-Toroanto 1955).
15. E.C. Stoner, E.P. Wohlfarth, Phil. Trans. Roy. Soc. **A240**, 599 (1948)
16. W. Schepper, A. Hütten, G. Reiss, J. Appl. Phys. **88**, 993 (2000)
17. H. Holloway, D.J. Kubinski, J. Appl. Phys., **83**, 2705 (1998)
18. D. Bozec, M. J. Walker, B. J. Hickey, M. A. Howson, N. Wiser, Phys. Rev. B **60** 3037 (1999)
19. S. Sanvito, C. J. Lambert, J. H. Jefferson, Phys. Rev. B **61**, 14225 (2000)
20. M. Ye. Zhuravlev, H. O. Lutz, A. V. Vedyayev, Phys. Rev. B **63**, 174409 (2001)
21. M. Ye. Zhuravlev, H. O. Lutz, A. V. Vedyayev, J. Phys. A **34**, 8383 (2001)

4 Interface Effects in Tunneling Magnetoresistive Systems

Branko Stahl[1,2], Horst Hahn[1], Holger Schmitt[1],
Mohammad Ghafari[1], and Sarbari Bhattacharya[1]

[1] Fachbereich Material- und Geowissenschaften,
Petersenstrasse 23, TU Darmstadt, 64287 Darmstadt, Germany
[2] Institut für Nanotechnologie, Forschungszentrum Karlsruhe,
76021 Karlsruhe, Germany

Abstract. In tunneling magneto resistance (TMR) devices the electronic and magnetic properties of the interfaces between the ferromagnetic metals and the insulating spacer layer influence the effective scattering potential for the conduction electrons. As this is especially relevant for structural and magnetic defects and inhomogeneities in the interface region, a non-destructive chemical, magnetic and electronic characterization with a sub-monolayer sensitivity is of importance.

In the present work, Mössbauer spectroscopy as a nuclear probe method was used to study the oxidation of the Fe-Al interface in an ultra high vacuum (UHV) system under successive oxidation steps. The formation of Fe-Al spinel and Fe oxide phases as well as a smoothening of the metal-insulator interface after annealing at around 230°C are identified. After a first pronounced reaction, the oxidation is observed to occur in constant amounts though the partial pressure and the time intervals varied significantly in the successive oxidation steps.

4.1 Interface Properties in Tunneling Magnetoresistance Devices

Tunneling magnetoresistance devices (TMR) are members of the family of magnetoresistive systems that are developed for non-volatile memory chips, spin electronic components and sensor applications like read heads for hard disks [1]. In general, the magnetoresistance of a trilayer consisting of two metallic ferromagnets with a separating insulating layer depends on the relative orientation of the magnetization vectors of the metallic layers. This phenomenon arises from the spin-polarization in the ferromagnetic metals and the spin conservation in the tunneling process through the electric insulator [2,3]. The principles of the electron transport in a trilayer arrangement were first studied by Juliere [4] a decade before the discovery of the now popular giant magnetoresistance effect (GMR) in purely metallic systems [5,6]. However, a breakthrough in the preparation of TMR layers with reproducible large resistance changes only came in 1995 with junctions consisting of CoFe/Al_2O_3/Co, stimulating the present activity in this field [7].

In the hypothetical case of a perfect metal insulator interface, i.e. epitaxial growth of the TMR trilayer system, the tunneling current would essentially be determined by the electronic band structure of the materials involved. Already in this case the crystallographic orientation of the ferromagnet with respect to the interface significantly influences the observable magnetoresistance effect [8]. In realistic multilayers the structural, electronic and magnetic imperfections lead to additional scattering of the conduction electrons. This accounts for the variety of TMR results in seemingly identical devices. It is important to note that minor changes in stoichiometry, crystallography and magnetic ordering on a sub-monolayer scale may already drastically influence the magnetoresistive properties [9]. For instance, the conservation of spin in the tunneling process may longer hold. In addition, the electronic states at the interface are modified by the local structure, thus giving rise to a lateral variation of the effective tunneling barrier for the TMR process. This is most obvious at unwanted pinholes in the nominally insulating system. The complexity of the problem is increased by the solid state chemistry at the interface during the synthesis and operation of the devices and the modification of the electronic band structure due to the interface. All these aspects are non-trivial and yet resist a detailed experimental as well as theoretical analysis.

Experimentally, the challenge is to find methods which allow a collection of detailed information of a buried interface while providing a sensitivity in the sub-monolayer range. Neutron diffraction as the work horse in structural and magnetic characterization lacks the necessary sensitivity for the given systems. Nuclear probe methods fortunately fulfill the above requirements of sensitivity at buried interfaces with atomic resolution [10]. In this work we will focus on Mössbauer spectroscopy although other methods like nuclear magnetic resonance (NMR) can provide similar or complementary information. The aim is not to give a complete survey of the efforts to characterize the interface properties in TMR devices. We will rather concentrate on the potential of nuclear methods such as the highly surface sensitive depth selective Mössbauer spectroscopy to demonstrate the importance of unconventional approaches for the study complex systems.

4.2 Depth Selective Mössbauer Spectroscopy

The principal idea of the non-destructive depth selective Mössbauer technique is simple and has first been applied by Bonchev using the isotope ^{119}Sn [11]. Due to the experimental and computational efforts necessary to guarantee an acceptable sensitivity and depth resolution this method found its way to a wider range of applications only recently. Major contributions are related to fields like the phase characterization after ion implantation [12,13], surface chemistry and corrosion studies [14,15] and near-surface magnetism [16–18]. The theoretical basis for the quantification of the depth profile was mainly provided by Liljequist [19,20].

In a standard Mössbauer experiment the re-absorption of γ-radiation in a second nucleus of the same type is studied as a function of the relative velocity of source and absorber. The Doppler modulation by the relative movement is necessary to compensate for the deviating energy shifts of the nuclear levels in both nuclei due to the different hyperfine interaction in the different atomic environments. Only an overlap of the resonance lines within the line width of the nuclear transition leads to a detectable absorption. Due to the high energy resolution that is typically smaller by several orders of magnitude with respect to the energy transfer to an isolated atom by the γ-photon, the chemical shift IS, electric quadrupole splitting QS and magnetic splitting B_{hf} are measurable. The binding to a crystal lattice prevents the energy transfer to the Mössbauer atom to a certain degree that is given as the recoil free fraction in form of the temperature dependent Debye-Waller factor. IS is directly correlated with the chemical state of the Mössbauer atom, QS with the electric field gradient due to the local charge distribution at the site of the nucleus in the lattice and B_{hf} gives information on the atomic magnetic moment. All three physical quantities provide valuable information to characterize chemical or magnetic phases. The sensitivity to thin films, small particles, defects and interfaces results from the fact that these quantities are probed directly at the Mössbauer nucleus. Therefore, the nuclear probe has to be placed at the appropriate sites of interest in the sample.

To acquire a certain degree of depth information the resonant absorption of the γ-radiation should not be observed directly by a detector behind the sample but instead by the detection of the subsequently emitted conversion and Auger electrons in backscattering geometry (Fig. 4.1). These electrons which occur in conjunction with the resonant absorption process in the sample have a finite range in the solid. Due to elastic and inelastic scattering the probability to leave the sample with a certain energy loss with respect to the initial kinetic energy depends on the depth of the Mössbauer absorption event. By discriminating on different electron energies when recording the Mössbauer spectra different depth weight functions are realized. A de-convolution of the experimental data by these depth weight functions provided by theoretical Monte Carlo simulations leads to the depth profile of the Mössbauer absorption, i.e. of the chemical and magnetic phases. Considering the limitations due to source strength and source-absorber geometry, an electron spectrometer with a highest possible transmission for a given relative energy resolution of 1 to 2 % must be used in these experiments. Additionally, the experiments have to be performed in ultra high vacuum to prevent adsorbates on thin films, especially at low temperatures. The most powerful

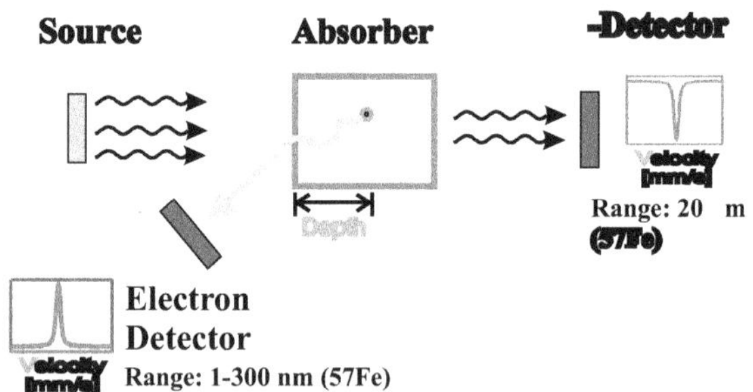

Fig. 4.1. Schematic sketch for the detection of γ-radiation and conversion electrons in a Mössbauer analysis.

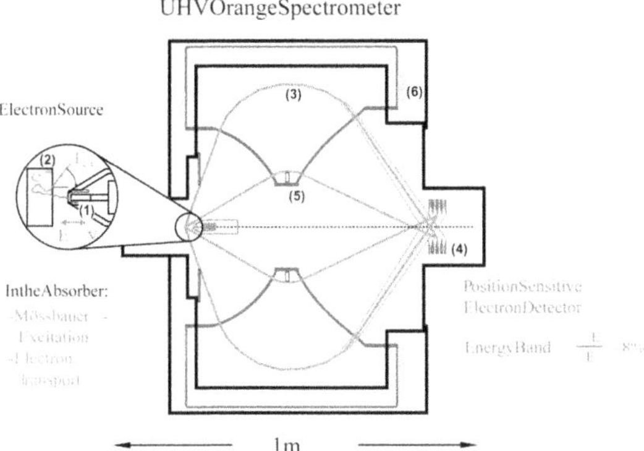

Fig. 4.2. Orange type magnetic electron spectrometer operating in ultra high vacuum. (1) Source, (2) absorber, (3) electron trajectory, (4) detector, (5) magnet coil, (6) water cooling.

instrument fulfilling these requirements is the orange type magnetic spectrometer [21] (Fig. 4.2). Depth selective Mössbauer experiments with ^{57}Fe have been performed on enriched as well as natural (2.2 % ^{57}Fe) samples with this instrument in a temperature range of 10 to 470 K.

4.3 Interface Magnetism

The magnetism at interfaces and surfaces is strongly modified compared to the bulk properties of the same material by the local change in electronic band structure [22–24]. This can influence the lattice parameters or even the crystal structure. The break of the translation symmetry may lead to a perpendicular magnetic anisotropy and to characteristic Friedel oscillations of the electronic wave functions. The atomic moment and charge density will vary as a function of distance to the interface. All three hyperfine parameters IS, QS, B_{hf} in a Mössbauer experiment are affected by these modifications. Using the two isotopes ^{56}Fe (no resonance effect) and ^{57}Fe in different atomic layers of a thin film structure allows to probe the hyperfine interaction at selected sites while other locations are suppressed. The chemical and magnetic properties are the same for both isotopes.

4.4 Solid State Reaction at the Fe-Al Interface under the Presence of Oxygen

As an example to demonstrate the potential of nuclear probe methods as a tool for a sample characterization on the nanometer scale, the oxidation of an Al/Fe/Al nanostructured trilayer is given in the following. Mössbauer spectroscopy as a nuclear probe method was used to study the oxidation of the top

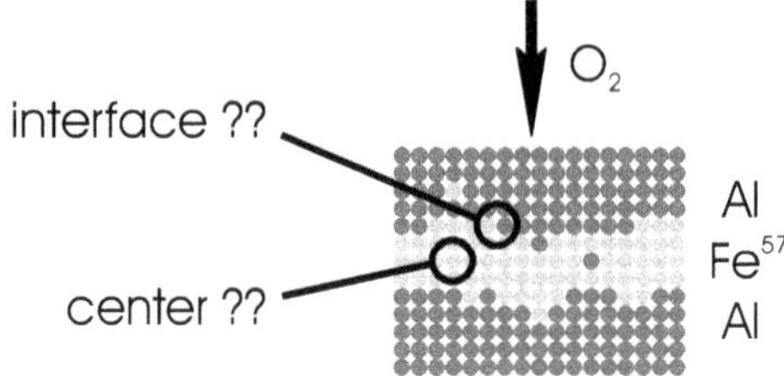

Fig. 4.3. To insure a highly sensitive sample characterization the iron film in the Al/Fe(1nm)/Al trilayer was enriched in the Mössbauer isotope ^{57}Fe to 95 %.

Fe-Al interface in an ultra high vacuum (UHV) system under successive oxidation steps (Fig. 4.3).

The functionality of tunneling magneto resistance structures depends on the structural, electronic and magnetic properties of a few atomic layers that are given by an electric insulator and the adjacent ferromagnetic metals. Usually, this layered structure is synthesized by a deposition of the first ferromagnetic layer on a smooth substrate that may previously be coated by a multilayer of magnetic and non-magnetic materials. The covering electric insulator can be produced by a deposition of a metal film that is oxidized in pure oxygen or even air with the partial pressure, the temperature, the duration and number of repeated steps as parameters. This solid state reaction may be assisted by various kinds of radiation, for instance ultra violet photons. Finally, the second ferromagnetic layer is deposited onto the insulator. This kind of synthesis combines relative simplicity with rather successful device properties. Therefore, the majority of fundamental and applied research activities still follow this route.

Mainly for applications, a clear correlation of the oxidation parameters and the magnetoresistive device properties is of great interest. Such a simple relation is hampered by the thinness of the insulating layer and the numerous influences like interface roughness, phase formation and separation, magnetic ordering, number and size of defect structures. Not surprisingly, at present the number of recipes to produce TMR structures is still increasing.

Sato et al. [25] studied the consequences of annealing on UHV prepared TMR junctions of the type Ni-Fe/Co/Al_2O_3/Co/Ni-Fe/Fe-Mn/Ni-Fe. By annealing steps for one hour up to a temperature of 300°C the magnetoresistance change could be improved from 15 to 23.5 % (Fig. 4.4, sample A). This is paralleled by a sharpening of the layer magnetization reversal as a function of applied field and an increase in the tunneling barrier height (Fig. 4.5). Above 300°C the TMR effect drops practically to zero.

Cardoso et al. [26] undertook annealing studies on $Co_{82}Fe_{18}/Al_2O_3/Co_{82}Fe_{18}$ structures and observed an increase in TMR effect from 27 to 39 % after annealing the sample at 290°C. Also in this case the effective barrier height increased from 2.15 eV to 2.43 eV accompanied by an effective thickness decrease of the insulation layer from 1.03 nm to 0.9 nm. Similar results are obtained by Sousa et al. [27]. Parkin [28] reports on TMR effects in a NiFe/Al_2O_3/NiFe structure as high as 40 % after heat treatment.

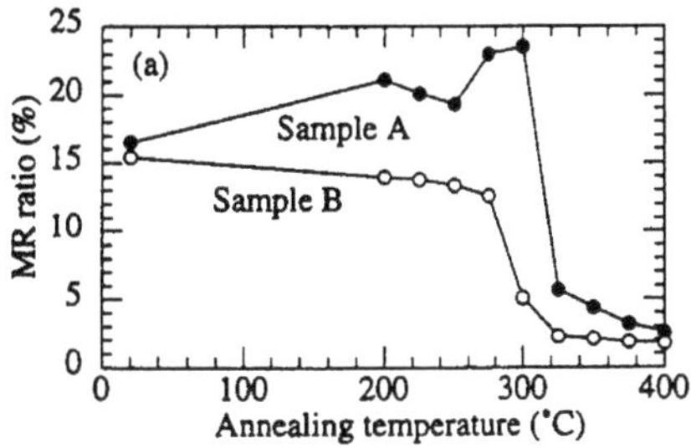

Fig. 4.4. Dependence of the magnetoresistance ratios on annealing temperature (courtesy ref. [25]).

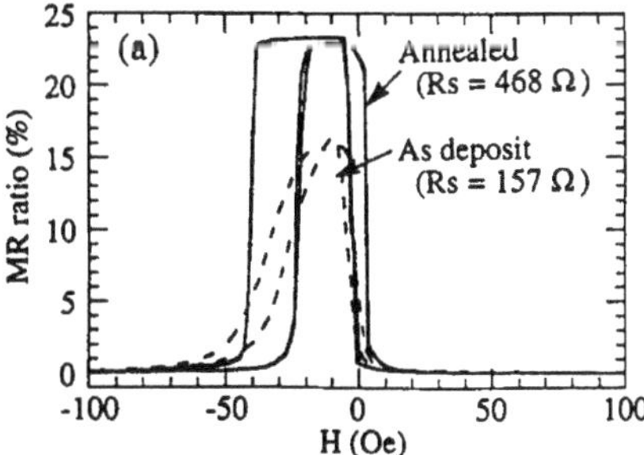

Fig. 4.5. Magnetoresistance curves before and after annealing at 275°C (courtesy ref. [25]).

In a study by Clark et al. [29] the structure of the Al_2O_3 barrier layer between Co and NiFe layers was measured by high resolution transmission electron microscopy (HRTEM). They discuss the change in TMR effect with respect to the thickness and roughness of the insulator. They observe a failure of the TMR device for the longest oxidation times.

What is needed at first hand to clarify the origin of the magnetoresistive properties is to non-destructively characterize the underlying structural and magnetic phase composition at the metal-insulator interface with high sensitivity for even small amounts in the sub-monolayer range. In this work we follow this aim by using a highly sensitive and depth selective Mössbauer technique that gives access to the desired properties via the hyperfine interaction of ^{57}Fe probe nuclei. For this purpose we study the Fe-Al interface as it is a most relevant model system for TMR structures. For the sake of rigor in the interpretation of the experimental data, in these first experiments a most simple and highly symmetric trilayer structure was used: (1) First, a 5 nm Al layer was evaporated on a clean oxidized Si substrate, (2) second, the ^{57}Fe layer and (3) third, the 2 nm (0.054 g/cm^2) top Al layer were deposited. The synthesis of complete TMR structures with a direct comparison of the magnetoresistive and mircostructural properties will be topic of future work. The Mössbauer experiments are done in the same UHV system as the preparation of the sample by mini-electron-beam evaporators.

Two samples with different thicknesses of the ^{57}Fe layer, namely 0.08 g/cm^2 $\approx$ 1.0 nm (sample I) and 0.36 g/cm^2 $\approx$ 4.5 nm (sample II), were prepared in UHV ($< 10^{-6}$ mbar during evaporation). The thinness of the first sample was chosen to give a fully (super-) paramagnetic signal in the Mössbauer analysis at room temperature while it guarantees a magnetically ordered state at low temperatures. This improves the phase analysis significantly as more constraints are present. In addition, the number of sub-spectra is reduced according to the lower number of different types of Fe sites. On the other hand, the thicker sample is of stronger relevance for TMR systems but one has to cope with a relatively weak fraction of the top interface in the whole resonant signal. Sample I was prepared at 200°C, sample II at room temperature. Mössbauer spectra were recorded directly after synthesis in a residual pressure of less than 10^{-9} mbar.

Fig. 4.6 shows the Mössbauer spectrum of sample I at 300 K in the as prepared state. The fully symmetric signal reflects the highly symmetrical structure of the trilayer system. Both Fe-Al interfaces seem to yield the same hyperfine interaction. The spectrum reveals a completely paramagnetic state of the iron probe atoms on a time scale of about 10^{-8} s. The line width is indicative of a distribution of isomer shifts (IS) and electric quadrupole splittings (QS) due to different Fe sites and Fe and Al co-ordinations in the layer. The value of the average isomer shift (relative to α-Fe) IS = 0.127(2) mm/s is characteristic for an Fe film between two Al layers. Fe-Al alloying or a diffusion of Fe into Al can be excluded as these known phases exhibit significantly larger isomer shift values and differ also in the other hyperfine parameters (Table 4.1). The electric field gradient results from the non-cubic symmetry at the Fe-Al interface. In addition, a tetragonal distortion of the Fe lattice cannot be excluded.

Sample I was oxidized in successive steps as follows: (1) 30 s pure O_2 , (2) 60 s O_2 , (3) 300 s O_2 , (4) 120 s O_2 , (5) 1800 s O_2 , (6) 1.5 h in air, (7) several times brought into air followed by evacuation. After each step 1-6 the sample

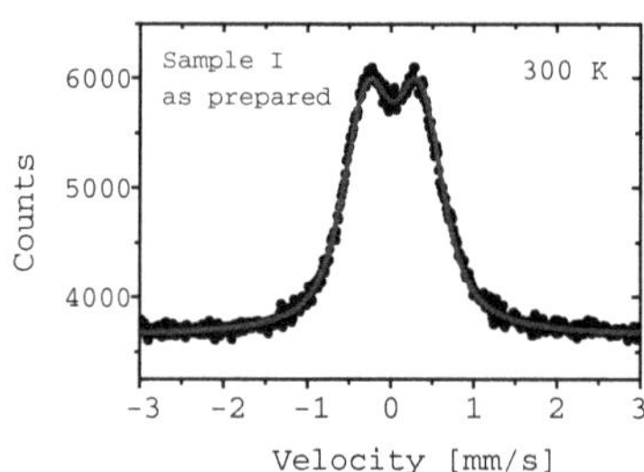

Fig. 4.6. Mössbauer spectrum of the as prepared sample I at room temperature.

Table 4.1. Hyperfine values of the as prepared sample I.

	line width [mm/s]	IS [mm/s]	QS [mm/s]	
sample I, 300 K	0.59(3)	0.127(2)	0.612(4)	as prepared

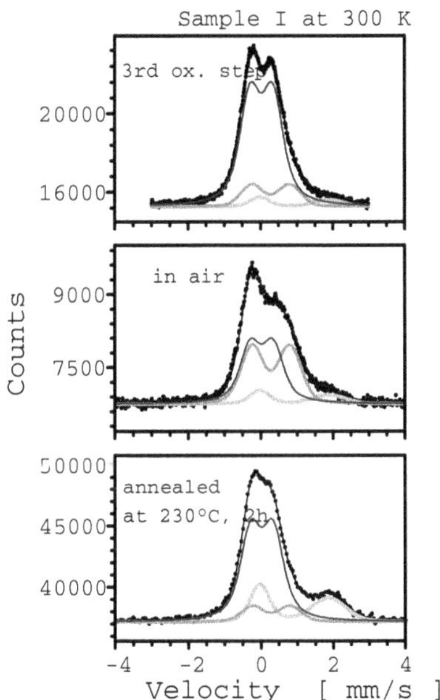

Fig. 4.7. Three Mössbauer spectra of sample I at room temperature at various stages of the oxidation and heat treatment.

Table 4.2. Hyperfine values of sample I after oxidation.

	line width [mm/s]	IS [mm/s]	QS [mm/s]	
component I	0.46(1)	0.374(3)	0.991(4)	interface zone
component II	0.20(1)	0.994(4)	1.84(1)	Fe-Al spinel

was evacuated to $< 10^{-9}$ mbar. Finally, the sample was heat treated in UHV ($< 6 \cdot 10^{-7}$ mbar) at 230°C and 330°C. Fig. 4.7 shows typical Mössbauer spectra of this series. Due to the presence of oxygen two additional components evolve in the paramagnetic spectra at room temperature. They are characterized by the hyperfine parameters in Table 4.2. The asymmetric line shape of component II is due to a correlation of the IS and QS distribution which is expected for a distribution of Fe sites at the Fe-Al-oxide interface zone.

The 9 Mössbauer spectra at room temperature of successive oxidation steps can be simultaneously fitted by the above mentioned three components. The oxidation of the Fe-Al interface follows a well defined route that increases the volume of the new phases with successive oxidation steps. The evolution of the relative fractions are plotted in Fig. 4.8. Surprisingly, these fractions do not develop in the same unsteady way as the oxygen treatment would suggest. Besides a fast initial oxidation of the top Al layer down to the Fe-Al interface, showing up in a first leap in Fig. 4.8, the increase of Fe sites with oxygen co-ordination follows a linear behavior as a function of oxygen cycling (O_2 inlet and evacuation

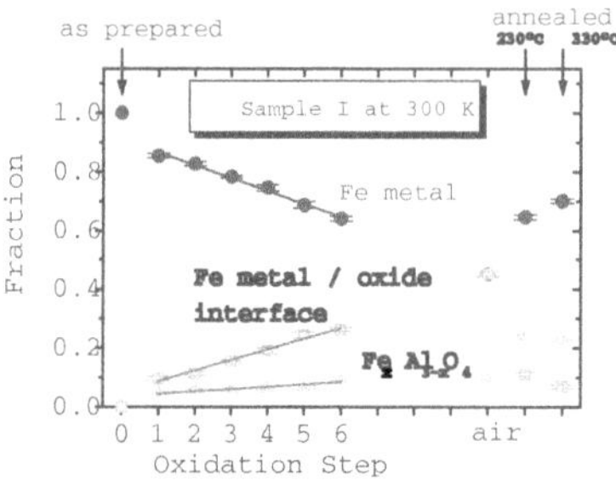

Fig. 4.8. Evolution of the relative fractions of the hyperfine components in Fig.4.7 as a function of oxidation cycles and annealing. Circles represent metallic Fe, triangles, the formation of an Fe-Al spinel phases in the top Al layer and squares, the presence of a distorted interface zone in between.

to 10^{-9} mbar). In this process the time interval, where oxygen is present, seems to play no crucial role. The same seems to hold for the O_2 partial pressure which was not precisely controlled in this experiment. Instead of a crudely increasing amount of oxygen containing Fe compounds the rather smooth behavior suggests some kind of passivation process that accompanies the oxygen cycles and that stops a further diffusion of O into the Fe layer. As the repeated evacuation of the sample area is crucial to this observation, the dissociation of O_2 on the surface of the sample might be hindered due to a built up of an adsorbate. In addition, structural changes and charge polarization in the top Al oxide layer due to the oxygen load may stop the oxygen diffusion as is the case for a free Al surface that gets passivated by an amorphous oxide coverage of 4 nm [30,31].

The annealing of the sample that followed after the sample was brought to air for several times (not precisely counted), reveals the metastable nature of the Fe-Al-O system at this stage. Component II of Table 4.2 increases significantly (Fig. 4.8). This is accompanied by a similar increase of the component that describes the as prepared metallic sample (Table 4.1). A further annealing at a higher temperature only weakly changes these observed fractions. The reason for the recover of the metallic Fe phase in the presence of the formation of a well defined Fe-Al-O phase lies in the separation of both constituents at the interface during annealing. In parallel, structural defects are eliminated. This also leads to a smoothening of the Fe-insulator interface as is reflected in a simplification of the Mössbauer spectrum, i.e. a nearly vanishing component I (Table 4.2). This last component is indicative for a distorted zone between the metallic Fe and the nearly stoichiometric Fe-Al-O phase.

According to ref. [32] (IS=1.0 mm/s, QS=1.59 mm/s) component II in Table 4.2 can be identified with the $Fe_xAl_{3-x}O_4$ spinel phase with x close to 1. This compound contains Fe in the 2+ state and is non-magnetic at room temperature. Component I in Table 4.2 is typical for the Fe^{3+} state and occurs as an intermediate phase to the formation of $FeAl_2O_4$, which is consistent with the preferential substitution of Fe^{3+} by Al. Its broad lines reveal a distribution of isomer shifts and quadrupole splittings. This is expected for a statistical distribution in the

Al and O content within an interface zone between the Fe metal and the Fe-Al spinel. This constitutes in effect a miscibility of the spinel $FeAl_2O_4$ with magnetite Fe_3O_4 and leads to an average sum formula $Fe_xAl_{3-x}O_4$ $(1 < x < 3)$ for this region.

This picture could be further clarified by Mössbauer measurements at low temperatures. Fig. 4.9 gives three typical spectra at different stages of the sample treatment. Again, the whole set of low temperature spectra can be described simultaneously with the indicated components. The correlations of the hyperfine parameters are plotted in Fig. 4.10 (closed symbols) together with those of the paramagnetic phases at room temperature. Numerical values are given in Table 4.3. The signal of the paramagnetic Fe-Al spinel component actually shows the typically strong temperature dependence of the electric quadrupole splitting which takes the value QS $= 2.68(4)$ mm/s at 10 K. This is depicted as an additional data point in the upper right part of Fig. 4.10. Fig. 4.11 shows the evolution of the absolute count rates of the low temperature components in addition to Fig. 4.8 (which shows the relative fractions).

Comparing the hyperfine parameters of Fig. 4.9 and the evolution of the relative and absolute intensities in Figs. 4.8 and 4.11, a correlation of the room and low temperature components can be given. The paramagnetic component that describes the as prepared sample corresponds well to the sextet S1 (Table 4.3) with an average hyperfine field of $\langle B_{hf} \rangle = 26.3(2)$ T and a sigma value for

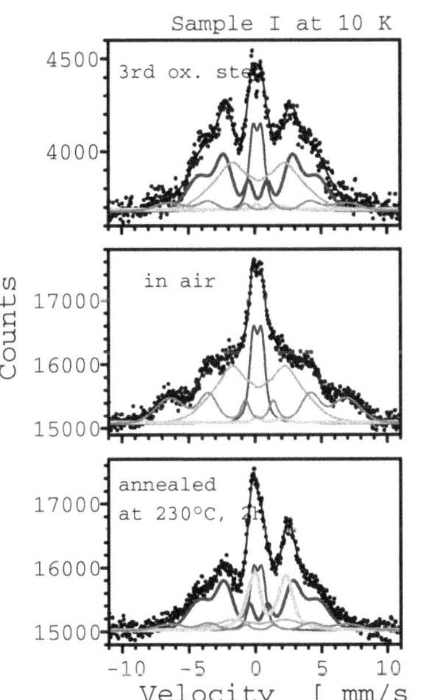

Fig. 4.9. Mössbauer spectra of sample I recorded at 10 K. They correspond to the spectra in Fig. 4.7 and show the complexity in magnetic ordering in the Fe-Al-O thin film system.

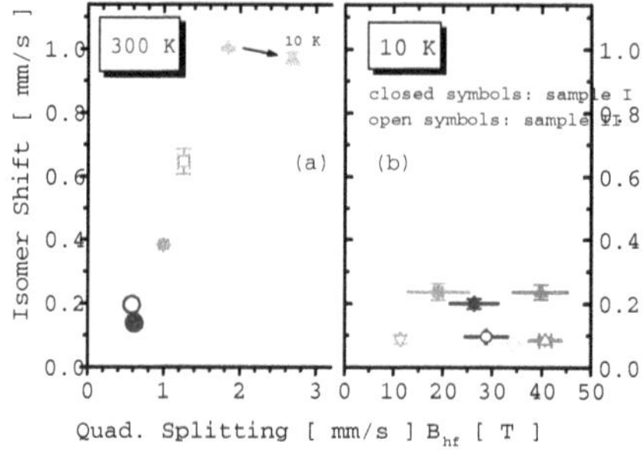

Fig. 4.10. Correlation of isomer shift and (a) electric quadrupole splitting, (b) magnetic hyperfine field. The 10 K values are corrected for a second order Doppler shift of 0.145(8) mm/s. Closed symbols: sample I, open symbols: sample II. The horizontal lines through the symbols in part (b) represent the sigma value of the gaussian hyperfine field distribution.

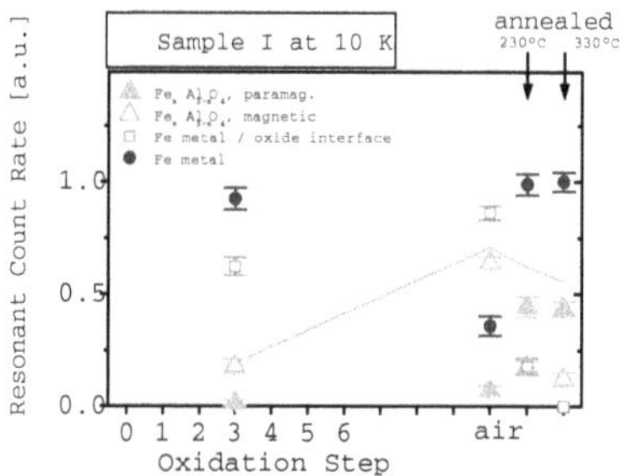

Fig. 4.11. Evolution of the absolute count rates of the hyperfine components in Fig. 4.9 as a function of oxidation cycles and annealing. Circles represent metallic Fe, triangles the formation of Fe-Al spinel phases in the top Al layer and squares the interface zone. The solid line gives the sum of the spinel phases.

Table 4.3. Hyperfine values of sample I after annealing.

	line width [mm/s]	IS [mm/s]	$\langle B_{hf} \rangle$ [T]	$\sigma(B_{hf})$ [T]	
S1	0.555(7)	0.19(2)	26.3(2)	4.9(2)	Fe metal
S2	0.62(2)	0.23(2)	39.8(3)	5.5(2)	Fe-Al spinel
S3	1.08(2)	0.23(3)	19.0(2)	6.2(3)	interface zone

the distribution of 4.9 T. These hyperfine parameters differ from those of bulk α-Fe because of the very thin structure of the trilayer in sample I, i.e. the influence of the top and bottom interfaces. The electronic band structure is modified by these adjacent layers. Part of the Fe metal signal is still paramagnetic at 10 K as can be seen in Fig. 4.9. The paramagnetic fraction stays constant through all steps of sample treatment. This indicates that it belongs to the bottom Fe-Al interface where -considering a certain degree of roughness- the Fe magnetic moment is suppressed for Fe atoms with predominant Al nearest neighbors.

The picture for the Fe-Al spinel phase is more complicated as a paramagnetic as well as magnetically ordered phase is present at low temperatures. This is due to a variability in Al content and structural defects. This is obvious when looking at the decrease of the magnetic fraction S2 (Table 4.3) that is nearly compensated by the increase of component II (Table 4.2) during annealing of the sample (Fig. 4.11). The $FeAl_2O_4$ phase is superparamagnetic at 10 K due to its nanocrystalline character. Magnetic ordering appears in regions with a higher Fe content and higher magnetocrystalline anisotropy. Nevertheless, magnetic ordering takes place only at very low temperatures as is depicted in Figs. 4.12 and 4.13 for sample I after the third and seventh oxidation state. Judging from the differences still present between the spectra at 10 and 20 K superparamagnetic relaxation cannot entirely be excluded even at 10 K. However, the size of the hyperfine field of component S2 characterizes this phase as ionic in nature. The average isomer shift is lower than in bulk Fe^{2+} and Fe^{3+} components, albeit the line with of 0.62 mm/s includes isomer shift values that are correspondingly larger. The reason lies in the extension of the electronic wave function of the metal layer into the first monolayer of the insulator where the small amount of ionic Fe is found. The increase in 4s electron density decreases the isomer shift value as well as the magnetic hyperfine field at these nuclei. In Fig. 4.11 the sum of the magnetic and paramagnetic ionic Fe signal is plotted as a line.

This interpretation is further supported by the intensity evolution of the third magnetic component S3. According to Figs. 4.8 and 4.11 it follows the same behavior as component I of Table 4.2. These components represent mainly

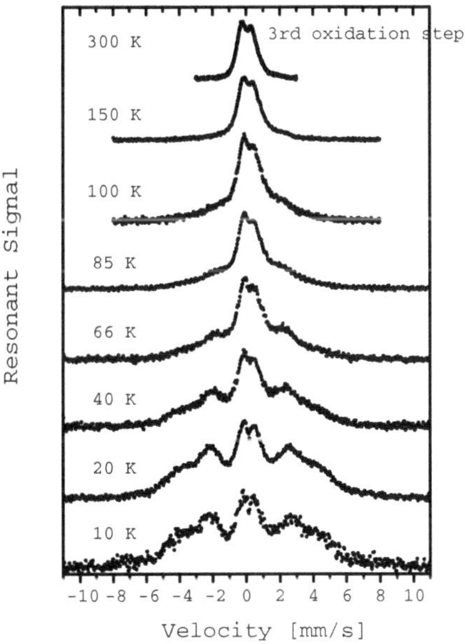

Fig. 4.12. Mössbauer spectra of sample I at various temperatures after the third oxidation step showing the superparamagnetic behavior of the nano-structured material.

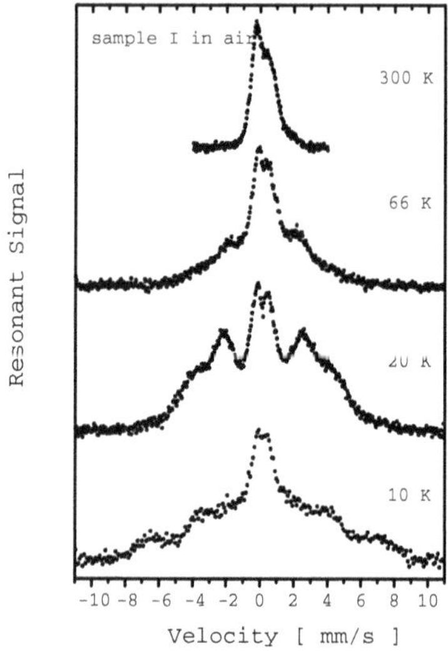

Fig. 4.13. Mössbauer spectra of sample I at various temperatures after exposing the sample several times to air.

metallic Fe in the metal-oxide interface zone. Due to defects, stresses and interface roughness the hyperfine field is suppressed in comparison to component S1. The annealing of the sample reduces these influences and leads to a significant increase of S1 paralleled by the equal drop in intensity of S3.

As expected, the second sample with a thickness of 4.5 nm shows magnetic ordering already at room temperature. Using the K conversion electrons of ^{57}Fe, this leads to the complication of a low signal to noise ratio for the components of the top Fe-Al or Fe-Al-O interface as it is superimposed by the strong signal of the metal Fe layer. To distinguish between the bottom and top interface in the trilayer structure, the use of low energy Auger electrons is most feasible. In the case of Fe, the ≈ 600 eV Auger electrons lead to a Mössbauer spectrum that completely suppresses the signal from the bottom interface region. The depth weight function for these low energy electrons is directly comparable with standard x-ray photo electron measurements. A future detailed analysis of the experimental results for sample II will reveal these features. In addition, the use of ^{56}Fe for the bottom part of the Fe layer could suppress the respective Mössbauer signal when using K conversion electrons. Fig. 4.14 shows the Mössbauer spectra of sample II in the as prepared state, after the seventh oxidation treatment and after the annealing at 125°C for 2 hours.

The spectrum of the as prepared sample is already quite complicated due to the modified electronic structure of the Fe metal in the presence of the two Fe-Al interfaces. In addition, a larger roughness is expected as the sample was prepared at room temperature. This is confirmed by a large amount of paramagnetic Fe (10.0(2) %) and of a low field component (7.7(2) %) with $\langle B_{hf} \rangle = 11.4(2)$ T. In sample I the purely paramagnetic fraction at 10 K amounts to 14(1) %, no

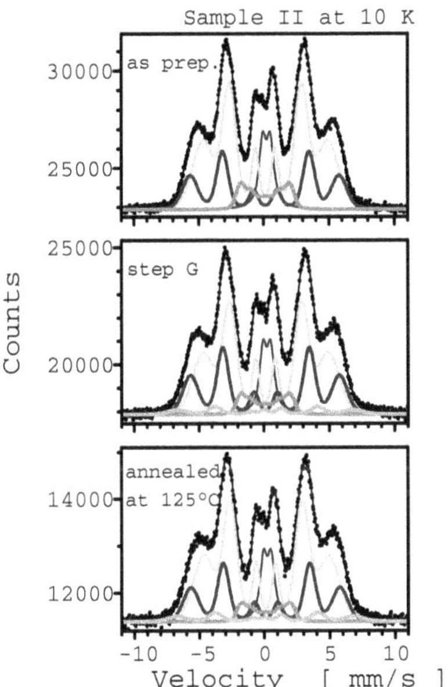

Fig. 4.14. Mössbauer spectra at 10 K of sample II at different stages of the oxygen and heat treatment.

pronounced low field component is observed. Considering the 4.5 times larger thickness of the Fe layer in sample II, the amount of Fe with predominantly Al nearest neighbors is larger by a factor of 5. The broad hyperfine field distribution of the remaining fraction in sample II has its origin in the Friedel oscillation of the electronic states near the interfaces. The hyperfine parameters of all these contributions are visualized in Fig. 4.10 as open symbols together with the closed symbols for sample I.

When oxidizing the sample similarly to the procedure for sample I, two additional resonant signals can be distinguished in the Mössbauer spectra. A very weak fraction has average hyperfine parameters that lie between those of superparamagnetic magnetite and Fe-Al spinel. A further refinement of this subspectrum will be possible by including the Mössbauer data that are based on the low energy Auger electrons. The second and stronger part is magnetically ordered and exhibits hyperfine fields that are similar to the ionic phase S2 in sample I. Both phases persist through the annealing process at 125°C.

As the second sample is significantly thicker than the first one, a depth selective Mössbauer experiment has revealed the difference in phase composition of the top and bottom interface. This is due to the observable energy loss of the 7.3 keV K conversion electrons of ^{57}Fe in a 4 nm Fe layer. In Fig. 4.15 the relative intensities of the various components are plotted as a function of electron energy. The respective values for the as prepared sample are given as open symbols at 7.22 keV. Roughly spoken, the values at 7.3 keV include a dominant signal from the top interface whereas the signal at 6.9 keV has a strong weight

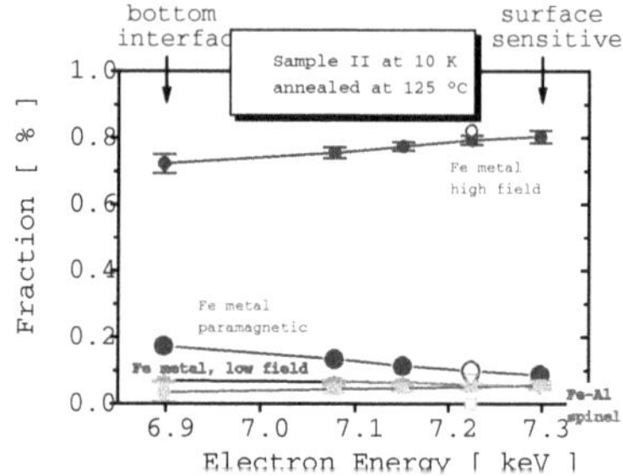

Fig. 4.15. Relative intensities of the components in Fig. 4.14 as a function of electron energy according to the DCEMS analysis. Open symbols are the data points for the as prepared sample. Closed symbols are the results for the annealed sample.

on the bottom interface. From this it is obvious that the Fe-Al spinel phase is found in the top interface. The increase of its intensity during oxidation is balanced by the decrease in intensity of the Fe metal phase, both, with low and high hyperfine field contributions. The paramagnetic signal from the Fe metal layer shows a significantly larger contribution from the bottom interface and has not changed compared to the as prepared state. This indicates a larger roughness of the bottom interface as the growth of the Fe layer on Al is different to the growth of the following Al layer on Fe.

In summary, within the statistical significance of the given experimental data, the oxidation in sample II follows the same route as in sample I. Differences due to a variation in interface roughness will be clarified in the not yet completed data analysis.

4.5 Discussion

The above results have to be viewed in the light of TMR characterizations of similar multilayer systems in the literature.

With regard to our analysis of the Fe-Al_2O_3 interface structural as well as magnetic properties vary strongly with oxidation and heat treatment. Both properties have to be considered as a source of the reported modification in magnetoresistance behavior due to various sample treatments. After the annealing of sample I, the smoothening of the interface that is accompanied by a reduction in the number of chemical phases and structural and magnetic defects or inhomogeneities improves the spin polarization and reduces scattering. This is in accordance with the observed increase in TMR effect in various systems for annealing temperatures of up to 300°C. The role of the magnetic properties of the insulator layer at the interface, i.e. the formation of Fe spinel compounds, is not yet clear. It is well known that magnetic insulators like NiO used as tunneling barriers in TMR structures lead to low changes in resistance with applied magnetic field [33–35]. This is especially true at higher temperatures, where non-conserved scattering with respect to spin moment can take place. One major difference of the present Fe-Al-O interface is the low thickness of the spinel phase. Therefore superparamagnetic relaxation on a time scale of 10^{-8} s is observed in the annealed state at room temperature. A clue on the influence of this phase on the TMR properties may be revealed in temperature dependent experiments where the hyperfine interaction is directly compared to the TMR signal. In general, to separate the influence of the changing magnetic ordering at the interface from the structural alterations on the TMR values is a difficult task and has to be considered in future work.

The observation of a worsening or failure of TMR structures for the longest oxidation times is a strong indication for the importance of the chemistry at the metal-insulator interface. The non-trivial oxidation behavior as plotted in Fig. 4.8 can obscure a systematic relationship between the oxidation procedure and magnetoresistive properties. Therefore it is necessary to have a high standard of interface characterization when conclusions are drawn with respect to the TMR properties. The XPS analysis of an Fe /Al_2O_3/CoFe structure by Mitsuzuka et al. [36] as a function of initial Al thickness on the Fe layer links the formation of Fe^{3+} for thin Al films to a drop in TMR effect. The possible role of Fe^{3+} in the region of a seemingly optimal Al thickness cannot be clarified due to the limits in sensitivity of the XPS method and the uncertainties in the background signal.

References

1. G.A. Prinz: Science **282** (1998) 1660
2. R. Meservey, P.M. Tedrow, P. Fulde: Phys. Rev. Lett. **25** (18) (1970) 1270-72
3. P.M. Tedrow, R. Meservey: Phys. Rev. Lett. **26** (4) (1971) 192-195
4. M. Julliere: Phys. Lett. **54**A (3) (1972) 225-226
5. P. Grünberg, R. Schreiber, Y. Pang, M.B. Brodsky, H. Sowers: Phys. Rev. Lett. **57** (1986) 2442; F. Saurenbach, U. Walz, L. Hinchey, P. Grünberg, W. Zinn: J. Appl. Phys. **62** (1988) 3473
6. M.N. Baibich, J.M. Broto, A. Fert, F. Nguyen Van Dau, F. Petroff, P. Eitenne, G. Creuzet, A. Friederich, J. Chazelas: Phys. Rev. Lett. **61** (1988) 2472
7. J.S. Moodera, L.R. Kinder, T.M. Wong, R. Meservey, Phys. Rev. Lett. **74** (16) (1995) 3272-3276
8. S. Yuasa, T. Sato, E. Tamura, Y. Suzuki, H. Yamamori, K. Ando, T. Katayama: Europhys. Lett. **52** (3) (2000) 344-350
9. J.M. De Teresa, A. Barthelemy, A. Fert, J.P. Contour, R. Lyonnet, F. Montaigne, P. Seneor, A. Vaures: Phys. Rev. Lett. **82** (1999) 4288-4291
10. Y.M. Tsipenyuk: *Nuclear methods in science and technology* (Institute of Physics Publishing, Bristol 1997)
11. T. Bonchev, A. Jordanov, A. Minkova: Nucl. Instr. Meth. **70** (1969) 36-40
12. S. Kruijer, W. Keune, M. Dobler, H. Reuther: Appl. Phys. Lett. **70** (1997) 2696
13. G. Walter, R. Heitzmann, D.M. Rück, B. Stahl, R. Gellert, O. Geiss, G. Klingelhöfer, E. Kankeleit: Nucl. Instr. Meth. B **113** (1996) 167-170
14. A.A. Novakova, A.P. Kuprin: Nucl. Instr. Meth. B (1993) 213-214
15. G. Belozerskii, C. Bohm, T. Ekdahl, D. Liljequist: Corros. Sci. **22** (9) (1982) 831
16. T. Yang, A. Krishnan, N. Benczer-Koller, G. Bayreuther: Phys.Rev.Lett. **48** (18) (1982)
17. A.S. Kamzin, L.A. Grigor'ev: Sov. Phys. Tech. Phys. **35** (1990) 840

18. B. Stahl, R. Gellert, M. Müller, E. Kankeleit, A. Kamzin: Phys. Rev. Lett., **84** (2000) 5632
19. D. Liljequist, T. Ekdahl, U. Baverstam: Nucl. Instr. Meth. **155** (1978) 529
20. D. Liljequist: Nucl. Instr. Meth. B **174** (2001) 351-360
21. B. Stahl, E. Kankeleit: Nucl. Instr. Meth. B **122** (1997) 149-161
22. A.J. Freeman, Ru-qian Wu: J. Mag. Mag. Mat. **100** (1991) 497-514
23. U. Gradmann: J. Mag. Mag. Mat. **100** (1991) 481-496
24. T. Shinjo: Surf. Sc. Rep. **12** (2) (1991)
25. M. Sato, H. Kikuchi, K. Kobayashi: IEEE Trans. Mag. **35** (5) (1999) 2946-2948
26. S. Cardoso, V. Gehanno, R. Ferreira, P. P. Freitas: IEEE Trans. Mag. **35** (5) (1999) 2952-2954
27. R.C. Sousa, Il Sun, V. Soares, P.P. Freitas, A. Kling: Appl. Phys. Lett. **73** (1998) 3288
28. S.S.P. Parkin: The 43th annual Conference of MMM proceedings, GA-**03** (1998) 255
29. T.E. Clark, F.B. Mancoff, S.X. Wang, B.M. Clemens, R. Sinclair: IEEE Trans. Mag. **35** (5) (1999) 2922-2924
30. T. Campbell, R. Kalia, A. Nakano, P. Vashishta, S. Ogata, S. Rodgers: Phys. Rev. Lett. **82** (24) (1999) 4866-4869
31. J.C. Sanchez-Lopez, A.R. Gonzalez-Elipe, A. Fernandez: J. Mat. Res. **13** (3) (1998) 703-710
32. E. Murad, J. H. Johnston: in *Mössbauer Spectroscopy Applied to Inorganic Chemistry*, Vol. 2 (ed. G.J. Long, Plenum Press, New York and London (1987))
33. S. Maekawa, U. Gäfvert: IEEE Trans. Magn. MAG-**18** (1982) 707-708
34. Y. Suezawa, Y. Gondo: Proc. Int. Symp. Phys. Magn. Mater, ed. M. Takahashi, S. Naekawa, Y. Gondo, N. Nose, World Sci. Singapore (1987) 303-306
35. J. Nowa, J. Rauluszkiewicz: J. Magn. Magn. Mater. **109** (1992) 79-90
36. T. Mitsuzuka, K. Matsuda et al.: J. Appl. Phys. **85** (8) (1999) 5807

5 Analysis of the Disturbing Influence of Stray Fields in Very Small MRAM Cells by Computer Simulation

Willi Schepper[1], Hitoshi Kubota[2], and Günter Reiss[1]

[1] University of Bielefeld, Department of Physics, Universitätsstraße 25, 33501 Bielefeld, Germany
[2] Dept. Appl. Physics, Tohoku University, 05 Aoba, Aramaki, Aoba-ku, Sendai, 980-8579 Japan

Abstract. We present computer simulations of sub-μ magnetic tunnel junctions. The devices with the small element sizes show a pronounced shift of the minor loop into the opposite direction as obtained for larger junctions, which can be interpreted by dipolar antiferromagnetic coupling. The devices show too an asymmetrical behavior in the magnetization reversal. The simulation calculations confirm both experimental results.

5.1 Introduction

Magnetic tunnel junctions (MTJ) are key devices for the development in magnetoelectronics [1] both for application as sensors or non-volatile memory (MRAM) devices [2]. Although large values of the tunneling magnetoresistance (TMR) were demonstrated [2–4], the influence of the geometry of the tunneling elements on the TMR is not yet understood. This is due to the complex interplay between magnetic anisotropies, domain splitting in the layers, new coupling effects across the tunneling barrier [5] and edge [6] or surface effects. One of the most interesting points is the magnetic and the related TMR behavior of MTJ's with tunneling areas well below 1 μ. Micromagnetic simulations with parameters adapted to the measurements can help answering the open questions.

CoFe(1.5)/Ru(0.9)/CoFe(2.2)/Al_2O_3(1.5)/NiFe(6) form the magnetic effective part of a multilayer [7,8]. The films similar to those discussed earlier[9] are deposited by dc magnetron sputtering on a Ta/Cu buffer and the stack was covered by a Ta(5 nm)/Au(30 nm) top layer in order to prevent oxidation. The film stack (thickness dimensions in nm) was patterned by e-beam lithography and Ar ion etching to rectangular junctions with different sizes and aspect ratios. Using secondary ion mass spectroscopy, the etching was stopped shortly after the Al signal vanished. Thus the patterned part of the films probably consists from the top down until the second layer of the artificial antiferromagnet (AAF) [CoFe(2.2 nm)], whereas the remaining films [Ru/CoFe(1.5 nm)] were still continuous. The etch stop turned out to be one of the crucial points of the preparation routine. The junctions were first magnetically saturated in order to have a well defined magnetization of the hard layer stack. Then, they were transferred to a customized conducting atomic force microscope (C-AFM) which allows topographic imaging and contacting the top electrode with a conducting diamond tip.

Two different types of programs are used for the phenomenological simulations: energy minimization [10,11] (model 1) and the local field method [12,13] (model 2). The commonly used algorithms in model 1 firstly generate the GMR/TMR contributions for only one easy axis ξ_j, then ξ_j is averaged in a subsequent step. This procedure is only an estimate, since spatial fluctuations of the anisotropy parameters don't enter into the minimization process. In model 2 the basic parameters (f- and af-coupling, anisotropy) of the curves can be changed at each grid point, a large advantage of a lattice model. Both models give the same result in the case of no spatial dependence of the anisotropy constants and periodic boundary conditions, this fact is essential for the confidence in the calculations. Otherwise they describe with the simplest parameter set a slightly different physical situation. In model 1 the multilayer consists of independent grains without any inter-grain coupling, in model 2 the grains normally are strongly coupled ferromagnetically. It is, of course, possible to simulate also weaker inter-grain coupling in model 2 [16]. The advantage of model 1 is, that it works very fast. Multilayers with up to 90 magnetic layers can be treated. The advantage of model 2 is its flexibility. The disadvantage of model 2 is, that it is very computer time consuming. Both programs have been developed to essential tools in understanding what happens physically within multilayer stacks and optimizing the devices.

Simulations on magnetic multilayer systems have been done for several years with both models. Investigations of the influence of pinholes on GMR characteristics and their temperature [14] dependence with model 2 have shown, that the presence of the biquadratic coupling can enhance the influence of the pinhole which implies a larger sensitivity of imperfect films to magnetic shorts between the layers [13]. For sensor applications, the dependence of the resistance on the external field should be mostly linear. Additional requirements concerning amplitude and saturation field depend strongly on the type of application. GMR-systems, where these characteristics can be modified in a relatively simple way would be therefore highly desirable. We reported on results obtained with model 1 on mixed systems, i.e. multilayers containing simultaneously the first and the second maximum of the af coupling. Magnetization as well as GMR can be numerically modeled for such complex layer sequences, where different types of coupling are present within one film system. Moreover, it is possible to tailor numerically the dependence of the resistance on the magnetic field by constructing film stacks with different sequences of strong and weak coupling. This procedure can be therefore very useful for the understanding of the magnetic and GMR properties of such complex film system and in designing optimized sensor materials [11]. It could be shown with model 1, that one parameter set is sufficient to explain the GMR characteristics for a multilayer series in very different coupling cases (mixed af coupling in the first and second af maximum by varying the spacer thickness) [16]. The experimental GMR characteristic show tails in the saturation region, this behaviour could be reduced to the influence of grains[16]. Model 1 was very successful in investigations of the influence of hysteresis [15] and layer thickness on the tailoring of GMR characteristics.

Single molecules e.g. in biological experiments can be detected by labeling with magnetic beads instead of the otherwise used dyes and by analyzing the influence of the beads on a GMR- or TMR-multilayer. The development of suitable sensors has been studied experimentally as well as theoretically by computer simulation in order to optimize the sensor parameters. Model 2 with the dipole fields of the beads as additional contributions to the local field $\boldsymbol{H}_{ijk}$ has been used in the simulations already very successfully. The bead fields produce vortex like arrangements within the spin ensemble of a magnetic layer and change the spin relaxation process and as a result also the GMR/TMR characteristic.

An overview of the tunnel magnetic resistance effect is given by Moodera and Mathon [1]. Recently the domain structures in an MTJ-element was calculated in a micromagnetic simulation using a hybrid finite element/boundary element method [17]. The equilibrium spin distribution was obtained from the solution of the Gilbert equation, the anisotropy term in the local field was neglected. The free layer ($500 \times 500\,\text{nm}^2$, t=10 nm) was placed in the middle of the pinned layer ($1 \times 1\mu^2$, t=3 nm), therefore no stray fields of the aaf can penetrate into the free layer. The numerical calculations show that the energy of different remanent states (C- and S-state) is rather similar. For a ferromagnet in contact with an antiferromagnet (AFM) a shift of the hysteresis loop along the magnetic field axis can occur which is called exchange bias. Monte Carlo simulations [18] with a heat-bath algorithm and single-spin flip methods for the simulation were used. The calculations show a strong dependence of the exchange bias on the degree of dilution and the thickness of the AFM layer in agreement with experimental observations.

5.2 Simulation by Energy Minimization

The energy of a multilayer structure, consisting of 3 layers, is given by:

$$
\begin{aligned}
E_j = \quad & \sum_{i=1}^{3} t_i [k_{ui} \sin^2(\alpha_i - \xi_{ij}) - H_e M_i \cos\alpha_i + \frac{1}{2} \sum_{\substack{j=1 \\ i \neq j}}^{3} H_{ij}^{(st)} M_i \cos(\alpha_i - \alpha_j)] \\
& + \sum_{i=1}^{2} [J_{li} \cos(\alpha_i - \alpha_{i+1}) + J_{qi} \cos^2(\alpha_i - \alpha_{i+1})]
\end{aligned}
\tag{5.1}
$$

j is the index for a subsequent step, in which the easy directions ξ_{ij} are varied, the energy E_j depends therefore upon j. J_l is the linear exchange constant and J_q the quadratic term. k_u is the uniaxial anisotropy constant, ξ_{ij} the easy direction, t_i the thickness of layer i. In principle there is a further energy term (i=j) in the second sum: the influence of the stray field of layer i on itself, it must not appear here because a constant in the energy equation doesn't change the position of the minimum and therefore doesn't determine the angles α_i. The Downhill-Simplex method was used to minimize this energy equation in order to obtain the equilibrium values of the angles α_i. For obtaining magnetization curves, the external field H_e is stepwise increased. At every new field H_e, the previously calculated angles α_i were used as a new initial distribution, but previously corrected by the method of largest descent. The GMR/TMR contributions are calculated for only one easy axis distribution ξ_{ij} within the region $-\pi/8 < \xi < \pi/8$ and then averaged in a subsequent step.

5.3 Simulation by the Local Field Method

$$\boldsymbol{H}_{i,j,k} = \boldsymbol{H}_e + \boldsymbol{H}_{i,j,k}^{(st)} + \boldsymbol{H}_{i,j,k}^{(un)} + \boldsymbol{H}_{i,j}^{(su)} + \sum_{i_l=1}^{5} J_f \boldsymbol{S}_{i_l} + \boldsymbol{S}_6 \begin{cases} J_f & \text{f} \\ J_{al} + 2J_{aq}|S_{i,j,k}||S_6|\cos(\alpha_{i,j,k} - \alpha_6) & \text{af} \end{cases} \tag{5.2}$$

At each grid location (i,j,k) within $1 \leq i,j,k \leq n_{i,j,k}$ the local field according to Eq.(5.2) has to be calculated. The angle of $\boldsymbol{H}_{i,j,k}$ supplies the new direction $\alpha_{i,j,k}$ of the spin. The spin is aligned until a local energy minimum is reached. The procedure is repeated until the corresponding magnetization changes ΔM drop below a limit of e.g. 10^{-6}. Surface anisotropy is used for the boundary conditions in the (i,j) or yz-plane of the layer. No contributions to the local field are considered above the top layer. The contributions to the local field correspond to the well-known references [6]. The formula for the uniaxial anisotropy is used in the simulations, the components of the anisotropy fields in the volume and at the surface are:

$$\boldsymbol{H}_{i,j,k}^{(un)} = \begin{pmatrix} H_y \\ H_z \end{pmatrix} = 2k_i^{(un)} \cos(\alpha_{i,j,k} - \xi_{ij}) \begin{pmatrix} \sin \xi_{ij} \\ \cos \xi_{ij} \end{pmatrix} \tag{5.3}$$

$$\boldsymbol{H}_{i,s}^{(su)} = \boldsymbol{H}_{s,j}^{(su)} = -2k_{su} \cos(\alpha_{i,j} - \xi_s) \begin{pmatrix} \sin \xi_s \\ \cos \xi_s \end{pmatrix}$$

The easy directions ξ_{ij} have been chosen randomly within the region $-\pi/8 < \xi < \pi/8$, ξ_s is the direction of the normal at the edge surfaces (s=1,n). The sign of $H_{i,s}^{(su)}$ is negative because the normal points into the strong direction. The surface anisotropy tries to bend the spins parallel to the edge surface.

5.4 Equation of Motion

The Landau-Lifshitz-Gilbert (LFG) equation set are partial differential equations

$$\frac{\partial \boldsymbol{S}}{\partial t} = -\boldsymbol{S} \times \boldsymbol{H}_{loc} - \alpha(\boldsymbol{S} \times (\boldsymbol{S} \times \boldsymbol{H}_{loc}))$$

often used for micromagnetic calculations with the damping constant α. In thermal equilibrium the local field $\boldsymbol{H}_{loc}$ points into the direction of $\boldsymbol{S}$. There is no more time change in the spin direction ($\partial \boldsymbol{S} = 0$) because the cross product $\boldsymbol{S} \times \boldsymbol{H}_{loc}$ on the right hand side of the LFG equation is zero. This condition is used immediately in the local field method described above [12,13]. The result of both methods must be the same.

5.5 Stray Fields, Magnetization and TMR

The dipole field H_s of a cubic body magnetized homogeneously along the length l=2c can be evaluated analytically [20]. First, we replace the outer edge faces of the element by magnetic poles with a strength of $M_s A$ (M_s: saturation magnetization, A: area of the edge face). Then, the related magnetic potential ψ_m and the interesting field components parallel to the layer surface H_z, H_y can be calculated as:

$$4\pi\psi_m/M_s = \int_{-a}^{a}\int_{-b}^{b}\left[\frac{1}{r(-z,x',y')} - \frac{1}{r(z,x',y')}\right]dy'dx'$$

$$r(z,x',y') = \sqrt{(x-x')^2+(y-y')^2+(z+c)^2}$$

$$\boldsymbol{H}^{(st)}(x,y,z) = \begin{pmatrix} H_y(x,y,z) \\ H_z(x,y,z) \end{pmatrix} = \begin{pmatrix} d\psi_m/dy \\ d\psi_m/dz \end{pmatrix}$$

(x', y') are the coordinates of a magnetic charge in the edge face. H_z is the field component in the direction of the external field H_e. $l = 2c$ is the length, $w = 2b$ is the width, $t = 2a$ the thickness of the layer. The stray field components can be evaluated into complicated formulas consisting of 8 arctan (H_z) or ln (H_y) terms:

$$\begin{aligned}
f_y &= \frac{1}{4\pi}\ln\frac{\sqrt{(x+a)^2+(y+b)^2+(z+c)^2}+x+a}{\sqrt{(y+b)^2+(z+c)^2}} \\
f_z &= \frac{1}{4\pi}\arctan\frac{(x+a)(y+b)}{(z+c)\sqrt{(x+a)^2+(y+b)^2+(z+c)^2}} \\
H_y/M_s &= f_y(x,y,-z)+f_y(-x,y,-z)+f_y(x,-y,z)+f_y(-x,-y,z) \\
&\quad - f_y(x,y,z) - f_y(-x,y,z) - f_y(x,-y,-z) - f_y(-x,-y,-z) \\
H_z/M_s &= f_z(-x,y,-z)+f_z(-x,y,z)+f_z(x,y,-z)+f_z(x,y,z) \\
&\quad + f_z(-x,-y,-z)+f_z(-x,-y,z)+f_z(x,-y,-z)+f_z(x,-y,z)
\end{aligned} \tag{5.4}$$

Fortunately the component H_z at the layer center ($y = z = 0$) can be considerably simplified in the case of thin layers ($x_i, t_i \ll w_i, l_i$):

$$H_z^{\approx}(x,0,0)/M_s = \frac{2t_i w_i}{\pi\, l_i\sqrt{w_i^2+l_i^2}} \rightarrow \frac{\sqrt{2}\, t_i}{\pi\, l_i}\quad for\ w_i = l_i \tag{5.5}$$

In this limit $H_z^{\approx}$ of Eq.(5.5) is in good agreement with the complete formula (Eq.(5.4)) for l>50 nm. H_z/M_s is proportional to the ratio of the layer thickness to the lateral dimensions $t/l \ll 1$. It is e.g. of the size 1.8% for a typical ratio

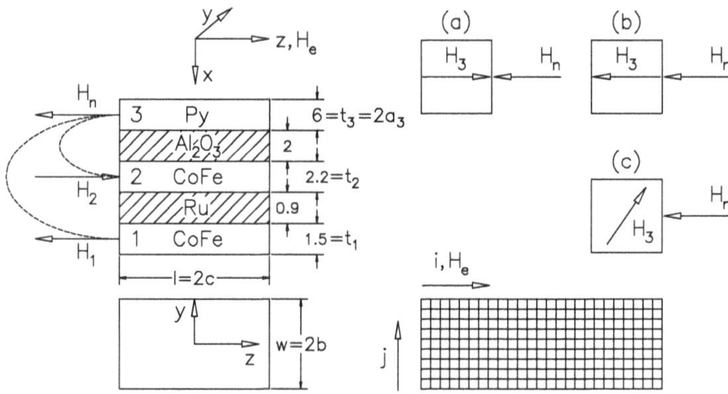

Fig. 5.1. Sketch of the MTJ element, coordinate system, AAF layer 1+2, sensor layer 3, direction of the stray field lines (dashed) of the AAF for the minor loop, spin direction in the sensor layer after magnetization with large positive (a) or negative (b) external fields H_e, transition case (c), dimensions l=w=50nm for the smallest element

t/l=4% (t=2 nm, l=50 nm), then it decreases to 0.18% for a quadratic device with a ten times larger length l=500 nm. It is remarkable, that in this approximation the stray field of layer i at the place of layer j is independent of the distance $x = d_{ij}$ between them. Layer 3 is influenced by two stray fields of the AAF with opposite directions according to Fig. 5.1. The field of Eq.(5.5) is only strictly valid above or below the center of the layer, it increases strongly towards the edge faces as Fig. 5.4 shows. Therefore the mean field value H_{zm} was introduced instead of $H_z(x, 0, 0)$, it can be quite big particularly for the larger devices.

$$H_{zm} = \frac{1}{twl} \int_{-a}^{a} \int_{-b}^{b} \int_{-c}^{c} H_z(x, y, z)\; dz\; dy\; dx \tag{5.6}$$

H_{zm} according to Eq.(5.6) decreases with increasing distance d_{ij} between the layers, especially for the smaller elements. It is remarkable, that devices with a larger dimension in the external field direction (aspect ratio $l/w = 2$, 100×50) have smaller stray field values as the quadratic devices ($l/w = 1$, 100×100) and could enable higher storage densities from this point of view. The stray fields can also be calculated using the dipole approximation method. The stray field of a magnetic dipole in polar coordinates is given by the well-known formula, which is also valid in the electrostatic case:

$$\boldsymbol{H}^{(st)} = \begin{pmatrix} H_r \\ H_\vartheta \end{pmatrix} = \frac{\mu_m}{4\pi r^3} \begin{pmatrix} 2\cos\vartheta \\ \sin\vartheta \end{pmatrix}$$

The spin points into the z-direction, the component H_φ is zero, $\mu_m = M_s V$ is the dipole moment. This formula must be firstly transformed into cartesian coordinates. Then the stray field at a given position (y,z) is a superposition of the field contributions of the whole spin ensemble:

$$\boldsymbol{H}_3^{(st)} = \begin{pmatrix} H_y \\ H_z \end{pmatrix} = \frac{M_{s3}\ l\ w\ t_3}{4\pi\ n_i\ n_j} \sum_{i=1}^{ni} \sum_{j=1}^{nj} \frac{1}{r^3} \begin{pmatrix} 3\cos\vartheta_{i,j} \sin\vartheta_{i,j} \sin\varphi_{i,j} \\ 3\cos^2\vartheta_{i,j} - 1 \end{pmatrix} \tag{5.7}$$

ϑ, φ are the angles in the polar coordinate system at one grid point i,j. Demagnetizing fields evaluated by Eq.(5.7) agree very well with Eq.(5.4) for layer distances $x > 4nm$. There is a problem in calculating the own stray field (x=0) because of divergence of the dipole field at its origin ($r = 0, 1/r^3 \to \infty$). An obvious possibility in order to avoid this difficulty is to omit the term i=j in the second sum. But in this case Eq.(5.7) overestimates also the stray field contributions of the nearest neighbors. According to our experience Eq.(5.4) should be applied in this case, the layer dimensions (a,b,c) in Eq.(5.4) must be replaced by the differential cell size of the lattice point (i,j,k). The differences between Eqs.(5.4), (5.7) are not only influenced on discretization errors [21].

The stray field components (H_y, H_z) in Eqs.(5.4), (5.7) are correct, if the spins point into the external field direction H_e ($\beta = 0$). If a spin during the relaxation process has been turned into the direction β, its components still have to be rotated on β correspondingly:

$$\begin{pmatrix} H'_y \\ H'_z \end{pmatrix} = \begin{pmatrix} \cos\beta & \sin\beta \\ -\sin\beta & \cos\beta \end{pmatrix} \begin{pmatrix} H_y \\ H_z \end{pmatrix} \tag{5.8}$$

If the angles $\alpha_{i,j,k}$ are known magnetization M and TMR $\Delta R/R$ are:

$$M = \frac{1}{n_i n_j n_k} \sum_{i=1}^{ni} \sum_{j=1}^{nj} \sum_{k=1}^{nk} \cos\alpha_{i,j,k} \to \Delta R/R = \frac{1}{2}(1 - M) \tag{5.9}$$

α are the angles of the spins in the yz-plane, $\alpha = 0$ corresponds to the direction of the external field H_e.

5.6 Results

The measured TMR curve in Fig. 5.8 shows the shift into the positive field direction and the remarkable asymmetry in the magnetization behavior.

The phenomenological model of energy minimization includes 3 layers [aaf (CoFe/Ru/CoFe), sensor Py] with layer thickness (1.5, 2.2, 6 nm) and M_s, a widening of the easy directions in the region of $\pm\pi/8$ to the external field, af-coupling (linear and quadratic) between the af-coupled layers 1-2. The center line of the hysteresis has been shifted by 10 Oe corresponding to a Néel-type f-coupling between layers (2-3) in agreement with measurements at the large devices ($600\times300\,\mathrm{nm}^2$). The other model parameters (anisotropy constants) were adjusted to the observed major and minor loops of the $600\times300\,\mathrm{nm}^2$ device. The appearance of domains in this case will further decrease the stray field influence.

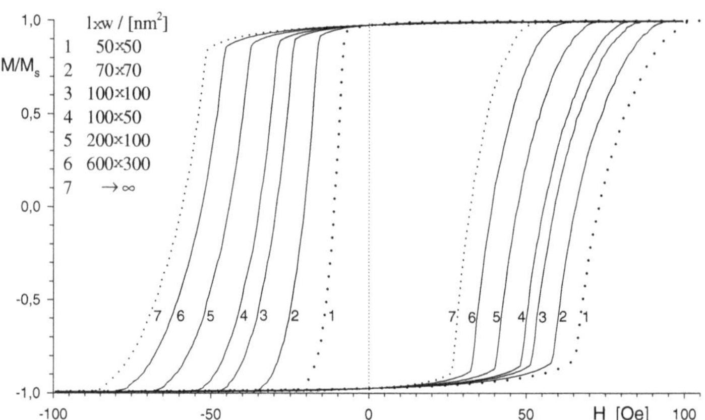

Fig. 5.2. Hysteresis curves $M(H_e)$ for different values of length l and width w, calculated by energy minimization due to Eqs.(5.1), (5.4), (5.6). Curve 7 represents the case without stray fields

Table 5.1. Stray fields H_z/M_s [%] of a cube $l \times w \times t$ in a distance of 5nm ($t = 2$nm) according to Eq.(5.4), Eq.(5.5) provides values with a relative deviation of 4.8% for the smallest and 0.04% for the largest device

$l \times w$[nm^2]	50×50	70×70	100×100	100×50	200×200	200×100	600×300
H_z/M_s [%]	1.713	1.254	0.889	0.561	0.449	0.284	0.095

The hysteresis curves (minor loops) have been calculated with Eqs.(5.1), (5.4), (5.6) by the method of energy minimization, including all stray field energies between the 3 layers, infinity means neglecting dipole coupling. The results in Fig. 5.2 show, that the shift in the minor loop is most pronounced in the case of the smallest device (50×50 nm^2). The shift decreases continuously towards the larger devices, which can be well understood by the stray field pre-factor in Eq.(5.5) (Table 5.1). The stray fields in devices with dimensions larger than 1μ are too small for shift values of 50 Oe, which have been reported earlier as results from micromagnetic calculations [22]. The results on minor loops show, that a significant improvement of the MRAM cell is attainable, if the thickness of the sensor layer 3 is reduced from 6 nm to 2 nm (Fig. 5.3).

Micromagnetic calculations with the local field method have been done according to Eqs.(5.2), (5.3). Eq.(5.4) is used for the stray field of the AAF, the own stray field of layer 3 was calculated by Eq.(5.7). First the calculations in Fig. 5.9 confirm the shift in the minor loops of the calculation by energy minimization (Fig. 5.2).

Our micromagnetic calculations show additionally magnetization patterns (Figs. 5.6, 5.7) and give insight into the magnetic reversal process. 4 patterns in Fig. 5.7 are calculated for the fields $H_e = 121.05(1), 10.52(2), 6.05(3), 5.79(6)$ $[Oe]$, the patterns 4,5 are non stationary states shortly before the spins completely

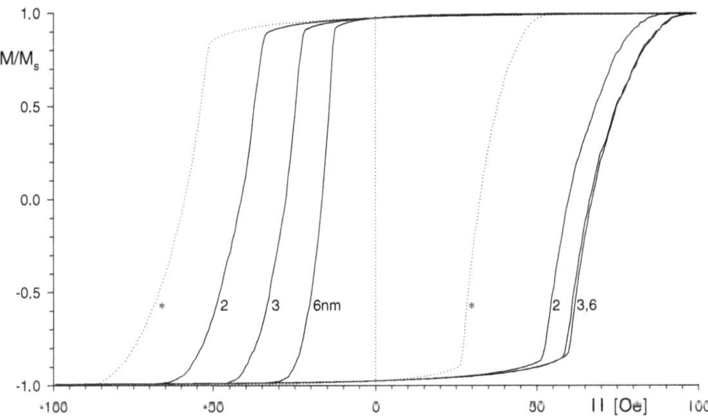

Fig. 5.3. Hysteresis curves $M(H_e)$ for the 50×50 nm^2-device but for different values of sensor layer thickness t_3, calculated by energy minimization due to Eqs.(5.1),(5.4),(5.6). The curve marked by $*$ agrees with the curve 7 in Fig. 5.2

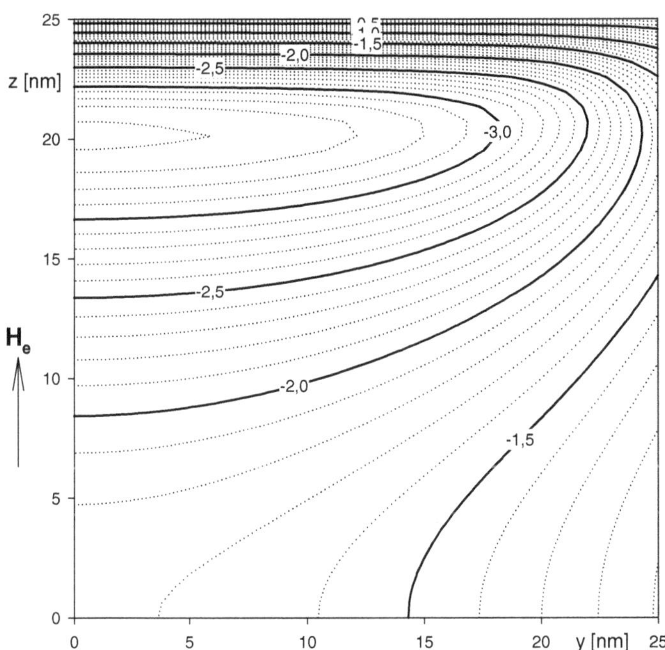

Fig. 5.4. Contour map of the stray field H_z/M_s [%] of a $50 \times 50\,\mathrm{nm}^2$ layer, 2 nm thick in a distance of 5 nm, $H_z/M_s(y,z)$ is always negative and shows a pronounced spatial dependence, the symmetry relations are $H_z(y,z) = H_z(-y,z) = H_z(y,-z)$, the value at the coordinate origin is $H_z/M_s(0,0)$=-1.71% (0,0) according to Eqs.(5.4,5.5), along the z-axis it falls to -3.42% (0,20.1 nm) and rises again to -0.28% (0.25 nm), along the y-axis it falls to -1.09% (25 nm,0)

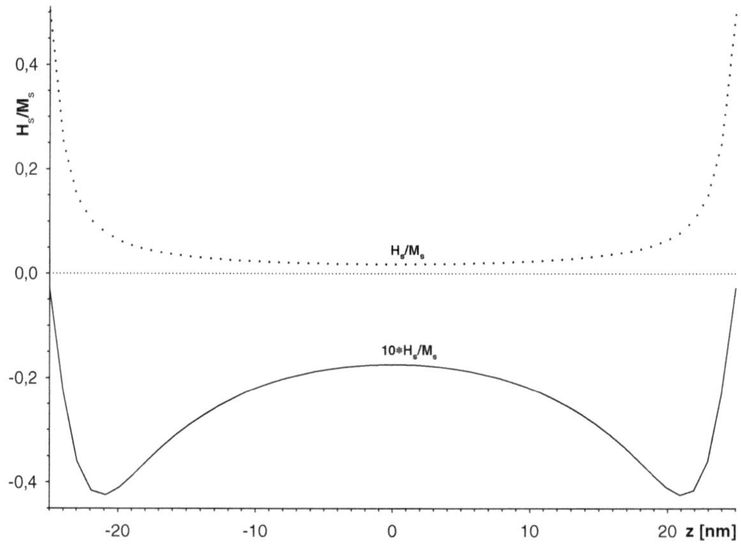

Fig. 5.5. Own (pointed) and foreign stray field of a $50 \times 50\,\mathrm{nm}^2$ layer at a distance of 4 nm (solid), layer thickness 2 nm, y=0

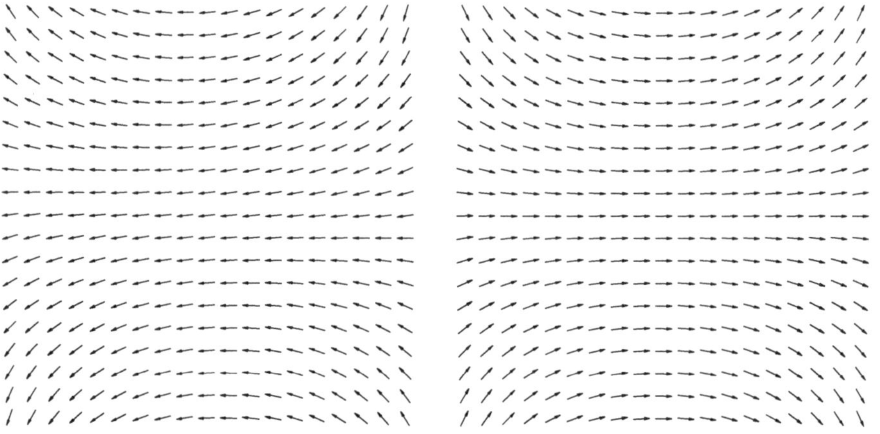

Fig. 5.6. Micromagnetic calculation in a 19×19 lattice, 2 spin patterns for a magnetization from the negative into the positive field direction, case (b) in Fig. 5.1

turn into the opposite direction, $M/M_s = 0.947(1), 0.816(2), 0.783(3), -0.970(6)$ are the corresponding magnetization values. H_e=121 Oe is a relatively high field, most of the spins are adjusted to the external field, the influence of the stray field components H_y can be recognized at the bended spins at the corners. In vector field 2 the external field is reduced to 10.52 Oe, therefore the influence of

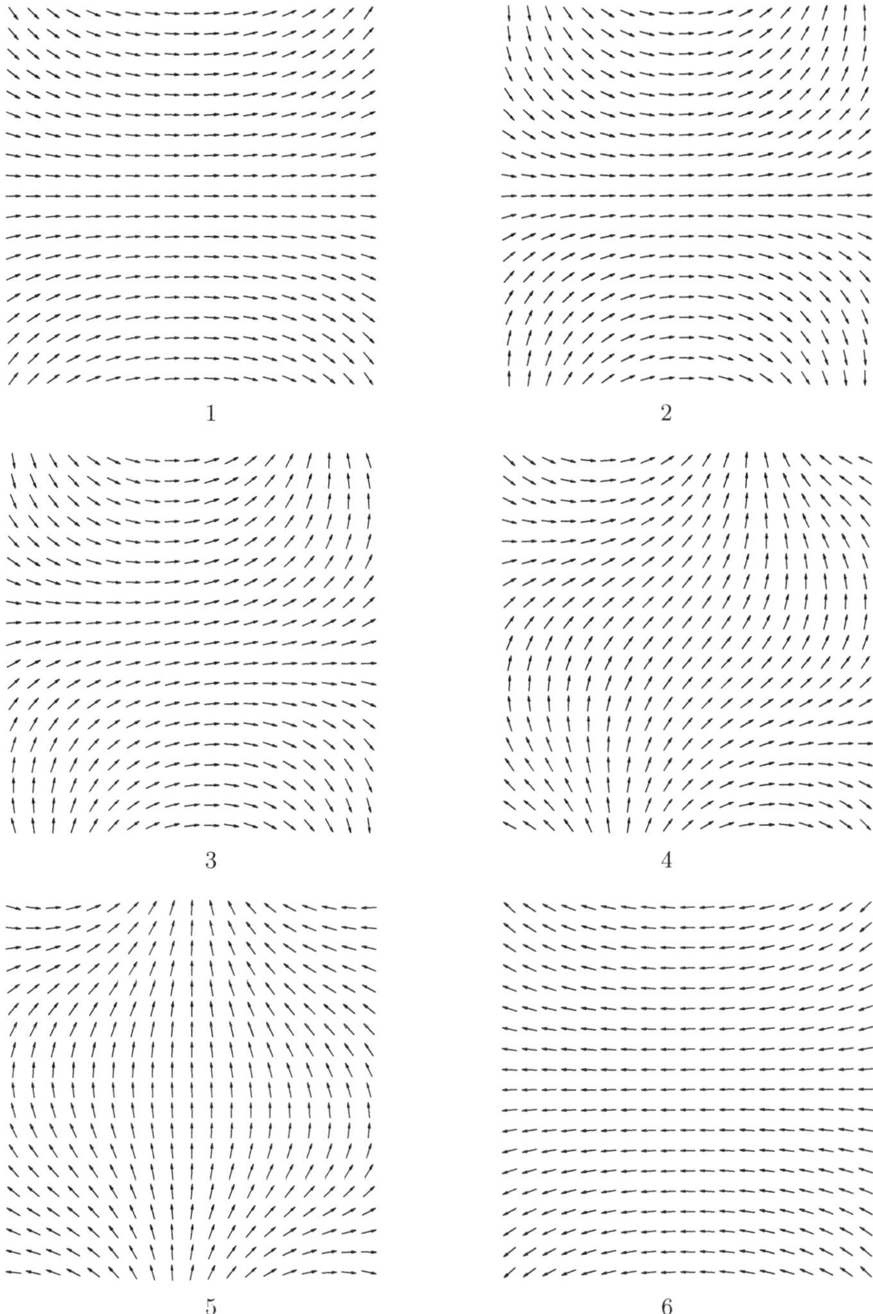

Fig. 5.7. Micromagnetic calculation in a 19×19 lattice, sequence of spin patterns for a magnetization from the positive into the negative field direction, case (a) in Fig. 5.1

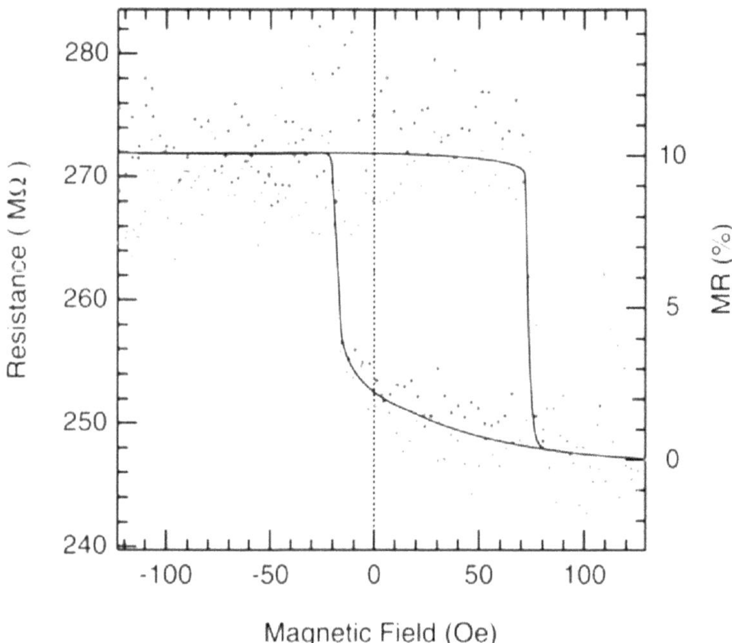

Fig. 5.8. Experimental curve for the 50×50 nm^2 element with the shift of the center line of the hysteresis and the asymmetry in the magnetization between the positive or negative field direction

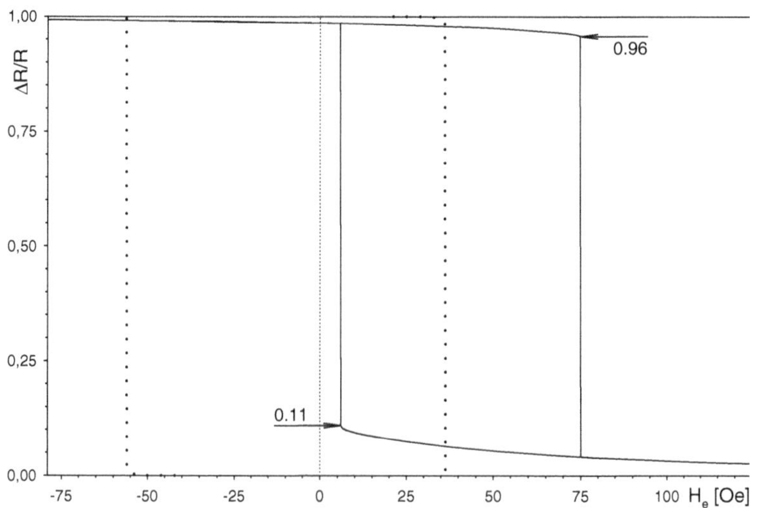

Fig. 5.9. $\Delta R/R(H_e)$, micromagnetic calculation in a 19×19 lattice with (solid) and without (pointed) stray fields

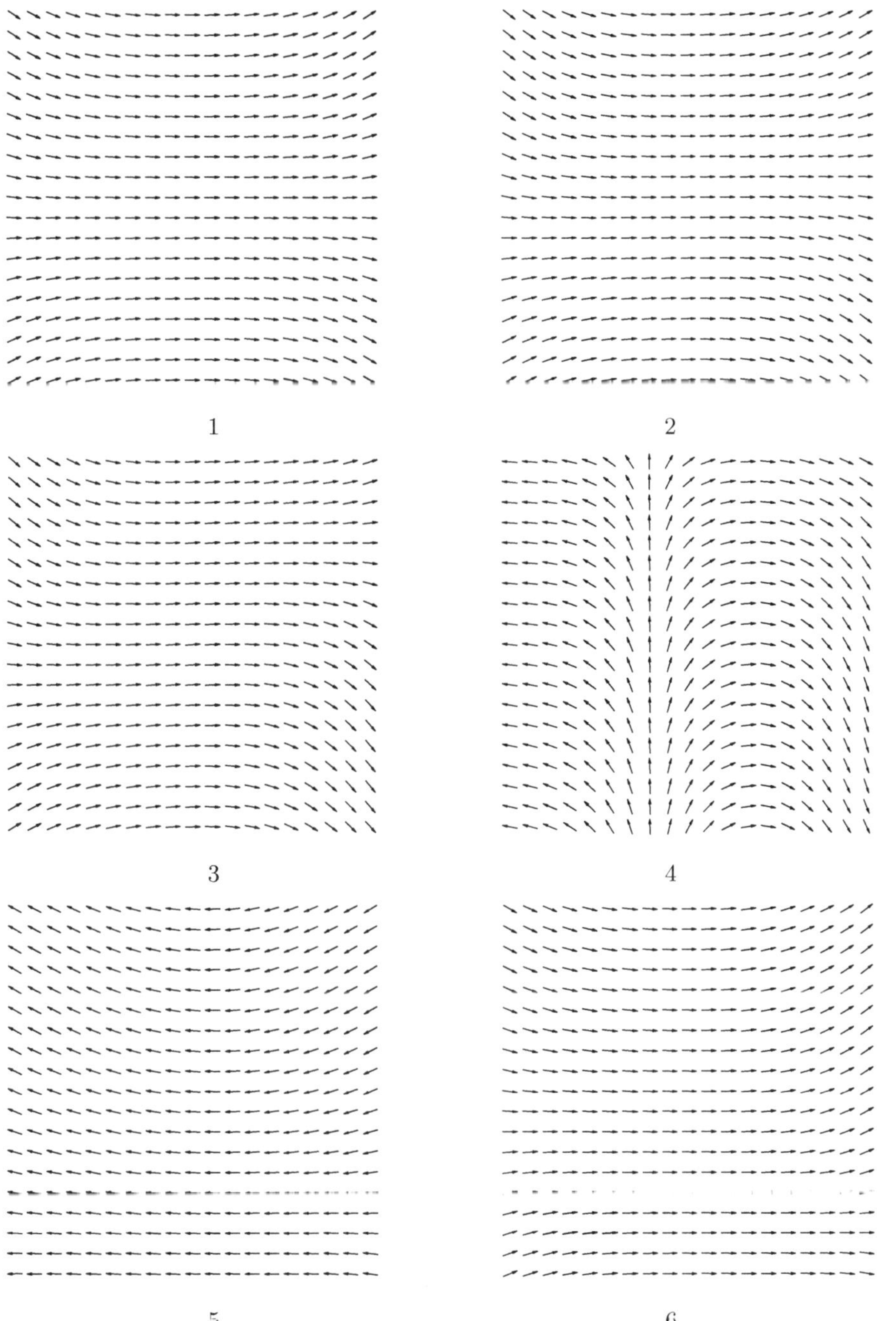

Fig. 5.10. Micromagnetic calculation in a 19×19 lattice, sequence of spin patterns for a magnetization from the positive into the negative field direction (1-4) or in the opposite direction (5,6)

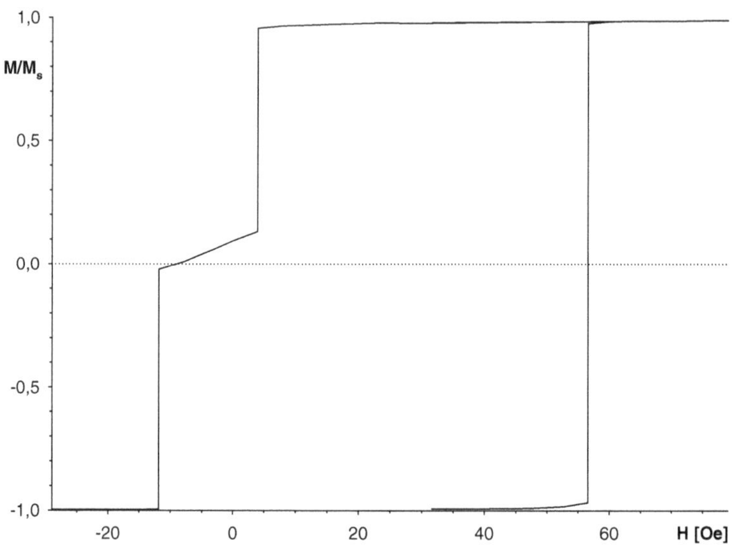

Fig. 5.11. $\Delta R/R(H_e)$, micromagnetic calculation in a 19×19 lattice, 150×150 nm^2

the stray field components H_z increases, the spins at the corners align already perpendicular to H_e, the spin distribution is still symmetrical to center line 10. Vector field 3 shows the situation at 6.05 Oe short before the transition. The center of the vortex in the upper half has shifted already one grid point to the left, the correspondig center in the lower half to the right, the spins in center line 10 turned up already. In the transition states 4 and 5 the parts of the vector field are growing at expense of others, which are lying favorably to the subsequent negative spin direction. Fig. 5.6 shows the vector fields for the magnetization into the opposite direction, just before (left, H_e = 74.74 Oe, M/M_s = −0.913) and after the jump (right, H_e = 75.00 Oe, M/M_s = 0.918).

The experimental curve in Fig. 5.8 shows a significant asymmetry between both branches of the TMR curve. This behavior could be confirmed by the micromagnetic calculations in Fig. 5.9. The M/M_s values just before the jump are -0.913 (1 in Fig. 5.6) and 0.783 (3 in Fig. 5.7), the corresponding TMR values are $\Delta R/R$=0.956=1-0.044(1) and 0.108(3). The difference between saturation and the $\Delta R/R$ values is 2.5 times larger in the case of Fig. 5.7 (case a in Fig.5.1). In this case the spin ensemble must bend at the 4 corners to minimize energy and therefore reaches saturation only slowly. In the case of Fig. 5.6 both fields point already into the same direction, case b in Fig. 5.1 is the state with the smaller energy. This asymmetry occurs in the micromagnetic simulation calculation only, if the contribution of the own stray field of layer 3 is taken into account, therefore it cannot be seen in the energy minimization. This asymmetry in the hysteresis and TMR curve can be even larger with smaller grid size, it is obviously induced by the edge spins. Fig. 5.5 shows that the magnitudes of both stray fields (own and foreign) are large short before the edge with the result that spin bending is very large in this region.

The asymmetry affect can also be found in the case of the larger devices ($150 \times 150\,\mathrm{nm}^2$) as is shown in Fig. 5.10. In this case the well-known C- and S-states appear in the micromagnetic calculations in the case of magnetization reversal towards the negative field values. Néel walls (4) can also appear in this case. These states never appear in the magnetization process in the opposite direction (5,6). In this case the jump of the spin ensemble starts always from a distribution with slightly bended spins at the edges to a similar state in the opposite direction. This is the reason for the observed asymmetry also for the larger devices. The following series of external fields belong to the vector fields: $H_e[Oe] = 31.58(1), 13.16(2), 5.26(3), 2.63(4), 55.26(5), 57.89(6)$. The corresponding magnetization curve is shown in Fig. 5.11. Néel walls appear in the region with magnetization values near zero.

Conclusion

The results of the minor loop calculations show the shift in the minor loop pronounced for the smallest device ($50 \times 50\,\mathrm{nm}^2$), it decreases continuously towards the larger devices, which can be well understood by the stray field pre-factor in Eq.(5.5). The results show too, that a significant improvement of the MRAM cell is attainable, if the thickness of the sensor layer 3 will be reduced from 6nm to 2 nm. The micromagnetic calculations confirm the shift in the minor loop, they show also the asymmetry in the magnetization behavior and give insight into magnetic reversal process. An ideal model TMR system would be desirable in order to adjust more correctly the parameters of simulation.

References

1. J.S. Moodera, G. Mathon, J. Magn. Magn. Mater. 200, 248 (1999)
2. S.S.P. Parkin, K.P. Roche, M.G. Samant, P.M. Rice, R.B. Beyers, R.E. Scheuerlein, E.J. O'Sullivan, S.L. Brown, J. Bucchigano, D.W. Abraham, Y. Lu, M. Rooks, P.L. Trouilloud, R.A. Wanner, W.J. Gallagher, J. Appl. Phys. 85, 5828 (1999)
3. Z.Z. Zhang, S. Cardoso, P.P. Freitas, P. Wei, N. Barradas, J.C. Soares Appl. Phys. Lett 78, 2911 (2001)
4. Y. Ando, H. Kubota, M. Hayashi, M. Kamijo, K. Yaoita, A. Chak Chung Yu, Xiu-Feng Han, T. Miyazaki, Jpn. J. Appl. Phys. 39, 5832 (2000)
5. M.R. McCartney, R.E. Dunin-Borkowski, M.R. Scheinfein, D.J. Smith, S. Gider, S.S.P. Parkin , Science 28, 1337 (1999)
6. A.Anguelouch, B.D. Schrag, G. Xiao, Y. Lu, P.L. Trouilloud, R.A. Wanner, W.J. Gallagher, S.S.P. Parkin, Appl. Phys. Lett. 76, 622 (2000)
7. H. Kubota, Switching of sub-μ MTJs probed by conducting AFM, MML'01, SDT-IV P, Aachen, June 24-29, (2001)
8. H.A.M. van den Berg, W. Clemens, G. Gieres, G. Rupp, M. Vieth, J. Wecker, S. Zoll, J. Magn. Magn. Mater. **165**, 524 (1997).

9. C. Tiusan, T. Dimopoulos, K. Ounadjela, M. Hehn, H.A.M. van den Berg, V. da Costa, Y. Henry, Phys. Rev. B 61, 580 (2000)
10. H. Holloway, D.J. Kubinski, J. Appl. Phys., **83**, 2705 (1998)
11. W. Schepper, A. Hütten, and G. Reiss, J. Appl. Phys. **88**, 993 (2000)
12. D.B. Fulghum and R.E. Camley, Phys. Rev. B, **52**, 13436 (1995).
13. W. Schepper, K. Diplas, and G. Reiss, J. Appl. Phys. **87**, 6597 (2000)
14. W. Schepper, K. Diplas, S. Heitmann, G. Reiss , 3D-Simulations of magnetic structures in af-coupled multilayers with pinholes - temperature dependence of the magnetization, International Conference on Magnetism ICM2000, Recife, Aug. 6-11, 4S-13, (2000).
15. W. Schepper, A. Hütten, T. Hempel, G. Reiss, Simulation of GMR- and magnetization characteristics in Co—Cu multilayers, International Conference on Magnetism ICM2000, Recife, Aug. 6-11, 4U-45, 2000.
16. W. Schepper, A. Hütten, K. Diplas, G. Reiss, Influence of grains in an af-coupled double layer on magnetization and magnetoresistance, 8th Joint MMM-Intermag Conference, San Antonio, Jan. 7-11, HQ-07 (2001).
17. T. Schrefl, J. Fidler, J.N. Chapman, K.J. Kirk, J. Appl. Phys. **89**, 7000 (2001)
18. U. Nowak, A. Misra, K.D. Usadel, J. Appl. Phys. **89**, 7269 (2001)
19. A. Hubert, R. Schäfer, Magnetic Domains, The Analysis of Magnetic Microstructures, Springer (1998)
20. R.I. Joseph, E. Schlömann, J. Appl. Phys., **36**, 1579 (1965)
21. X. H. Huang, M. Pardavi-Horvath, IEEE Trans. Magn., **32**, 4180 (1996)
22. J. O. Oti, S. E. Russek, IEEE Trans. Magn., **33**, 3298 (1997)

6 Stability of Magnetic Tunnel Junctions

Günter Reiss, Hubert Brückl, Jan Schmalhorst, and Andy Thomas

University of Bielefeld, Dept. of Physics, P.O. Box 100 131, 33501 Bielefeld, Germany

Abstract. Understanding and improving the properties of magnetic tunnel junctions are of basic importance for the further development of magnetoelectronics. Here we summarize results on the stability of these devices against thermal, magnetic and electrostatic influences with special emphasis on the physical reasons for the degradation. Critical issues like interdiffusion, magnetic degradation of the hard layer and properties of very small junction systems will be addressed and possible solutions of the related problems suggested.

6.1 Introduction

Magnetic tunnel junctions (MTJ) are key components for the development in magnetoelectronics [1] both for application as sensors or non-volatile memory (MRAM) [2] as well as for, e.g., magnetic logic devices. Although large values of the tunneling magnetoresistance (TMR) were demonstrated [2], [3], [4], it seems to be not yet safe to predict a broad turnover of the RAM market from dynamic random access memory to MRAM.

Besides purely technological issues of the production process, this uncertainty is mainly due to aspects of the physical stability of the tunneling elements. The devices should be stable against thermal treatment up to 400°C and withstand about 10^{17} magnetization cycles without degradation of the TMR signal. Moreover, dielectric stability at tunneling voltages of some 100 mV is required for a large number of years. All these features have to be fulfilled for small elements with anticipated sizes of 200 nm×200 nm and below [5].

In this article, we discuss results obtained for the magnetic [6], the dielectric [7] and the thermal [8] stability of magnetic tunnel junctions consisting typically of different antiferromagnetically coupled trilayers as hard electrode, Al_2O_3 as insulator and a soft magnetic layer. The discussion will expand to elements with sizes down to 50 nm×50 nm.

6.2 Experimental

The magnetically hard electrode of our MTJs consists of an artificial antiferromagnet (AAF) (either Co/Cu/Co or CoFe/Ru/CoFe) in the first (Co/Cu) or second (CoFe/Ru) coupling maximum, which is an alternative or an addition to an exchange-biased system (antiferromagnetic pinning)[5,6]. The layer stacks are deposited in a dc magnetron sputtering system with a base pressure of $5 \cdot 10^{-8}$

mbar on the native oxide of Si(100) wafers. The magnetically soft bilayers (Fe/Co or $Ni_{81}Fe_{19}$/Co) have no intentionally induced magnetic anisotropy, but exhibit a shape anisotropy due to their elongated form in case of non quadratic junctions. The TMR defined as $(R_{max} - R_{min})/R_{min}$ is measured by two- and four-probe DC technique. The Al_2O_3 was prepared from Al films by plasma oxidation. For patterning, e-beam lithography with subsequent Ar ion etching was used.

6.3 Thermal Stability

The temperature stability of the junctions is a crucial point for their applicability due to the process requirements. Especially during the backend passivation of an MRAM chip (often with Si_3N_4), temperatures up to $400°C$ occur. Although low temperature passivation processes (down to $200°C$) are available, it will be very difficult to introduce these modifications in a RAM production facility.

6.3.1 Experimental – Thermal Stability

In order to obtain the influence of the magnetic film thickness, three AAF systems were prepared: Co(1 nm)/Cu(1 nm)/Co(1.5 nm) and (currently under investigation) CoFe(1 nm)/Ru(0.8 nm)/CoFe(1.5 nm). The samples were isochronally annealed in a high vacuum chamber with a base pressure of $5 \cdot 10^{-5}$ mbar up to 750 K for 1 h. The structural properties of the MTJs were investigated by Auger depth profiling in a Scanning Auger Microscope Phi 660 (Physical Instruments, SAM), atomic force microscopy (AFM), and scanning electron microscopy (SEM). In the SAM, the samples are continuously rotated during sputtering with a 500 eV Ar ion beam for optimum depth resolution. The Auger electrons are excited by a 10 keV/100 nA electron beam. The analyzed area has a diameter of about 70 μm. For all types of samples, similar structural changes are found after annealing.

6.3.2 Results – Thermal Stability

Resistance and TMR signal of more than 100 MTJs are measured before and after annealing. The as prepared MTJs have shown an average resistance of 50 k$\Omega\mu$m^2 . The resistance increases by annealing up to 530 K and then decreases monotonically. In Fig. 6.1, this behavior is shown for Co/Cu/Co-AAF (note that a similar dependence is found for Co/Ru/Co as hard layer system).

This behavior is independent of the area of the MTJs. The large statistical fluctuations at 670 K can be explained by a recrystallization of the layers and the formation of large crystallites (discussed below).

With a modified layer stacking and CoFe/Ru/CoFe as AAF, this remarkable stability of the tunneling barrier itself can be further improved. Depending on the thickness of the Al_2O_3, stable barriers up to about $500°C$ could be obtained.

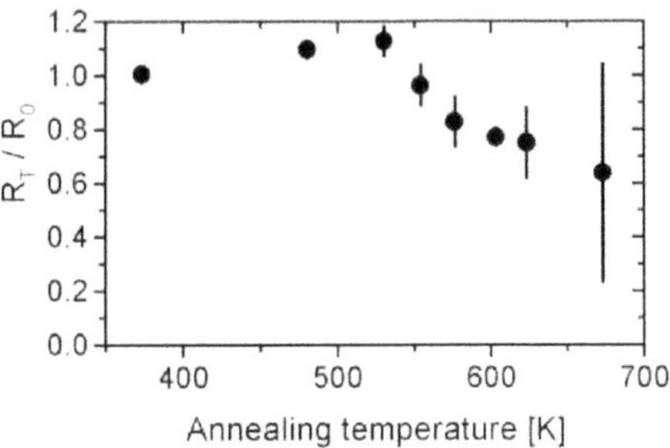

Fig. 6.1. Dependence of the normalized resistance R_T/R_0 on the annealing temperature (R_0 is the original resistance of the junction before annealing and R_T is the resistance of the MTJ after annealing), measured at room temperature.

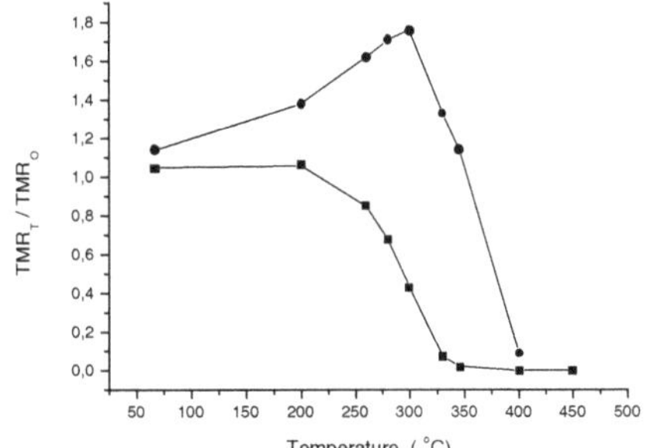

Fig. 6.2. Dependence of the normalized tunneling magnetoresistance TMR_T/TMR_0 (Squares: Co/Cu/Co-AAF, Circles: Co/Ru/Co – AAF) on the annealing temperature (TMR_0 is the original tunneling magnetoresistance of the junction before annealing and TMR_T is the tunneling magnetoresistance of the MTJ after annealing), measured at room temperature.

The TMR signals $\Delta R/R$ extracted from minor loops with a field range of typically ± 4 - 5 kA/m, which is well beyond the saturation field of the soft-magnetic Co/Fe bilayer are shown in Fig. 6.2 both for a Co/Cu/Co- and for a CoFe/Ru/CoFe-AAF.

For the Co/Cu/Co system, $\Delta R/R$ increases slightly (2 % at 370 K and 1 % at 210°C) by annealing up to 210°C and, then, decreases monotonically between 260°C and 330°C to only 1.4 % of the original value. Again, this behavior was found to be independent on the junction area. It should be noted, that the TMR vanishes at around 350°C although the junctions still show a tunneling behavior, i.e. are still operative. The same is true for the case of CoFe/Ru/CoFe-AAF's, where we find a strong increase of the TMR up to an annealing temperature of $\approx$ 350°C and a strong decrease for larger T_A. The increase of the TMR is attributed to an improvement of the barrier, which is also evident by an improvement of the dielectric stability discussed later.

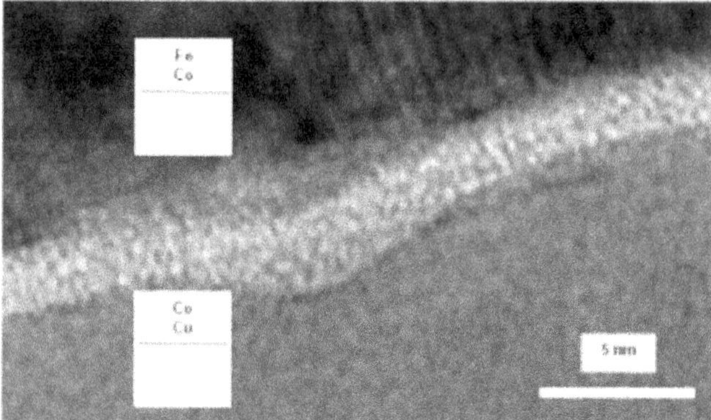

Fig. 6.3. TEM cross section after annealing at 400°C for one hour. There are no cracks or thickness changes of the barrier (light region) in comparison to the as prepared state.

The questions arising from these data are the obvious difference between the behavior of the tunneling resistance and the TMR signal upon annealing and the reasons for the degradation of the respective properties. For the degradation of the tunneling behavior, we usually found crystal growth and cracking of the Al_2O_3 as already mentioned. The first question is a more subtle one. Fig. 6.3 shows a TEM cross section of a TMR junction annealed at around 400°C. From this image, no hints for a degradation can be found.

Therefore, Auger depth profiling was performed prior to and after annealing (Fig. 6.4, note that the sensitivity of the Auger depth profile analysis is restricted to about 1.2 nm).

In Fig. 6.4, the AAF is visible as a small shoulder both in the Co as well as in the Cu intensity on the right hand side of the barrier in the as prepared state. The results showed for $T_A \approx 300°C$ a degradation of the AAF evidenced by the vanishing shoulder and a broadening of the Auger signals at the interfaces. Additional first arrival Auger experiments on the AAF itself done in situ during annealing showed, that Cu appears at this temperature at the surface of the top Co-layer. Thus a Cu diffusion through the top layer of the AAF to the interface with the Al_2O_3 barrier seems to occur at this annealing temperature.

This would be a possible reason for the decrease of the TMR, because Cu was reported to suppress spin dependent tunneling very effectively. The tunneling characteristic itself, however, contradicts this assumption, because the asymmetry of the dj_{tun}/dU_{tun} curve evolving at this annealing temperature has the wrong sign with respect to the work functions of Cu and Co.

Following the Auger results, however, there is no doubt that Cu diffuses to the barrier. In order to remove this inconsistency, we were lead to assume a semiconducting spinell-type $CuAl_2O_4$ layer forming at the interface between Co and Al_2O_3. For this system, quantummechanical calculations of the spin dependent tunneling were done using a transfer matrix approach in order to check the physical consistency of this assumption. In Fig. 6.5, we first show the experimental results for the dj_{tun}/dU_{tun} measurements.

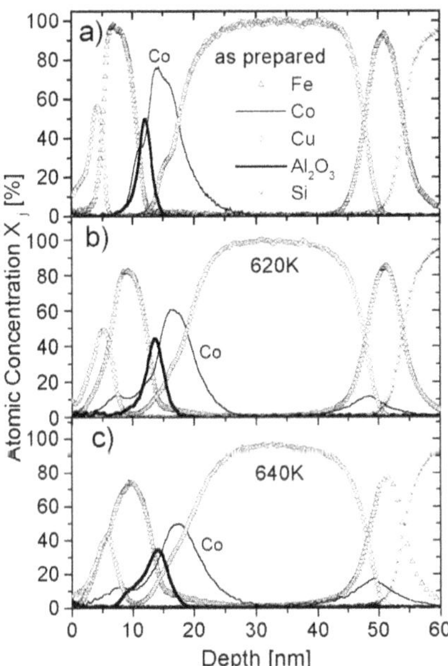

Fig. 6.4. Auger depth profiles of as prepared samples (a) and samples, which are annealed at 620 K (b) and 640 K (c) for 60 min. Only the profiles of Fe, Co, Cu, Al_2O_3 and Si are shown for clarity. The measured Auger intensities $I_j(t)$ (defined as peak to peak heights of the differential spectrum) of the different components j are converted into the atomic concentrations $X_j(t)$ by using relative sensitivity factors S_j. The sensitivity factors have been corrected with respect to the backscattering term14. Target factor analysis is used for removing some of the noise associated with the spectra, for deconvoluting overlapped Auger lines and for extracting chemical information. The calibration of the depth scale is based on the assumption, that the change of the sputtering rate dz/dt is proportional to the atomic concentration: $dz/dt = \sum_j X_j(t) \cdot \rho_j, (\rho_j$ is the sputtering rate of the pure component j).

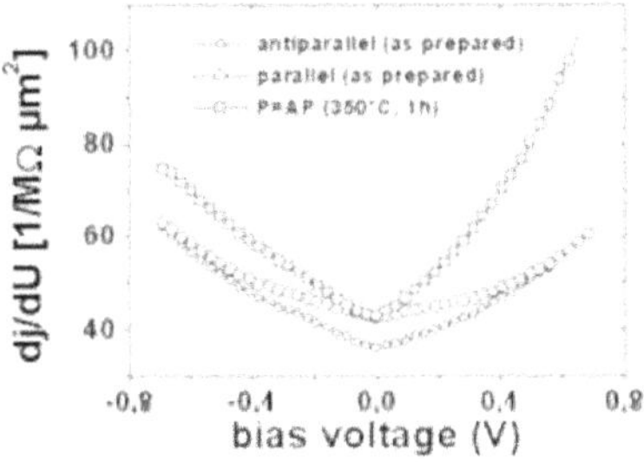

Fig. 6.5. Experimental results for the as prepared and annealed state. The TMR nearly vanishes above 350°C, i.e. the parallel and antiparallel curve are equal.

As already seen, the TMR vanishes for the Co/Cu/Co-AAF at $T_A \approx 350^oC$, i.e. there is no difference between the parallel- and antiparallel-state dj_{tun}/dU_{tun}-curve. At the same time, however, an asymmetry evolves which has to be at-

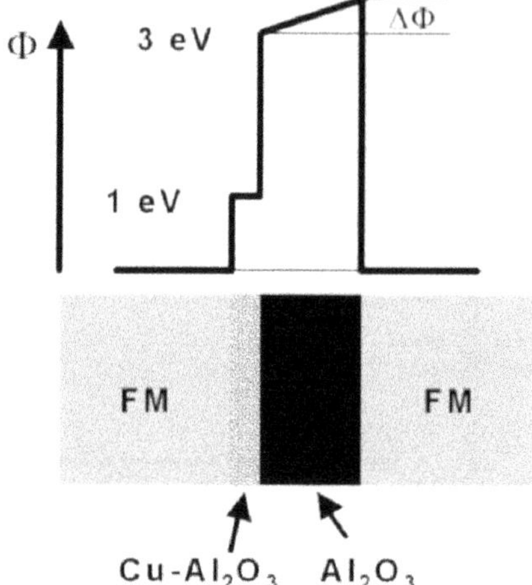

Fig. 6.6. Model of the step-like barrier with an intermixing zone at one of the barrier faces. $\Delta\Phi$ symbolizes a work function difference between the ferromagnetic electrodes (FM).

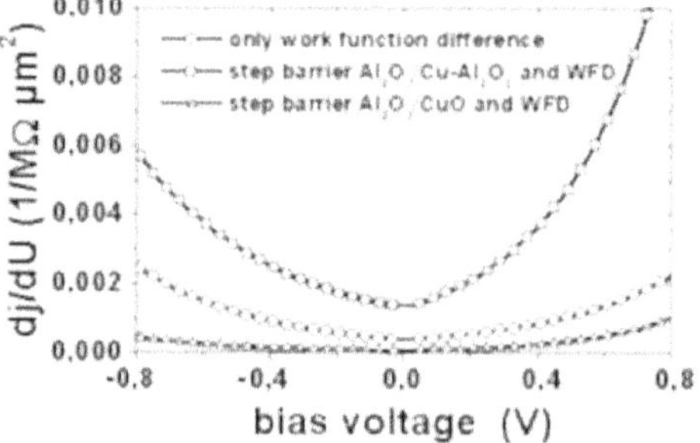

Fig. 6.7. Numerical calculation of a step-like barrier without image potential correction as sketched in Fig. 6.3. The Cu-Al_2O_3 zone is supposed to be 0.2 nm thick with a barrier height of 1 eV, about one third of Al_2O_3. The total barrier thickness is 1.5 nm.

tributed to a change of the potentials at the barrier. In order to explain these results consistently, we assume a step like barrier shown in Fig. 6.6.

Using this potential for the calculations, the results shown in Fig. 6.7 were obtained for different scenarios of material composition and distribution at the interface.

As can be stated by a comparison of Figs. 6.5 and 6.7, a very good agreement was obtained only for the case of a Al_2O_3/Cu-Al_2O_3 system including the related work function change, were the Cu-Al_2O_3 mixture at the interface is only about one or two monolayers thick and has semiconducting properties known from, e.g. $Cu_2Al_2O_4$.

At temperatures larger than around 400°C, additional diffusion processes can be found, depending on the layer sequence and the corresponding thickness. This will be subject of further investigations.

6.3.3 Conclusions – Thermal Stability

The good correspondence of the calculated with the measured curves and the results of the Auger experiments therefore give clear evidence that the diffusion of Cu to the interface between hard electrode and tunneling barrier degrades the TMR signal at a temperature $T_A \approx 300°C$ for the Co/Cu/Co-AAF.

For the CoFe/Ru/CoFe-AAF, we found similar results for larger annealing temperatures than for Co/Cu/Co ($T_A \approx 400°C$), i.e. Cu diffusing in this case from the wiring towards the barrier. The better stability of the Ru-containing AAF can then be simply explained by the larger distance from the Cu-conduction lines to the barrier and the very small diffusion of Cu through Ru.

Thus a suppression of the Cu diffusion will be one of the major tasks for any TMR element, because the conducting lines in a possible MRAM cell will most probably consist of Cu. With regard to the huge amount of knowledge about Cu diffusion barriers already gained from the Cu-DRAM and CPU development, this will probably not be a serious obstacle for MRAM's.

6.4 Magnetic Stability

Magnetic stability means that the magnetization state of the hard layer and the TMR amplitude are not affected by switching the magnetization of the soft electrode (e.g. minor loops). Some groups investigated the magnetization state of hard electrodes in tunnel junctions and spin valves, consisting of either hard magnets (e.g. $Co_{75}Pt_{12}Cr_{13}$) or exchange-biased layers (e.g. MnFe/Co). They reported that the magnetization of the hard magnets is unaffected by rotating cycles, but unstable if the soft layers of Co or CoFe are switched unidirectionally, whereas the magnetization in the exchange-biased system is stable even after switching cycles. This instability of the magnetization in the hard layer induced by cycling the soft layer can be attributed to the large fringing fields of the magnetic domain walls of the soft layer.

In Fig. 6.8, we show a micromagnetic simulation of a model system consisting of a soft permalloy electrode (3 nm thick), a 1.2 nm nonmagnetic gap and a 3 nm thick Co (hard) electrode. In this simulation, we have included some roughness at the interfaces.

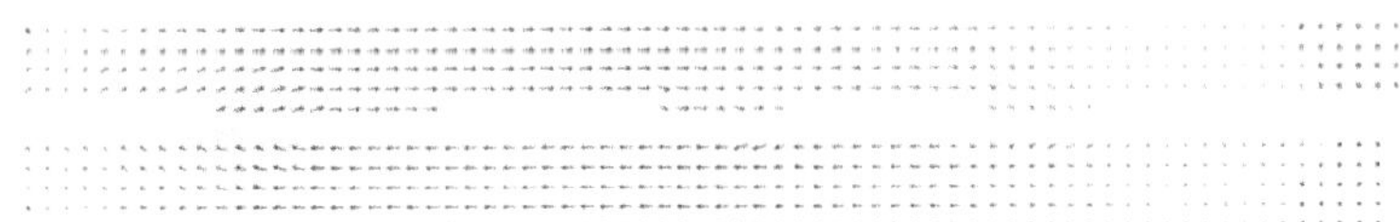

Fig. 6.8. Numerical micromagnetic simulation of the magnetization in a permalloy/Al_2O_3/Co system during switching the element. The starting point was complete saturation. The figure shows the magnetization at the coercive field. The arrows indicate the direction of the local magnetization. The grey scale is proportional to the out-of-plane component, indicating coupling of the magnetic layers.

The critical coupling mechanism is indicated in Fig. 6.8 by the correlated dark regions at the interfaces of the magnetic layers and the nonmagnetic spacer. This can give rise to a "magnetic fatigue" of the hard layer, because the stray fields of the soft layer's domains are large enough to induce domain splitting.

Here we discuss investigations of the magnetization state in MTJs with an artificial antiferromagnet as pinning layer. The TMR amplitude is measured after switching and rotating cycles of the soft electrode. Stable switching is reached using two perpendicular and time-shifted field pulses, demonstrating a feasibility of writing a MRAM bit by addressing the cell via two perpendicular current lines.

6.4.1 Experimental – Magnetic Stability

Prior to each measurement, the junctions were saturated in a 360 kA/m field along the easy axis of the rectangular elements. The initial TMR_0 is measured in a 4 kA/m rotating field (360° angle scan). Then, the field cycling of rotating (360° in a fixed field H_{max}) or switching ($+H_{max} \rightarrow -H_{max} \rightarrow +H_{max}$ in the initial saturation direction) is started. The cycle frequency was some Hz or 50 Hz for switching, 0.3 Hz for rotation and 3.6 Hz for double pulse switching. Finally, the TMR is measured in a $4\,kA/m$ rotating field again.

6.4.2 Results – Magnetic Stability

Rotating and Switching Cycles. Fig. 6.9 compares the dependence of the normalized TMR on H_{max} for switching and rotating of the soft electrode at a fixed cycle number n = 1.2×10^4.

The TMR for both soft electrodes is unaffected by rotating at 4 kA/m; for larger fields, the TMR decreases slightly. In case of switching, the TMR starts to decay immediately, with a more pronounced signal loss for Fe(Co).

The same result can be found at higher cycle numbers with a fixed field of 4 kA/m (Fig. 6.10). The TMR of Fe(Co) soft electrodes decreases more rapidly than the TMR of Py(Co) electrodes. In addition to the strong TMR decrease, for

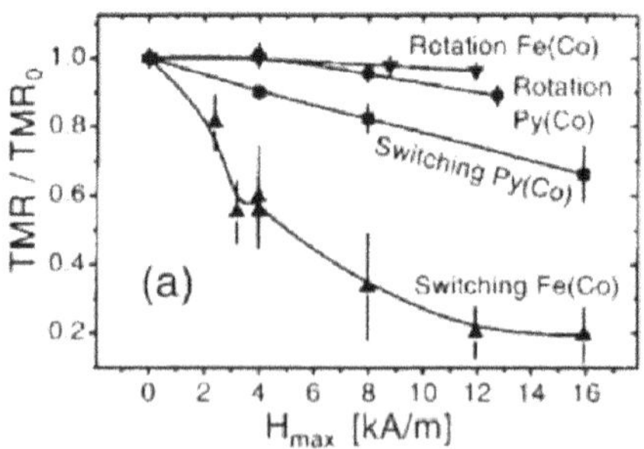

Fig. 6.9. Field dependence of the TMR signal (normalized to the initial value TMR_0) after 1.2×10^4 cycles of switching ($\pm H_{max}$ varying field) and rotating (H_{max} fixed field) at samples with two different magnetically soft electrodes: Fe(Co) and Py(Co).

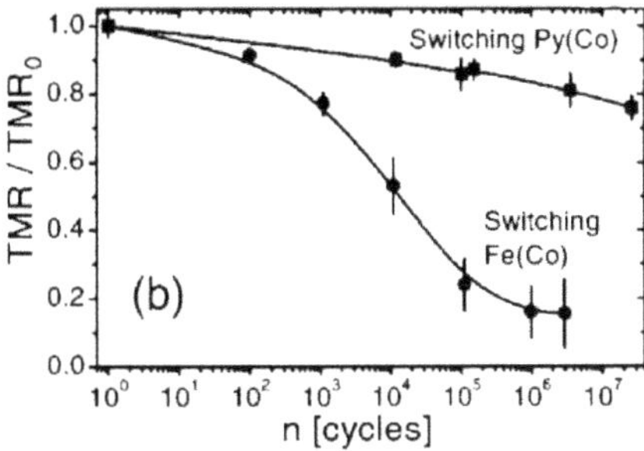

Fig. 6.10. Normalized TMR in dependence of the switching cycle number n in a maximum field of H_{max}=4 kA/m.

Fe(Co) a directional change (up to 180°) of the hard electrode's magnetization can be found after $n = 10^5$ cycles by angle scans of TMR at 4 kA/m. For Py(Co) however, the magnetization direction of the hard electrode remains stable. The stronger degradation of junctions with Fe(Co) can be understood by the larger domain wall width in Py(Co) compared to Fe(Co), which is related to a smaller fringing field.

A degradation of the AAF due to the external field can be excluded from the fact that both elements show no TMR decrease in a rotating field larger than the saturation fields of the soft electrodes. It seems reasonable to expect that the TMR stays constant at fields above 4 kA/m because the magnetization of the soft bilayers and the Fe buffer layer are completely saturated. In contrast, switching of large elements usually includes nucleation and movement of domain walls. Thus these experimental results strongly point to a degradation of the AAF by the fringing fields of domain walls in the soft electrode.

A further decrease of the TMR is observed, however, for both soft bilayers regardless of rotating or switching at H_{max}¿4 kA/m. Therefore, an intrinsic instability of the AAF must be taken into account. We expect, that this effect can be avoided by additionally coupling the AAF magnetically to a natural antiferromagnet. Corresponding experiments, which will be reported elsewhere, seem to support this assumption.

Double Pulse Switching. In a magnetic memory cell, the applied fields for switching the soft electrode should be kept as small as possible; additionally, the suppression of domain walls is essential to circumvent any fatigue. It would be, however, very complicated to apply a rotating field to an MRAM cell. A field sequence of a first pulse perpendicular to the original magnetization direction and a second one opposite to it, however, can produce first a rotation of the magnetization out of the easy axis and then a completion of the reversal by further rotation into the opposite (easy axis) direction.

Using Py(Co) we investigated whether this double pulse switching leads to an apparent "coherent rotation" of the magnetization of the soft bilayer, leaving the TMR signal unaffected after a large cycle number. The double pulse cycling

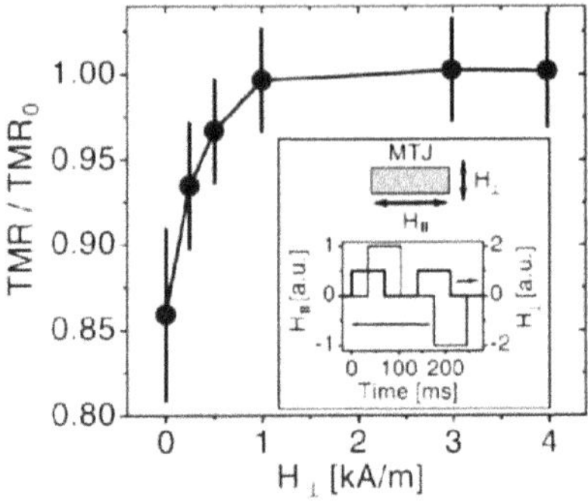

Fig. 6.11. Normalized TMR of MTJs with a Py(Co) soft layer in dependence of an applied perpendicular field pulse $H_\perp$ at a constant cycle number 10^5 and parallel field $H_\parallel$=4 kA/m. The inset shows the chronological order of the applied parallel ($H_\parallel$) and perpendicular ($H_\perp$) field cycles.

is done by periodically switching the fields $H_\parallel$ and $H_\perp$ (with $H_\perp$ first) with a phase shift of 90° (inset of Fig. 6.11).

While $H_\parallel$=4 kA/m is kept constant, the decay of the TMR during cycling reduces with increasing $H_\perp$ (Fig. 6.3), becoming completely suppressed at $H_\perp \gg$ 1 kA/m like in the case of pure rotation. The demagnetization of the AAF, which was induced by domain walls and identified as the main source to degrade the TMR signal, can thus be suppressed by tilting the magnetization of the soft bilayer Py(Co) with a field $H_\perp$ of only 1 kA/m, considerably smaller than the saturation field. This procedure obviously avoids domain walls in Py(Co) and favors a magnetization reversal similar to a "coherent rotation".

6.4.3 Conclusions – Magnetic Stability

The long-term TMR behavior of tunnel junctions with an artificial antiferromagnet as part of the hard layer depends critically on the magnetic stability of the magnetically hard electrodes. Whereas uniaxial switching of the soft electrodes degrades the AAF due to domain wall fringing fields, cycling with fields up to 4 kA/m avoids the decay of the TMR signal.

Simulating this rotation by a double pulse switching using perpendicular fields $H_\perp \gg$ 1 kA/m, showed that the degradation of the AAF can be also circumvented by this procedure, which can be simply realized in an MRAM architecture with two perpendicular (word- and bit-) lines. A further improvement of the magnetic stability should be possible by additional pinning of the AAF with a natural antiferromagnet.

6.5 Dielectric Stability

The integration of MTJs in the microelectronic process technology requires additional features like stability against dielectric (i.e. electrostatic) breakdown, which directly influences the lifetime of the devices. Breakdown voltages above 1 V are reported for "good" junctions only in the as prepared state. Here, we report on the evolution of the TMR and the voltage stability during annealing, and try to shed light on some physical processes leading to a degradation of the voltage stability and the barrier itself.

6.5.1 Experimental – Dielectric Stability

The as prepared MTJs show an area resistance of about $23\,\mathrm{k}\Omega\mu\mathrm{m}^2$ and a TMR signal of 18% in the minor loop at room temperature. In order to investigate the breakdown in dependence of thermally induced changes in the barrier, the samples are isochronally annealed up to 410°C for 1 h in a high vacuum furnace with a base pressure below $5 \cdot 10^{-5}$ mbar.

The dielectric breakdown of the MTJs is studied by increasing the voltage in 1mV steps with a constant rate of dU/dt=23 mVs^{-1} to a maximum of 12 V referred to the magnetically hard electrode. This is one of the standard methods to evaluate the lifetime.

6.5.2 Results – Dielectric Stability

Resistance and TMR signal of about 450 MTJs are measured before and after annealing. The area resistance of the MTJs changed upon annealing (see chapter 3) independent of the junction size. The dependence of the differential conductance $G_d = dI/dU$ on the tunneling voltage U shows the typical parabolic shape; fitting this dependence with Brinkman's model reveals an increasing asymmetry of the tunneling barrier with increasing T_A up to 380°C. Above 380°C the resistance drops strongly to $1\,\mathrm{k}\Omega\mu\mathrm{m}^2$ at 410°C and G_d becomes linear, i.e., shows ohmic behavior. Between 300°C and 350°C, the TMR drops down to 1% as already discussed within the context of thermal stability.

Dielectric Breakdown. In most cases, the breakdown of samples in the as prepared state occurs by a sudden large increase of the current at the breakdown voltage U_{BD} (Fig. 6.12). Upon annealing, both breakdown voltages indicated in Fig. 6.12 show a slight increase, which can again be explained by healing effects in the barrier. The breakdown itself originates from the formation of a highly conducting localized path, shunting the remaining tunneling resistance.

This interpretation is supported by the observation that the typical resistance R_0^{BD} after breakdown at zero bias is between $2\,\Omega$ and $5\,\Omega$, independent of the junction size. To show this experimentally, the bias current (maximum 100 mA) was limited additionally by a serial resistor R_{pre}. Then the resistance after the breakdown was $R_0^{BD} = 11 \pm 3\,\Omega$ for $R_{pre} = 220\,\Omega$ and $R_0^{BD} = 26 \pm 9\,\Omega$ for $R_{pre} = 330\,\Omega$. This evidences a thermal process because R_{pre} limits the current and hence the power for the thermally formed local path through the barrier.

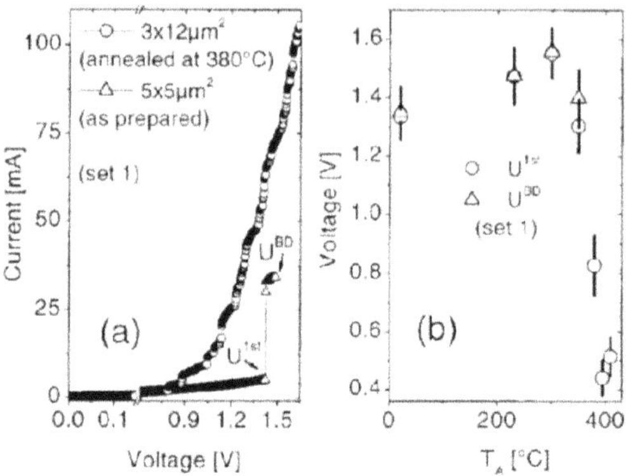

Fig. 6.12. a: Current / voltage characteristics of typical dielectric breakdown experiments for as prepared (triangles) and annealed junctions (circles; 380°C for 60 min). b: Dependence of the size-averaged mean breakdown voltages U_{BD} and U_{1st} on the annealing temperature.

Breakdown Statistics. Oepts et al. [9] analyzed their measurements of the junction breakdown with different models proposed earlier for SiO_2. They stated that only the assumption of a field induced displacement of ions in the barrier (E model) can explain the observed breakdown effects. Our data of the dielectric breakdown experiments are thus discussed using this model. In breakdown experiments, identical junctions are stressed by a time independent or time dependent voltage (or current) under the same conditions. If $F(t)$ represents the fraction of junctions breaking down after a time t, the breakdown probability density can be defined as $p(t) = (dF/dt)/(1-F)$. For the E model, $p(t)$ is given by $p(t) = A\ \exp(E(t)d_B/B)$, where $E(t) = U(t)/d_B$ is the time dependent electric field. If no explicit time dependence is assumed for $p(t)$ (no wear out), A is proportional to the junction area S_J and depends on the activation energy ΔH for dielectric breakdown: $A \propto S_J \cdot \exp(- \triangle H/kT)$. An external electric field $E(t)$ lowers the activation energy, which is described by the field accelerating parameter 1/B. For experiments with $dU/dt = constant$, $F(t)$ is given within the E model by

$$F(t) = 1 - \exp(-p(t)B(dU/dt)^{-1} + AB(dU/dt)^{-1}) \tag{6.1}$$

The maximum of dF/dt is found at

$$U_{max} = B\ \ln((dU/dt)/AB) \tag{6.2}$$

and corresponds to the most frequently found U^{BD} of the MTJ ensemble. For applying the E model, we measured U^{1st} and U^{BD} for an ensemble of 266 as prepared MTJs with a size of 100 μm^2 and $dU/dt = 23\,mVs^{-1}$. One should note that fitting Eq.(6.1) to experimental results is only reasonable for a large number of nearly identical MTJs, because the fitting parameters are extremely sensitive to the shape of $F(t)$.

The E model fit with Eq.(6.1) yields the parameters $A = (1.79 \pm 0.09) \cdot 10^{-11}\text{s}^{-1}$ and $B = (55.8 \pm 0.1)\,\text{mV}$. Note that no breakdown is found below 1 V. These values differ from those reported by Oepts et al. ($A = 6.3 \cdot 10^{-17}\text{s}^{-1}$ and $B = 35\,\text{mV}$), who used Eq.(6.2) for fitting the breakdown voltages of five junctions with a size of $2500\,\mu m^2$ measured at five ramp speeds dU/dt between 10^{-3} and $3\,\text{mVs}^{-1}$. If we extrapolate our data to a junction size of $2500\,\mu\text{m}^2$ and the data of Oepts et al. to a larger ramp speed of $23\,\text{mVs}^{-1}$, the breakdown voltages would be about the same, namely U^{BD}=1.2 V with 1.0 nm Al for ours and $U^{BD} = 1.3$ V with 1.8 nm Al for Oepts et al. Within the E model however, one would expect that the breakdown voltage increases linearly with the barrier thickness. We therefore suggest that this dependence is compensated by the low barrier height (1.0 eV) of their junctions.

MTJ Lifetime. For the application of MTJs in MRAM, a mean lifetime of about 30 yrs is necessary for a typical operating voltage of, e.g., 300 mV. The extrapolation of the junction lifetime to low voltage from the high voltage data, however, depends strongly on the employed model, similar to electromigration testing. Additionally, it is not clear that the junction degradation at low voltage is dominated by the same physical process as at high voltages.

For a time independent breakdown probability density p (e.g., the E model with $U = const : p = Aexp(U/B)$) the mean lifetime $\tau_{1/2}$ of a junction can be defined as $\tau_{1/2} = ln(2)/p$. With the parameters of our MTJs, the lifetime of $400\,\mu\text{m}^2$ junctions at 1000 mV is estimated to 160 s within the E model. This value, however, is much lower than the measured lifetime of several hours for this junction size, i.e., the real lifetime seems to be underestimated.

Nevertheless, the estimated lifetime for a $1\mu m^2$ junction at a typical operating voltage of 300 mV in realistic MRAMs results in about 600 yrs. On should additionally note, that the typical breakdown changes drastically to a multistep behavior for $T_A = 350°\text{C}$, although the area resistance remains nearly the same as for the as prepared samples. Within the E model, a significantly lower breakdown voltage is explained by a smaller activation energy and/or an increased field accelerating factor 1/B. This implies that the breakdown process for MTJs annealed at $T_A >350°\text{C}$ is triggered by defect formation in the barrier. A plausible mechanism is the diffusion of electrode atoms into the barrier. It should be emphasized that only a very small number of defects is necessary to dominate the breakdown process.

6.5.3 Conclusions – Dielectric Stability

In conclusion, $Co/Al_2O_3/Co$ tunnel barriers show a typical breakdown voltage between 1.3 and 1.5 V and their dielectric stability remains almost unchanged upon annealing up to 350°C. At higher temperatures a significant change in the breakdown mechanism is found. Instead of only a few large current steps a quasicontinuous current increase occurs during ramping the voltage, which can be explained by diffusion of electrode atoms into the barrier. A size dependence of the breakdown voltage can be understood by the statistical nature of the breakdown. From the E model, a lifetime of 600 yrs can be deduced for a junction area of 1 μm^2 at 300 mV operating voltage. For applications of MTJs in MRAMs, neither the lifetime nor the voltage of first current jumps $U_{1st} > 1\,V$ seems to be a limiting factor.

6.6 Ultrasmall Tunnel Junctions

One of the most interesting points is the magnetic and the related TMR behavior of MTJ's with tunneling areas well below 1 μm^2. For, e.g., MRAM applications, the anticipated size of the MTJ's is in the range of 0.2 μm×0.2 μm. Further downscaling to areas below 0.1 μm×0.1 μm will be inevitably necessary for the success of these new devices. There is, however, also interesting physics involved: As the MTJ becomes smaller and smaller, will there be still domains? Or will there be generally only some bending of the magnetization to S- or C-states as found by micromagnetic simulations. Or will the behavior be dominated by edge effects? In order to contribute to answering these questions, we customized an atomic force microscope (AFM) in order to characterize junctions as small as possible. In this section, we discuss briefly the instrumentation and show results for tunneling junctions with sizes ranging from about 1 μm×1 μm down to 0.05 μm×0.05 μm.

6.6.1 Experimental – Ultrasmall Tunnel Junctions

Using secondary ion mass spectroscopy, the Ar ion beam etching was stopped shortly after the Al signal vanished. Thus the patterned part of the films consisted in this case of a top CoFe/Al2O3(1.5 nm)/NiFe(6 nm)/Ta(5 nm)/Au(30 nm) layer stack, whereas the remaining base film was still continuous. The etch stop turned out to be one of the crucial points of the preparation routine.

The samples were subsequently characterized in two ways: On large junctions, conventional major and minor loops were taken in order to obtain information about the mean value of the TMR of the film stack. The small junctions where first magnetically saturated in order to have a well defined magnetization of the hard layer stack. Then, they were transferred to the customized conducting atomic force microscope (C-AFM) which allows topographic imaging and contacting the top electrode with a conducting diamond tip. Although the measurement seems to be straightforward, special care has to be taken in order to prevent electrostatic damaging of the MTJ's. In the C-AFM, we performed two types of measurements: Simultaneous topographic and conductance imaging by applying a constant ac voltage to the tip and measuring the current in lock-in technique. Secondly, we placed the tip on top of selected MTJ's and took TMR curves at some mV bias voltage. Due to the restricted field range of the Helmholtz coils surrounding the C-AFM, only minor loops could be taken. The field direction was always (anti-) parallel to the fixed magnetization of the hard layer along the long axes of the elements.

6.6.2 Results – Ultrasmall Tunnel Junctions

Scanning MTJ Arrays. In Figs. 6.13 and 6.14, we show a topographic image of an array of MTJ's (6.13) and the corresponding conductance (6.14). We did not obtain any current at areas between the MTJ's, because the AAF was exposed to air and oxidized in these regions.

On the Au covered elements, however, a stable current was obtained except on spots, where some e-beam resist remained. The elements appearing bright in Fig. 6.14 have high and the darker ones low conductance, respectively. For the elements in Fig. 6.14 with an area of $0.5\,\mu\text{m}\times 1.5\,\mu\text{m}$ one obtains a resistance of about $700\,\text{k}\Omega$ in good correspondence with the mean area resistance product of around $2\,\text{M}\Omega\mu\text{m}^2$ measured on $50\,\mu\text{m}\times 50\,\mu\text{m}$ large samples. The comparison of

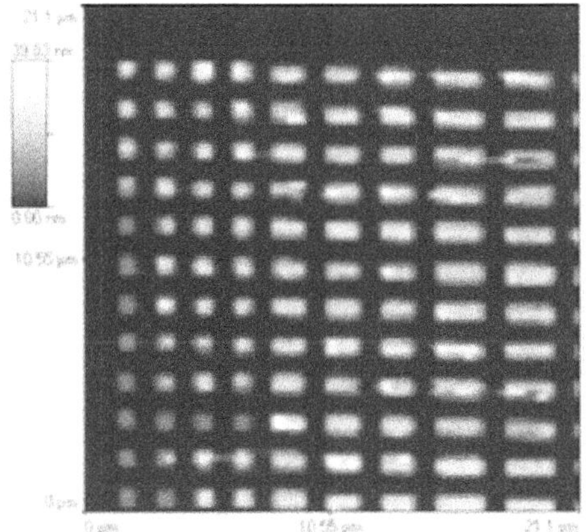

Fig. 6.13. Topographic image of an array of magnetic tunnel junctions with varying size and aspect ratio

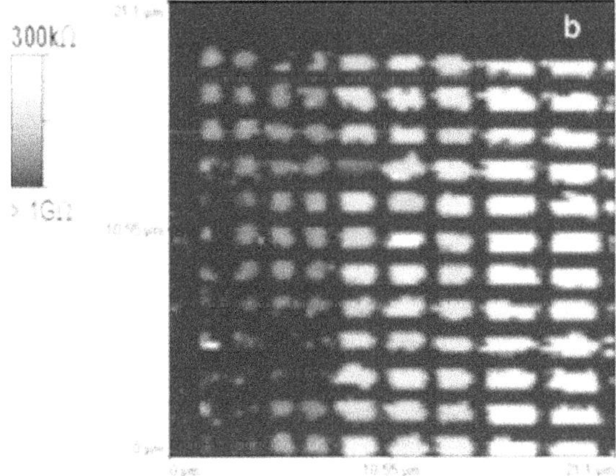

Fig. 6.14. The corresponding conductance image for some mV bias voltage. In the middle of the image, there is an element which shows much more conductance than expected.

Figs. 6.13 and 6.14 moreover gives information about defect MTJ's, e.g., that some of the smaller junctions in the lower left corner do not show any conductance. One of the larger elements (indicated in Fig. 6.14) has a conductance which is too high, probably due to a damaged barrier. Most of the MTJ's, however, show a conductance which corresponds very well with the expected values. For taking TMR curves, selected MTJ's were contacted with the C-AFM tip. In order to keep a stable electric contact to the Au on top of the elements during the TMR measurement, the force had to be increased to some 100 nN. This, however, was done in the static mode, i.e. without lateral movement of the tip. Subsequently, minor loops were taken as already described.

TMR Curves with C-AFM. Fig. 6.15 shows an example for the smallest MTJ's we were capable of measuring up to now. The area of this junction is approximately $0.05\,\mu\mathrm{m}\times 0.05\,\mu\mathrm{m}$. The noise in the resistance measurement is due to the large resistance of the element, which becomes comparable to the input resistance of the lock-in amplifier. Improvement of the experimental setup is currently on the way.

As a remarkable result, however, we would like to emphasize first, that these smallest MTJ's are still operative, i.e. show a well defined minor loop, a TMR amplitude between 10% and 17% and no kinks or steps. Such features were frequently found for larger MTJ's, most pronounced for areas around $1\,\mu m^2$ and can be related to magnetic domains in the soft electrode.

On larger elements (down to $0.1\,\mu\mathrm{m}\times 0.1\,\mu\mathrm{m}$), a TMR amplitude of around 19% was usually found which is close to the value of 20% of the $50\,\mu\mathrm{m}$ MTJ's. The larger range of the TMR for the smallest MTJ's can not yet be completely explained. One reason could be an edge contamination by metal deposition during the Ar ion etching which has a larger influence for smaller junctions. Again, however, the largest TMR measured on small junctions is close to the mean value of large elements.

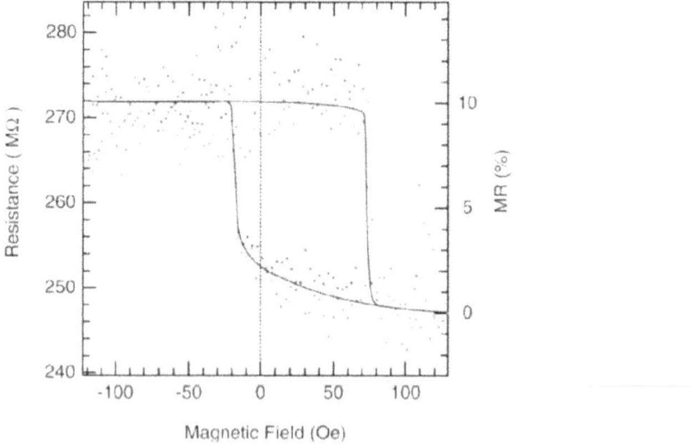

Fig. 6.15. TMR minor loops of an element with a size of $0.05\,\mu m \times 0.05\,\mu m$. The figure shows in two subsequent measurements in order to demonstrate the reproducibility.

Magnetic Peculiarities. Measuring smaller and smaller MTJ's, we additionally found an increasing shift of the minor loops from the field direction corresponding to a Néel-type ferromagnetic coupling of about H_N=10 Oe in the direction related with antiferromagnetic coupling of the electrodes adjacent to the barrier. In Fig. 6.15, this amounts to 25 Oe in positive field direction. The largest value we found on such elements was 58 Oe.

This can be explained by a dipole coupling across the edges, the influence of which increases as the MTJ's get smaller. The maximum field related with this dipole coupling can be analytically evaluated: First, we replace the outer edge faces of the element by magnetic poles with a strength of $M_s \cdot A$ (M_s: saturation magnetization, A: area of the elements edge face). Then, the related potential and the field H_D in the near field region can be calculated. Although this gives usually complicated integral equations, the fact that the film thickness is much smaller than the lateral dimensions allows for a considerable simplification of the results. For an element magnetized homogeneously along the length L, this approximation gives for the net field component parallel to the surface:

$$H_S = \frac{2M_s tW}{\pi\left(L^2+4x^2\right)\sqrt{1+\frac{W^2+4x^2}{L^2}}} - H_N \tag{6.3}$$

with t, W and L the thickness, width and length, respectively, and x the distance from the central point of the surface, i.e. x corresponds to the thickness of the barrier. A comparison with numerical simulations showed, that this simple formula holds within a few percent accuracy up to a thickness t and distance x of about one tenth of L and W. In case of quadratic elements (W=L) and for $x \ll L$ (i.e. the tunneling barrier much thinner than the in plane size of the element), Eq.(6.3) reduces to:

$$H_S = \frac{\sqrt{2}M_s t}{\pi L} - H_N \tag{6.4}$$

i.e. in this case an apparent 1/L-dependence of H_D appears as already found by other authors [10]. This dependence, however, is not valid in the general case of rectangular elements with $L \neq W$. It should be noted, that H_S does not vary strongly as a function of x in the near field region ($x \ll L, W$), i.e. H_S is almost constant over the thickness of the counter electrode (i.e. NiFe in our case). Moreover, the 'out of plane' components are much smaller than the parallel component in this region. As found by other authors with simulations and as given in Eq.(6.3), the dipole coupling field varies in a more complicated manner as a function of the element's width.

In Fig. 6.16, we show the net field H_S at a location 1.5nm away from the middle of the surface of a 2.5 nm thick CoFe electrode as a function of the element's length for different widths and experimental results.

For elements with an area of 0.05 μm×0.05 μm, Eq.(6.3) gives a maximum possible net coupling field of around 50 Oe. Thus the maximum field shift for the smallest elements should be in this range in agreement with the measured

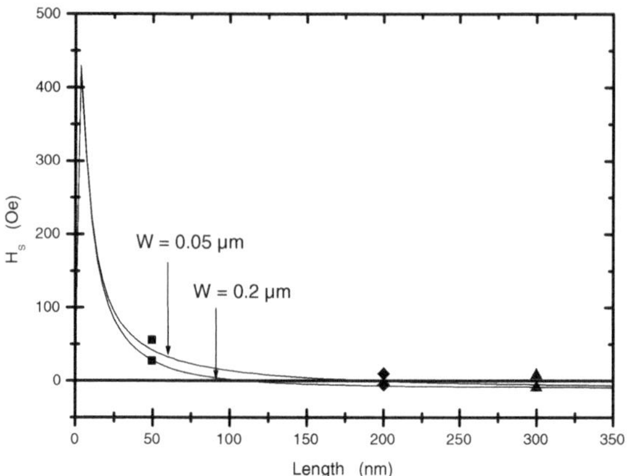

Fig. 6.16. The calculated net shift field at a distance of 1.5 nm from the center of a ferromagnetic electrode in dependence of the length for t=3 nm and W=0.05 μm and 0.2 μm (lines) and experimental results for W=L=0.05 μm (circles), L=0.2 μm, W=0.1 μm (diamonds) and L=0.3 μm, W=0.1 μm (triangles).

results. For elements with lateral dimension around 0.2 μm, we found net shifts around 0 Oe. This also corresponds to the analytic expression, because the dipole field at this size should be around 10 Oe and compensates the Néel coupling. Finally, it should be noted, that the influence of the dipole coupling will become considerably smaller as soon as the whole AAF is patterned. In this case, the effective field will be approximately

$$H_S^{eff} = \frac{2(t_1 M_s^1 - t_2 M_s^2) W}{\pi \left(L^2 + 4x^2\right) \sqrt{1 + \frac{W^2 + 4x^2}{L^2}}} - H_N \tag{6.5}$$

with $t_{1/2}$ the thickness of the ferromagnetic films of the AAF and $M_s^{1/2}$ the corresponding saturation magnetizations, respectively. Thus the same arguments leading to the well known stiffness of the AAF against external fields will also hold for the influence of the dipole coupling.

6.6.3 Conclusions – Ultrasmall Tunnel Junctions

In summary, we have shown in this last section, that C-AFM is an appropriate tool for the characterization of ultrasmall MTJ's. The results show, that a simultaneous topography and conductance scan can give a fast overview on the quality of the MTJ's. Minor loops of selected elements furthermore demonstrated, that tunneling elements with an area of 0.05 μm×0.05 μm are still operative, but can show a wide distribution of TMR signal heights and a large antiferromagnetic field-shift of the minor loop. These results can be explained by taking into account the dipolar coupling of the ferromagnetic electrodes which can be described by a simple analytic formula.

6.7 Summary

We reported on a variety of aspects of the stability of magnetic tunnel junctions, all of which are related with subtle physical mechanisms of diffusion, breakdown mechanisms and magnetic coupling phenomena. It turned out, that with respect to thermal stability Cu diffusion to the barrier's interfaces seems to be one of the most crucial points, because Cu will always be present within a MRAM layout at least in the electrical on chip wiring. Cu diffusion barriers and a Cu free AAF, however, seem to be an appropriate solution of this problem.

The magnetic switching of the soft electrode turned out to be another critical process due to the influence of the fringing fields of domain walls with the hard layer. In an MRAM layout, however, there will be most probably two perpendicular current (i.e. word- and bit-) lines. Thus timely correlated double pulses can be used to induce a coherent rotation of the magnetization of the soft layer, avoiding the magnetic fatigue of the hard one.

Dielectric breakdown occurs at voltages above 1V for a barrier thickness of about 1.2 nm. Thus the dielectric stability does not yet seem to be a serious obstacle for the currently anticipated MRAM cells. For very small elements, which may be important for the future development, however, the resistance and yet the barrier thickness have to decrease. Thus, although ultrasmall TMR junctions have been shown to be still operative, stability remains a critical issue for MRAM as well as for other applications in, e.g. logic devices.

Acknowledgments

The authors are indebted to J.Wecker and G.Gieres (Siemens AG Erlangen) for samples and valuable discussions and to the German Ministry of Education and Research (BMBF) for financial support.

References

1. J. S. Moodera, L. R. Kinder, T. M. Wong and R. Meservey, Phys. Rev. Lett. **74**, 3273 (1995)
2. S. S. P. Parkin, K. P. Roche, M. G. Samant, P. M. Rice, R. B. Beyers, R. E. Scheuerlein, E. J. O'Sullivan, S. L. Brown, J. Bucchigano, D. W. Abraham, Y. Lu, M. Rooks, P. L. Trouilloud, R.A. Wanner and W. J. Gallagher, J. Appl. Phys. **85**, 5828 (1999)
3. R.C. Sousa, J. J. Sun, V. Soares, P. P. Freitas, A. Kling, M. F. da Silva and J. C. Soares, Appl. Phys. Lett. **73**, 3288 (1998)
4. M. Sato, H. Kikuchi and K. Kobayashi, J. Appl. Phys. **83**, 6691 (1998)
5. T. Palucka , MRS Bulletin **25**, 4 (2000)
6. J.Schmalhorst, H.Brückl, and G.Reiss, Appl. Phys. Lett. **77**, 3456 (2000)
7. J.Schmalhorst, H.Brückl, M.Justus, A.Thomas, G.Reiss, M.Vieth, G.Gieres, and J.Wecker, J. Appl. Phys. **89**, 586 (2001)
8. J. Schmalhorst, H. Brückl, G. Reiss, G. Gieres, M. Vieth and J. Wecker, J. Appl. Phys., **87**, 5191 (2000)

9. W. Oepts, D. B. de Mooij, V. Zieren, H. J. Verhagen, R. Coehoorn, and W. J. M. de Jonge, J. Magn. Magn. Mater. **198-199**, 164 (1999)
10. A.Anguelouch, B.D. Schrag, G. Xiao, Y. Lu, P.L. Trouilloud, R.A. Wanner, and W.J. Gallagher, S.S.P. Parkin, Appl. Phys. Lett. **76**, 622 (2000)

7 Electrodeposition of Multilayered Nanostructures

Mürsel Alper

Department of Physics, Science and Art Faculty, Uludag University, 16059, Görükle, Bursa, Turkey

Abstract. Electrodeposition is one of the techniques used to produce nanostructured magnetic multilayers as well as single thin films. We have developed an experimental system to produce nanostructure materials such as metal/metal multilayers and multilayer nanowires by electrodeposition. The properties of such structures may depend on many electrochemical parameters such as the electrolyte concentration, the electrolyte pH, the deposition potentials of metals and the current efficiency. The growth mechanisms and the structural and magnetotransport properties of ferromagnetic multilayers grown by electrodeposition from a single electrolyte are studied.

7.1 Introduction

Over the last ten years magnetic multilayers have been one of the subjects of continued interest in solid state physics because of both scientific and technological applications. They consist of alternating ultrathin layers of a ferromagnetic and a non-ferromagnetic component. The ferromagnetic layers in a multilayer are a ferromagnetic transition metal such as Fe, Ni, Co or their alloys. These ferromagnetic layers are separated by non-ferromagnetic or noble-metal spacer layers such as Cr, Mn, Cu, Ag etc. When the layer thicknesses are less than the mean free path of an electron, magnetic multilayers may exhibit novel properties significantly different from those of bulk metals. The most remarkable property of a magnetic multilayer is giant magnetoresistance (GMR), which was first discovered in Fe/Cr multilayers grown by molecular beam epitaxy (MBE) in 1988 [1]. Since the discovery of the GMR, a large effort has been spent on the synthesis and characterization of magnetic multilayers such as Fe/Cr, Co/Cu, Co/Ag and NiCo/Cu and the GMR effect was observed in many magnetic multilayers [2–5]. Such structures have generally been produced using sputtering or molecular beam epitaxy (MBE) which require vacuum system.

There are various sputtering systems used to prepare the magnetic multilayers, for example, d.c., r.f., and ion beam sputtering. In these techniques, the different target materials are used for each constituent to be deposited, such as Cu and Co. During sputtering the target materials are bombarded by ionising particle beams such as Ar^+ and electrons from a heated filament under vacuum conditions. The sputtered atoms ejected from the target are directed to a substrate, which can be kept at a desired temperature and collected there to form a film. The layers of different materials is successively grown as the substrate, in turn, passes by each target. A computer can be used to control the process.

The individual layer thicknesses may be determined by the deposition rate of the sputtered atoms. During deposition, the hot particles emitted from the target also impinge the substrate and the growing film, and the gas ions used in the bombardment may also be entrapped in the deposit, such effects may destroy the uniformity and smoothness of the film. The sputtered samples, therefore, have a less perfect crystal structure and more rough interfaces than MBE. However, the sputtering systems are widely used to grow multilayered films since they are lower cost, faster and less complicated than MBE [6–11].

The other techniques used to produce multilayered structures are thermal evaporation (including effusions cells) and electron beam heating. In the electron beam heating, the materials to be deposited are evaporated by a beam of high-energy (greater than 3 keV) electrons. In the thermal evaporation, the materials to be evaporated are placed in resistively heated crucibles and the resulting vapour within the crucible is emitted as a beam of atoms (or molecules). In the case of growing epitaxial films using the effusion cells, this process is referred to as molecular beam epitaxy (MBE). In such a system, the multilayers are deposited on a suitable substrate in an ultrahigh vacuum chamber with pressure as low as 10^{-11} mbar. Each material to be deposited is evaporated from a separate effusion cell. The atomic (or molecular) beams from a small hole of the effusion cell travel to the substrate, which is kept at an appropriate temperature. It is possible to obtain a stable atomic (or molecular) flux simply by controlling the temperature of the effusion cells, which can individually be controlled by a computer. During deposition, the shutters over the cells are alternately opened and closed by the computer, depending on the desired layer thickness. Therefore, the multilayer growth is performed by switching the corresponding atomic (or molecular) beams on and off. The growth rate of the layers is kept low to provide good epitaxy and monitored using devices such as quartz crystal oscillators and mass spectrometers. Since the substrate temperature and the growth rate can be controlled, the sharp interfaces between the layers are obtained and hence a perfect crystal. A MBE system can also be furnished with in situ characterization tools such as Reflection High Energy Electron Diffraction (RHEED), which gives information about the crystallographic structure of the multilayered film and the oxide of the surface during deposition [6–8,12].

However, it is possible to grow magnetic multilayered nanostructures by electrodeposition. In comparison to other methods, the electrodeposition technique has some advantages, for example, simple, high deposition rate and low cost. It is also a room temperature and pressure technology. Although the technique requires a conductor as substrate it is possible to employ semiconducting substrates or insulating substrates with evaporated or sputtered seed metallic layers as thin as 10 nm.

In the production of the layered structures, electrochemical deposition was first used to deposit alternating layers of pure Cu and Ni from two separate electrolytes by Blum [13]. Later, Brenner was able to produce Cu/Bi alloy multilayers from a single electrolyte and succeeded in growing the multilayers with layer thicknesses as thin as 300 nm [14]. After the multilayers with ultrathin (a few nm) layer thickness were electrodeposited from a single electrolyte containing two metal ions [15], the technique has been used more frequently for the production of multilayered films, and high quality multilayers with the layer thicknesses 1-2 nm have been grown epitaxially from a sulphamate bath [16,17]. It was reported that the electrodeposited multilayers also exhibited significant GMR, which was first observed in CoNiCu/Cu multilayers [18]. These multilayers were grown on (100) textured polycrystalline Cu substrates. After growth, the films were required to be removed from their substrates for magnetotransport measurements because of the conducting Cu short-circuit. This might be a disadvantage for the electrodeposition process. Therefore, recently several multilayers have been electrodeposited on semiconductors such as Si and GaAs and it was found that they exhibit GMR [19–21]. The room temperature GMR values measured in the current in the film plane (CIP) has generally remained below 30% for electrodeposited multilayers [22–29]. However, a GMR magnitude of 55% has been reported for an electrodeposited Co/Cu multilayers at room temperature [30], but this result has not yet been confirmed by other investigators studied on electrodeposited multilayers.

Also, the electrodeposition technique has successfully been used to fabricate granular alloys [31,32] and multilayer nanowires [33–37] exhibiting GMR as well as conventional multilayers. Magnetic multilayered nanowires give a good way to measure the resistance in the current perpendicular to the film plane (CPP). The CPP geometry is more interesting than the CIP, because the CPP-GMR is considerably larger than the CIP-GMR obtained for the same material [38] and because it may be used to determine the spin diffusion lengths [39,40]. The measured room temperature GMR value is larger than 50% for CoNi/Cu multilayered nanowires electrodeposited in aluminum oxide membranes [35,36]. Therefore CPP-GMR in multilayered nanowires is the promising systems for sensor applications. For such nanostructure materials, the electrodeposition is a production technique that is easy to apply and inexpensive.

Previous studies show that many physical properties of electrodeposited conventional and nanowired multilayers, including GMR, may depend on several factors such as layer thicknesses, the composition of metals in the film, the electrolyte pH value and the crystallographic orientation. The studies on conventional and nanowire multilayers grown by electrodeposition were reviewed by several authors [41–43].

In this study, we describe the experimental system developed to produce nanostructure materials such as multilayers and nanowires by electrodeposition in our department. The synthesis and structural characterizations of ferromagnetic multilayers such as Ni/Cu and CoNi/Cu grown on (100) textured polycrystalline and (100) single crystal copper (Cu) substrates are considered. The magnetotransport properties of electrodeposited multilayers are also summarized.

7.2 Fundamentals of Electrodeposition

Electrodeposition of a metal takes place by the reduction of its ions at the cathode of an electrochemical cell which contains metal ions. The ions receive electrons from the cathode and therefore are reduced to the metal. The reaction occurring at the cathode can be represented as

$$M^{n+} + ne^- \to M \tag{7.1}$$

where M^{n+} and M denote a metal ion and metal atom respectively, e^- indicates an electron and n is the number of electrons transferred per atom (the valance of the metal). Each metal has a reduction potential which depends on the concentration of its ions in solution. In solutions containing metal ions of more than one species, the ions of the more noble metals such as Ag and Cu are usually reduced in preference to those of the less noble metals such as Ni and Co. Hence the more noble metals deposit preferentially so that their concentrations in the deposit may be greater than those of the less noble metals, even though their concentrations in the solution are lower. However, in solutions containing one or more of the three metals of the iron group (Fe, Ni and Co), anomalous codeposition is generally observed, that is, the less noble metal deposits preferentially [14].

During electrodeposition, the amount of the material deposited can be monitored using Faraday's law, because the charge passed (q) is proportional to the number of moles (N) of metal deposited and is given by

$$q = NnF \tag{7.2}$$

where F is Faraday's constant (96487 Coulomb/mole). The number of moles deposited N can be found by dividing the mass m of deposited metal by its relative atomic mass M_a. Hence the thickness t of the deposited film can be calculated from:

$$t = \left(\frac{qM_a}{A\rho Fn}\right) \tag{7.3}$$

where A is the surface area of the film and ρ is the bulk density of the deposited metal. The most important parameter affecting the thickness of a film is the current efficiency, which may be defined as the ratio of the actual amount of metal deposited to that expected theoretically according to (7.3). For a given metal, the layer thickness calculated in this way is called the nominal thickness of that metal. Since the current efficiency may be smaller than 100%, in practice in order to find the real layer thickness it is necessary to determine the current efficiency for each metal deposition.

The parameters that play particularly important roles in the growth of an electrodeposited film are the deposition potentials, the pH of the electrolyte, the current density and the electrolyte temperature. In addition, the concentration of ions of the metal being deposited in the electrolyte, the presence of additives or impurities, and whether the current is continuous or pulsed are important variables in electrodeposition. All these factors can control whether epitaxial growth takes place or not. Some parameters, especially potential and concentration of metal ions may need to be varied until the optimum conditions to obtain high quality films are found [44,45].

In electrodeposition of multilayers, two techniques are used: One is the dual-path technique, the other is the single-bath technique. In the dual-bath technique, a conducting substrate is immersed into two different electrolytes alternately while controlling the applied voltage or the current. During the growth process, the sample must be transferred rapidly from one solution to the other. Although pure layers of metals can be obtained, the dual-bath technique is not usually preferred because the sample is exposed to atmospheric contamination during transfer between solutions and because the apparatus needed is quite complicated [41,46].

In the single-bath technique, it is possible to deposit multilayers from a single electrolyte by changing the voltage and current density. If there are several species of ion in an electrolyte, they will be reduced at different potentials since each species has a different standard electrode potential. In a solution containing ions of metals A and B, if it is assumed that metal A is more noble than metal B, in order to deposit metal A the applied potential is kept at a more negative value than the reduction potential of A, but more positive than that of B. A more negative potential than the reduction potential of B is then applied for the deposition of B. During deposition of B, material A will codeposit with B so that a multilayer deposited from a single electrolyte consists of alternating layers of pure A and A-B alloy, but the A content in the A-B alloy layers can be minimized by reducing the concentration of A in the electrolyte [15,16,42,47].

7.3 Experimental Methods

Fig. 7.1 shows the schematic diagram of an experimental system developed for the electrodeposition technique. The system is the similar one to those in previous studies [16,17,42,47]. It consists of a potentiostat with three electrodes, a computer, a digital to analog converter (DAC), an analog to digital converter (ADC) and an electrochemical cell. The potentiostat (EGG Model 362) keeps the working electrode (cathode) potential fixed at the selected value with respect to the reference electrode (R.E.). The DAC and ADC are used to interface the computer and the potentiostat. The computer controls the process and also integrates the current to obtain the charge passed between the anode and the cathode of the cell.

Film growth in electrochemical deposition depends significantly on the choice of substrate material and its preparation. For this work, polycrystalline copper plates or single crystal (100) copper disks are used as substrates. The polishing of the substrates is the first step in the growth of a multilayer. Therefore, first, one face of each substrate is mechanically polished using emery paper, and then covered with electroplating tape, except for a circular area (a few cm^2) of the polished face. The uncovered area, which is exposed to the electrolyte, is

Fig. 7.1. Schematic diagram of experimental system used for electrodeposition

electropolished in a glass dish containing a 50% (w/o) orthophosphoric acid electrolyte at a suitable voltage vs. a counter electrode, for example, carbon rods. After electropolishing, the substrate is immersed, in turn, in 25% H_3PO_4, 10% H_3PO_4, distilled water, 10% H_2SO_4 and distilled water.

As soon as the polishing procedure is completed, the substrate is placed immediately in a electrolyte used to deposit desired metals. Ni/Cu (or CoNi/Cu) multilayers are grown from a single electrolyte containing Ni^{2+} and Cu^{2+} (or Ni^{2+}, Cu^{2+} and Co^{2+}) ions using the potentiostatic deposition method. During deposition, the electrolyte temperature is generally kept at about 30°C. The deposition is carried out at the cathode of a cell with three electrodes and monitored by the personal computer (PC) with our own software. A platinum (Pt) sheet with an area of 16 cm^2 is used as counter electrode (C.E.), which is the anode terminal of the cell. The reference electrode is a saturated calomel electrode (SCE). Copper substrates are connected to the working electrode (cathode) terminal of the cell. For both Ni/Cu and CoNi/Cu multilayers, the deposition of copper layers can be made at the cathode potentials such as -0.2 V or -0.25 V with respect to the SCE, while Ni layers (or NiCo layers) can be deposited at the cathode potentials such as -1.6 V, -1.7 V or -1.8 V with respect to the SCE. At the Cu deposition voltage, only Cu which is the more noble metal will be deposited, while at the Ni deposition voltage both Cu and Ni (or CoNi) will be deposited at rates limited by their concentrations, and hence the Ni (or CoNi) layers always contain Cu as well. For this reason, the Cu layers are made of pure Cu while ferromagnetic layers will be an alloy containing Cu. The NiCu/Cu or CoNiCu/Cu multilayers (for brevity, referred to as Ni/Cu and CoNi/Cu respectively) can be grown by altering the cathode potential between two values chosen for Ni (CoNi) and Cu deposition. When the charge passed between the anode and the cathode reaches the value corresponding to the desired thickness of multilayer component being deposited, the computer changes the working electrode potential to start depositing the other component. The charge amount required for the desired layer thickness is calculated in according to (7.3) by assuming 100% current efficiency for each metal and using bulk densities.

However, since the hydrogen evolution leads to a decrease in the current efficiency, in particular at low pH values and high deposition potentials, the current efficiency for each metal should be determined before beginning to grow multilayers systematically. This can be done by plotting the modulation wavelength (repeat distance) of a series of multilayers (which can be obtained from x-ray diffraction patterns) as a function of nominal layer thickness of one component while keeping the nominal layer thickness of the other component at a fixed value. Since the more noble metal (Cu layers) is deposited at lower potentials, its current efficiency may be assumed close to 100%, but for the less noble metals (Ni and NiCo layers) deposited at high potentials it is generally less than 100%. The current efficiency for these metal layers was found to vary with the electrolyte pH value between 80% and 50% [47,48].

After growth, the multilayers can be stripped from their Cu substrates electrochemically using a chromic acid solution, and mounted on an insulator such as glass, which is required to make some measurements such as magnetoresistance.

7.4 Electrochemical Characterization

The solution prepared for the deposition of Ni/Cu multilayers was characterized by the method of cyclic voltammetry. The cyclic voltammogram curve of the solution is shown in Fig. 7.2. The scan was performed on a Pt wire electrode with a diameter of 0.5 mm at 30°C and between -1 V and +1 V with respect to the saturated calomel reference electrode (SCE). A platinum foil electrode served as counter electrode. The rate of change of applied potential was 20 mV/s. This current-voltage curve provides us to choose the potentials to be applied for the deposition of metals. As seen in Fig. 7.2, there are two strong peaks on the cyclic voltammogram curve. One is at -0.3 V vs. SCE in the cathodic side and corresponds to Cu deposition, the other is in the anodic part, at about +0.15 V

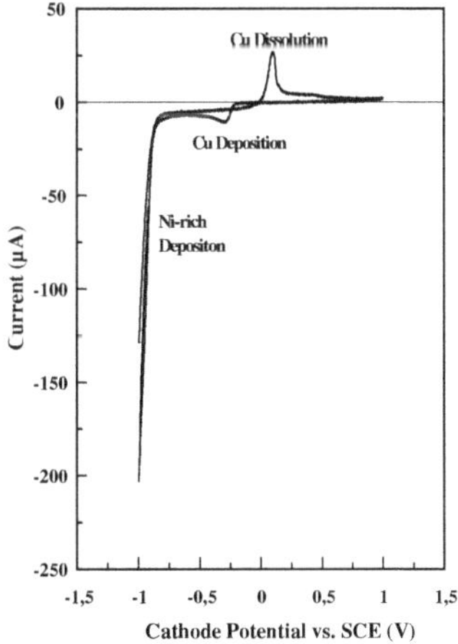

Fig. 7.2. Cyclic voltammetry of the solution used to deposit Ni/Cu multilayers

vs. SCE and is due to Cu dissolution. In the potential range between -0.4 V and -0.8 V, a current plateau with low current occurs. This is probably arising from diffusion limited Cu deposition. Although a large cathodic current starts flowing at about -0.8 V, presumably corresponding to deposition of Cu and Ni in the electrolyte and possibly also H_2 generation, this potential gives black deposits if used to grow multilayer. To grow "high quality multilayers" that are metallic in appearance, NiCu alloy (Ni-rich) layers must be deposited at the potentials higher than -1.5 V vs.SCE.

Fig. 7.3 shows the variation of current with time during the growth of a multilayer with 100[Ni(2.0 nm)/Cu(2.0 nm)] grown on a (100) textured polycrystalline Cu substrate. Similar current transients have also been observed for multilayers grown on (100) single crystal Cu substrates. These curves show similarity to those reported for multilayers such as Co/Cu [27,28], NiCo/Cu [48] and NiFe/Cu [29]. In cathodic part, the high current pulses correspond to the deposition of Ni layers while the low current pulses to the Cu layers. When the cathode potential is switched from the Ni deposition voltage to the Cu deposition voltage, a large anodic current is observed. It is thought that this anodic peak may be due to redissolution of less noble metal and/or to the oxidation of absorbed or occluded hydrogen during the cathodic deposition of ferromagnetic layer [28,48]. For the first few Cu layers, the Cu deposition current is large because the initial Cu concentration close to the surface of the electrode is large. After the Cu layer is deposited, up on switching to the nickel deposition potential the increase in the current is due to the high concentration of Ni in the electrolyte. The Ni deposition current is almost constant for all layers, but the Cu deposition current only reaches a steady value of about 3 mA after depositing a few layers. Note that the Cu deposition time varies with the Cu current, indicating the effect of the computer control maintaining a constant layer thickness.

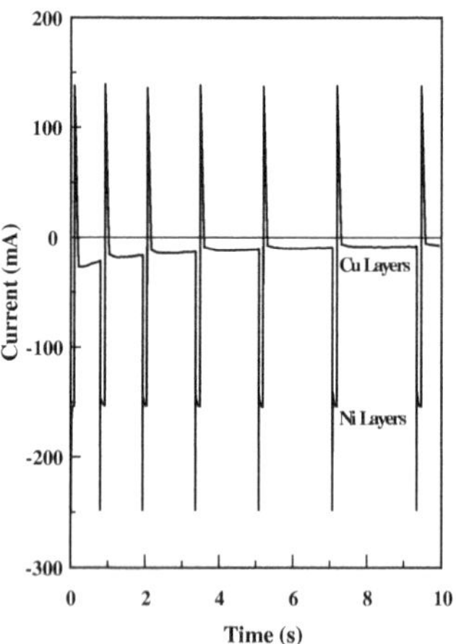

Fig. 7.3. Current transients for the first few layers during the growth of a Ni/Cu multilayer

7.5 Structural Characterization

X-ray diffraction (XRD) is still one of the most widely used techniques to determine the atomic structure of crystalline materials. It is also a non-destructive technique and does not affect the properties of a sample. This technique can also be used to study the structure of multilayers as it gives information about the modulation wavelength, strain, how sharp the interfaces are, and how periodically the individual layers are stacked [49,50].

Our samples were structurally characterized in the large angle range (40° - 60°) using the XRD with Cu K_α radiation (wavelength 0.15406 nm) on their substrates and after removing them. The XRD pattern of a multilayer with 150[Ni(2.0 nm)/Cu(2.0 nm)] grown on a single crystal Cu (100) is shown in Fig. 7.4. In this figure, the peak labelled as S(200) at about 50.5° belongs to the substrate, and the peak labelled as ML(200) corresponds to the mean spacing of the atomic planes within the multilayer itself and is called the main Bragg peak of the multilayer. These reflections from (200) planes of the multilayer and the Cu(100) substrate indicate that the multilayer structure is oriented in the [100] direction as its substrate. The satellite peaks up to third order, which are labelled as (+1), (-1), (+2) (-2) and so on, arising from the constructive and destructive interference of the X-rays scattered by alternate layers of the multilayer, are apparent in the XRD pattern. This shows that the multilayer has a good periodicity. The modulation wavelength (Λ) can be directly calculated from the positions of two satellite peaks having the same order using the relation [16,51],

$$\Lambda = \frac{n\lambda}{\sin\theta_{+n} - \sin\theta_{-n}} \tag{7.4}$$

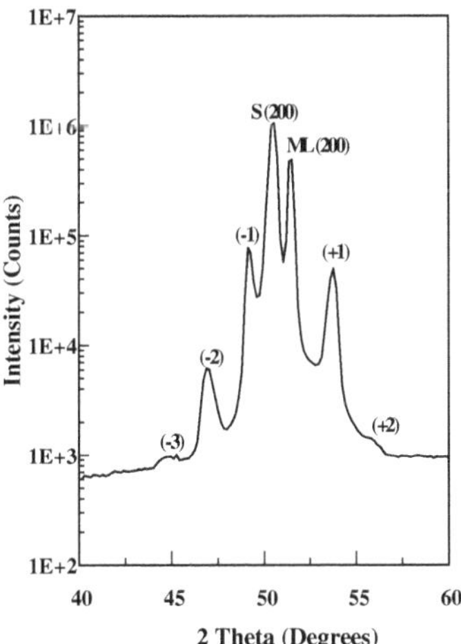

Fig. 7.4. XRD pattern of a multilayer with 150[Ni(2.0 nm)/Cu(2.0 nm)] grown on a single crystal Cu(100) substrate

where $\lambda = 0.15406\,\text{nm}$ for Cu K_α radiation, and $-n$ and $+n$ refer to adjacent satellite peaks of the same order around the main Bragg peak. The modulation wavelength was calculated to be $(4.28\pm0.32)\,\text{nm}$ using (7.4) from the first order satellites peaks of the XRD pattern in Fig. 7.4, which agrees with the bilayer thickness of the multilayer predicted by considering the current efficiency. Similar results were also obtained for multilayers with different individual layer thicknesses and bilayer numbers, showing that superlattice growth is highly reproducible.

Fig. 7.5 shows the XRD pattern of a multilayer with 50[Ni(3.0nm)/Cu(2.0nm)] grown on a (100) textured polycrystalline Cu substrate. The substrate has been removed to clearly observe satellite peaks. The main Bragg peaks of the multilayer, which are ML(111) and ML(200) peaks and correspond to the mean interatomic spacing in the multilayer, are observed. The ML(200) peak is stronger than the ML(111) peak. This shows that the multilayer has the same strong (100) texture as its polycrystalline Cu substrate. The first order satellite peaks are apparent on both low angle and high angle side of the ML(200) Bragg peak. The modulation wavelength was found to be $(5.34\pm0.37)\,\text{nm}$ using (7.4), which agrees with the bilayer thickness of 5 nm of the given multilayer. It was also observed that in CoNi/Cu multilayers the periodic structure was formed although the ferromagnetic CoNi layers contain Cu as much as 20% [48].

7.6 Magnetoresistance

Magnetoresistance (MR) is defined as a change in the electrical resistance of a material when it is placed in an external magnetic field. The resistance can increase (positive MR) or decrease (negative MR) with increasing magnetic field.

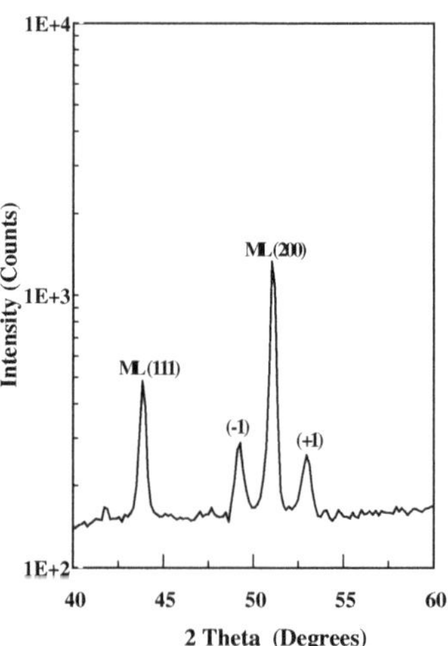

Fig. 7.5. XRD pattern of a multilayer with 50[Ni(3.0 nm)/Cu(2.0 nm)] grown on a (100) textured polycrystalline Cu substrate, after removing its substrate

The magnitude and sign of the magnetoresistance are different for different metals. For a given metal, the measured MR may depend not only on the strength of the magnetic field, but also on the direction of the magnetic field with respect to the current flowing through the material.

In order to make the magnetoresistance measurements four point contacts are applied to the corners of a square of film as shown in Fig. 7.6. This geometry is known as the van der Pauw (VDP) method [52]. When the film is oriented as shown in Fig. 7.6, R_L is defined as the potential difference $V_D - V_C$ between contact D and contact C per unit current flowing from contact A to contact B and is the "longitudinal" magnetoresistance (magnetic field parallel to the current in the film plane). Similarly, R_T is defined as the ratio of the potential difference $V_D - V_A$ between contact D and contact A to the current flowing from contact C to contact B, and is the'transverse' magnetoresistance (magnetic field perpendicular to the current in the film plane).

In ferromagnetic single layer thin films such as Ni, Fe, Co and their alloys, the resistance increases with increasing magnetic field when the field is parallel to the electric current, but decreases with increasing field when the field is perpendicular to the current. This effect is called anisotropic magnetoresistance (AMR) and is caused by the spin-orbit interaction [53]. In order to observe this effect single layer Ni films of various thickness in the μm range were electroplated on a (100)textured polycrystalline Cu substrates from a nickel sulphamate bath, which contains a small amount of Co^{2+} ($\sim$ 2%) in addition to Ni^{2+} ions. After deposition the films were stripped from their substrates and mounted on glass, and then magnetoresistance measurements (longitudinal and transverse) were made in the film plane by applying a magnetic field as shown in Fig. 7.6. As an example, the room temperature magnetoresistance curve of a Ni single layer

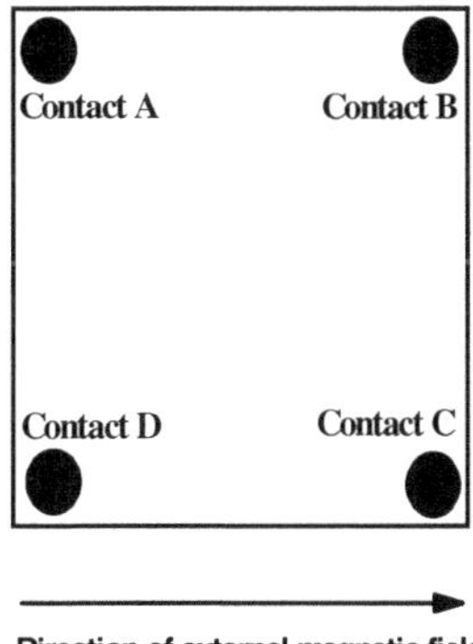

Fig. 7.6. Van der Pauw (VDP) geometry used for magnetoresistance measurements of a square of film. Longitudinal magnetoresistance (magnetic field parallel to current) is measured by passing the current between contact C and contact D, and measuring the voltage between contact A and contact B, while transverse magnetoresistance (magnetic field perpendicular to current) is measured by measuring the potential difference between contact A and contact D per unit current flowing from contact C to contact B

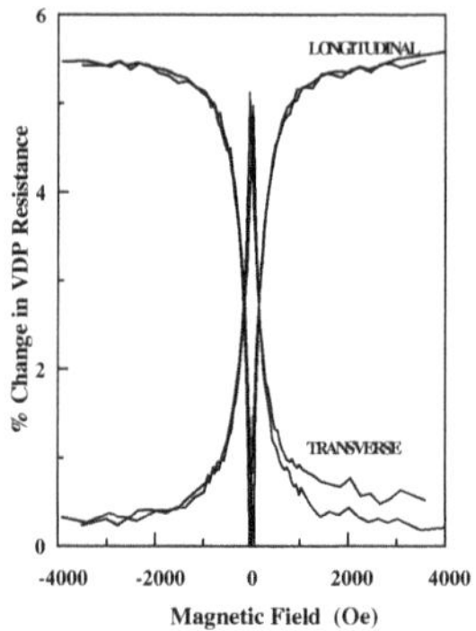

Fig. 7.7. Magnetoresistance curves for an electrodeposited single Ni film with a thickness of $1\,\mu m$

film with a thickness of $1\,\mu m$ is shown in Fig. 7.7. The percentage changes in the van der Pauw (VDP) resistances R_L (longitudinal MR) and R_T (transverse MR) ($\Delta R/R$) were calculated using the relation:

$$\frac{\Delta R}{R}(H) = \left[\frac{R(H) - R_{\min}}{R_{\min}}\right] \times 100 \tag{7.5}$$

where $R(H)$ is the resistance in any magnetic field and R_{min} indicates the minimum resistance in transverse or longitudinal magnetoresistance measurements, which correspond to that in maximum magnetic field and that in coercive magnetic field respectively. As shown in this figure the longitudinal magnetoresistance increases with increasing magnetic field whereas the transverse magnetoresistance decreases, indicating that the film has the AMR effect. Both the longitudinal and transverse MR saturate in under 2000 Oersted (Oe) applied field. The maximum percentage change (AMR magnitude) in longitudinal and transverse MR is about 5-6 % and the same for all single Ni films having different thicknesses between $1\,\mu m$ and $5\,\mu m$, which indicates that the AMR magnitude is independent of film thickness. This is unsurprising because AMR is a bulk effect. It can be seen that there are two peaks on each magnetoresistance curve as a result of hysteresis, one corresponding to increasing field and the other to decreasing field. The positions of these peaks on the MR curves correspond closely to the coercive field on the hysteresis loops.

Giant magnetoresistance observed in ferromagnetic multilayers is defined as a very large change in electrical resistance in an external applied magnetic field. In such structures, the resistance decreases with increasing magnetic field when the field is both parallel and perpendicular to the current. Since this decrease is so large, the MR in ferromagnetic multilayers is widely referred to as giant magnetoresistance (GMR). The feature that distinguishes GMR from AMR is not just its magnitude, but also its being negative in all field directions: longitudinal, transverse and perpendicular. The resistance of a ferromagnetic multilayer to a current flowing parallel to the layers depends on the magnetic configuration between ferromagnetic layers. In the ferromagnetic configuration, that is, when all magnetic moments of the ferromagnetic layers are aligned parallel the resistance is minimum, while in the antiferromagnetic configuration when neighboring moments are antiparallel the resistance is maximum. The change in the resistance may reduce by more than 100% between two configurations. This sensitivity of resistance to an external magnetic field, known as GMR, is attributed to the spin depending scattering of electron which occurs within the ferromagnetic layers and at the interfaces between the ferromagnetic and spacer layers [1].

Although the first multilayers exhibiting GMR were grown by MBE [1], it was reported that sputtered multilayers also exhibited GMR [2] and that the GMR magnitude could be larger in sputtered samples than in samples grown by MBE. After these observations GMR has attracted increasing interest. A number of groups has started studying multilayers grown by electrodeposition. In 1993, the electrodeposited CoNiCu/Cu multilayers have been proven to exhibit significant GMR [18]. These multilayers were grown on (100) textured polycrystalline Cu substrates from a single electrolyte containing Cu^{2+}, Co^{2+} and Ni^{2+} ions. After removing their substrates, magnetotransport measurements were carried out at room temperature using the VDP geometry as described in Fig. 7.6. Fig. 7.8 shows the magnetoresistance curves of a multilayer with 80[CoNiCu(1.5 nm)/Cu(0.8 nm)]. In this figure, the percentage change in the VDP resistance as a function of the applied field was calculated according to:

$$\Delta R/R(H) = \left[\frac{R(H) - R(H_{\max})}{R(H_{\max})}\right] \times 100 \tag{7.6}$$

where $R(H)$ is the resistance in a particular applied magnetic field and $R(H_{max})$ is the resistance in the maximum applied magnetic field which is 8 kOe in our case. It is obvious from Fig. 7.8 that both the transverse and the longitudinal VDP resistance decrease with increasing magnetic field. In this multilayer, the magnitudes of the observed GMR for transverse and longitudinal measurements are 26% and 19% respectively, in a magnetic field of 8kOe at room temperature.

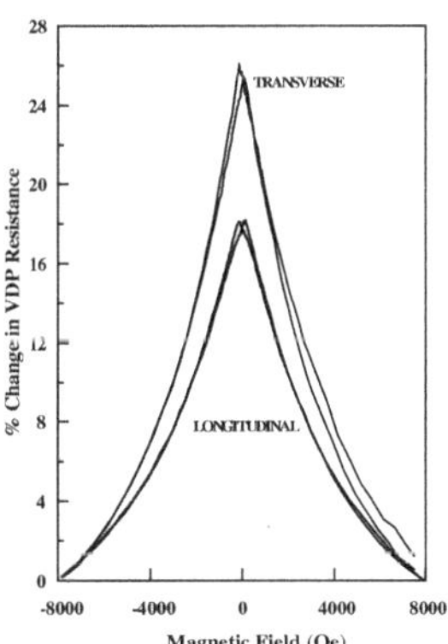

Fig. 7.8. Magnetoresistance curves for a multilayer with 80[CoNiCu(1.5 nm)/Cu(0.8 nm)]

A magnetic field of 8 kOe is insufficient to saturate the magnetoresistance. It may be possible to obtain the larger GMR values in higher magnetic fields. The observed magnetoresistance is largely due to the multilayer structure of the film. As seen in Fig. 7.7, AMR generally causes the transverse resistance to decrease and the longitudinal resistance to increase as the applied magnetic field increases. It is clear that AMR also contributes to the magnetoresistance of the multilayers because the magnitude of the transverse GMR is larger than that of the longitudinal GMR. The magnetoresistance curves of electrodeposited multilayers are very similar to those grown by other systems [1–5,22,23] and have two peaks: one of them being observed at a small negative magnetic field when the applied field is decreasing and the other at a small positive field when the applied field is increasing. The positions of these peaks corresponding to maximum resistance are close to the coercive field of the sample where the net magnetization passes through zero [54].

The GMR magnitude of electrodeposited multilayers can significantly change with parameters such as layer thicknesses and electrolyte pH. The details of how these parameters affect GMR were published in elsewhere [48,55]. One of the most important factors affecting the GMR of the electrodeposited multilayers is the electrolyte pH. A series of CoNiCu/Cu multilayers were grown from the electrolyte with different pH values, and then their magnetotransport properties were investigated. The multilayers with the larger Cu thicknesses than 1.5 nm show GMR when they are grown both from the electrolyte with high pH (3.3) and the electrolyte with low pH(1.8). In contrast, for the smaller Cu thicknesses (0.8 nm), the multilayers grown from the electrolyte with the low pH (1.8) exhibit GMR while the multilayers from the electrolyte with high pH (3.3) have predominantly AMR although their XRD measurements showed the formation of a periodic structure [48].

Figs. 7.9 and 7.10 show the magnetoresistance curves for the multilayers with 107[CoNi(2.0 nm)/Cu(0.8 nm)] grown at low pH (1.8) and high pH (3.3) respectively. The percentage changes in the magnetoresistance were calculated using (7.5) for the sample grown at high pH and (7.6) for the sample grown at low pH. It is clear from these figures that for the multilayers grown from the electrolyte with the low pH, both the longitudinal and transverse magnetoresistance are negative (see Fig. 7.10), but for the films grown at high pH there is a region where anisotropic magnetoresistance is dominant (see Fig. 7.9). If the Fig. 7.9 is compared to the MR curve of the single Ni film in Fig. 7.7, the maximum percentage changes in the resistance of single Ni film are almost the same both for the transverse and longitudinal MRs, while in the multilayer the maximum percentage change in the transverse MR is larger than the longitudinal MR. This is because, as seen in Fig. 7.9, in the low magnetic field region the longitudinal MR increases with increasing magnetic field whereas in the high magnetic field it decreases as the magnetic field increases. This indicates that the multilayer structure has a tendency to exhibit GMR even though grown at high pH.

The dependence of GMR on the electrolyte pH was studied in detail for multilayers with the same thickness, 130[CoNiCu(1.5 nm)/Cu(0.8 nm)], grown

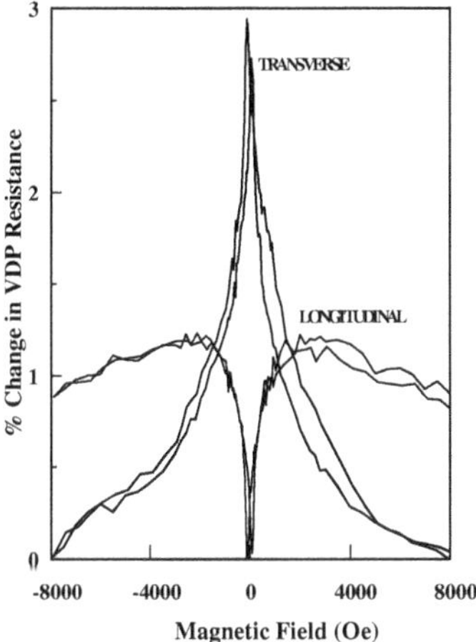

Fig. 7.9. Magnetoresistance curves for a multilayer with 107[CoNiCu(2.0 nm)/Cu(0.8 nm)] grown from the solution with a high pH value (3.3), (From [48])

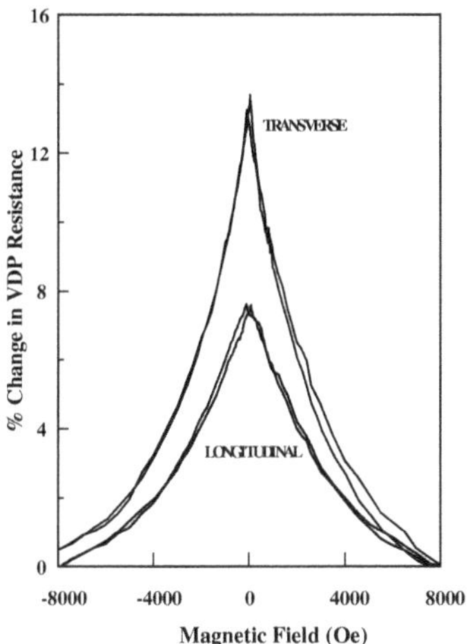

Fig. 7.10. Magnetoresistance curves for a multilayer with 107[CoNiCu(2.0 nm)/Cu(0.8 nm)] grown from the solution with a low pH value (1.8), (From [48])

at different electrolyte pH values. Fig. 7.11 shows the variation of the maximum percentage changes in transverse and longitudinal VDP resistances (GMR magnitudes) as a function of the electrolyte pH. As seen from this figure, the maximum percentage changes both in the transverse and in the longitudinal magnetoresistance, that is, the GMR magnitudes increase as the electrolyte pH decreases. The large increase in GMR with decreasing electrolyte pH is attributed to the decrease in the Cu content or the increase in the Co content of the NiCo alloy layers in the films grown at lower pH, but may also be a consequence of

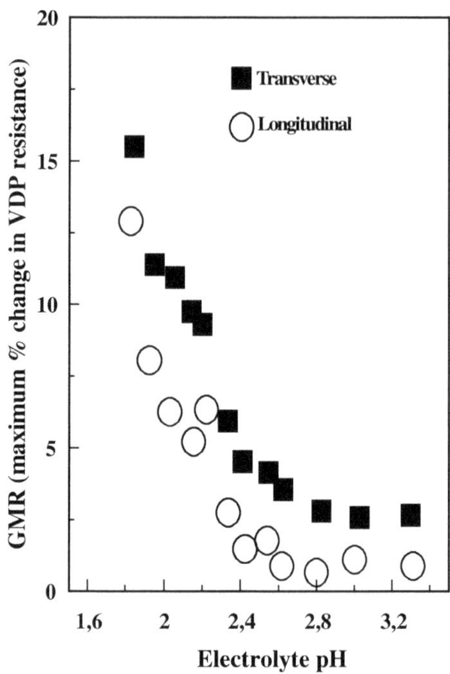

Fig. 7.11. Variation of the GMR magnitude with the electrolyte pH for a series of multilayers with 130[CoNiCu(2.0 nm)/Cu(0.8 nm)], (From [48])

films prepared at low pH having better structural quality or different growth mode, because the current transients for ferromagnetic layers are different for the electrolytes with low and high pH [48]. These questions require further work.

7.7 Conclusions

Although the electrodeposition technique has been used in many fields for a long time, it is a new technique in the production of nanostructure materials. We have developed an experimental technique to synthesis nanostructure materials electrochemically. In this technique, the ferromagnetic multilayers can be grown from a single electrolyte containing their ions. The XRD measurements obtained from the multilayers show that they have the same crystal structures as their substrates. The observation of the satellite peaks up to third order demonstrates that high quality multilayers could be produced by electrodeposition. Electrodeposited multilayers can also have GMR values comparable to those of the multilayers grown by other techniques. The GMR in electrodeposit multilayers strongly depends on the electrolyte pH. It is not still completely understood how the electrolyte pH affects the structural, magnetic and magnetotransport properties of electrodeposited multilayers and hence further work is required.

Acknowledgements

The financial support of the Scientific and Technical Research Council of Turkey (TUBITAK) under Grant no. TBAG-1771 for the electrodeposition technique in our department is gratefully acknowledged. I would like to thank Dr. W. Schwarzacher from University of Bristol (U.K) for his valuable help.

References

1. M.N. Baibich, J.M. Broto, A. Fert, F. Nguyen Van Dau, F. Petroff, P. Etienne, G. Creuset, A. Friederich, and J. Chazelas: Phys. Rev. Lett. **61**, 2472 (1988)
2. S.S.P. Parkin, Z. G. Li, and D. J. Smith: Appl. Phys. Lett. **58**, 2710 (1991)
3. D. H. Mosca, F. Petroff, A. Fert, P. A. Schroeder, W. P. Pratt, Jr., and R. Laloee: J. Magn. Magn. Mater. **94**, L1 (1991)
4. F. Petroff, A. Barthelemy, A. Hamzic, A. Fert, P. Etienne, S. Lequen, and G. Creuzet: J. Magn. Magn. Mater. **93**, 95 (1991)
5. X. Bian, J. O. Ström-Olsen, Z. Altounian, Y. Huai, and R.W. Cochrane: Apply. Phys. Lett. **62**, 3525 (1993).
6. T. W. Jr. Barbee: In:*Synthetic Modulated Structure*, Edited by L. L. Chang and B. C. Giessen, Academic Press, London, 1985 p.313.
7. B. Y. Jin, and J. B. Ketterson: Advances in Physics. **38**, 189 (1989).
8. L. M. Falicov, D. T. Pierce, S.D. Bader, R. Gronsky, K. B. Hathaway, H. J. Hopster, D. N. Lambeth, S. S. P. Parkin, G. Prinz, M. Saloman, I. K. Schuller, and R. H. Victora: J. Mater. Reports, **5**, 1299 (1990)
9. H. J. de Wit: Rep. Prog. Phys., **55**, 113 (1992)
10. R. J. Highmore, W. C. Shih, R. E. Somekh, and J. E. Evetts: J. Magn. Magn. Mater.,**116**, 249, (1992)
11. M. Naoe, N. Matsushita, and S. Nakagawa: J. Magn. Magn. Mater., **134**, 395 (1994)
12. H. Ibach, and H. Lüth: *Solid State Physics,* Springer-Verlag, Berlin, 1991, p. 324.
13. W. Blum: Trans. Amer. Electrochem. Soc. **40**, 307 (1921)
14. A. Brenner: *Electrodeposition of Alloys*, (Academic Press, New York 1963)
15. J. Yahalom and O. Zadok: J. Mater. Sci. **22**, 499 (1987)
16. D.S. Lashmore and M. P. Dariel: J. Electrochem. Soc. **135**, 1218 (1988)
17. M. Alper, P. S. Aplin, K. Attenborough, D. J. Dingley, R. Hart, S.J. Lane, D.S. Lashmore, and W. Schwarzacher: J. Magn. Magn. Mater. **126**, 8 (1993)
18. M. Alper, K. Attenborough, R. Hart, S.J. Lane, D.S. Lashmore, C. Younes, and W. Schwarzacher: Appl. Phys. Lett. **63**, 2144 (1993)
19. A. P. O'Keeffe, O. I. Kasyutich, W. Schwarzacher, L. S. de Oliveria, and A. A. Pasa: Appl. Phys. Lett. **73**, 1003 (1998)
20. A. A. Pasa, and W. Schwarzacher: Phys. Stat. solid. (a), **173**, 73 (1999)
21. O.I. Kasyutich, W. Schwarzacher, V.M. Fedosyuk, P. A. Laskarzhevskiy, and A. I. Masliy: J. Electrochem. Soc. **147**, 2964 (2000)
22. S. Z. Hua, D. S. Lashmore, L. Salamanca-Riba, W. Schwarzacher, L. J. Schwarzendruber, R. D. McMichael, L. H. Bennett, and, R. Hart: J. Appl. Phys. **76**, 6519 (1994)
23. S.K.J. Lenczowski, C. Schonenberger, M.A.M. Gijs, and W.J.M. de Jonge: J.Magn. Magn. Mater. **148**, 455 (1995)
24. L. Peter, A. Cziraki, L. Pogany, Z. Kupay, I. Bakonyi, M. Uhlemann, M. Herrich, B. Arnold, T. Bauer, and K. Wetzig: J. Electrochem. Soc. **143**, C168, (2001)

25. I. Bakonyi, E. Toth-Kadar, T. Becsei, J. Toth, T. Tarnoczi, A. Cziraki, I. Gerocs, G. Nabiyouni, W. Schwarzacher: J. Magn. Magn. Mater. **156**, 347 (1996)
26. Y. Jyoko, S. Kashiwabara, and Y. Hayashi: J. Electrochem. Soc. **144**, L5, (1997)
27. P. Nallet, E. Chassaing, M.G. Walls, and M.J. Hytch: J. Appl. Phys. **79**, 6884 (1996)
28. E. Chassaing, A. Morrone, and J. E. Schmidt: J. Electrochem. Soc. **146**, 1794 (1999)
29. E. Chassaing, P. Nallet, and M.F. Trichet: J. Electrochem. Soc. **143**, L99 (1996).
30. K.D. Bird, and M. Schlesinger: J. Electrochem. Soc. **142**, L65 (1995)
31. V. M. Fedosyuk, O. I. Kasyutich, D. Ravinder, and H. J. Blythe: J. Magn. Magn. Mater. **156**, 345 (1996)
32. V. M. Fedosyuk, O. I. Kasyutich, and W. Schwarzacher: J. Magn. Magn. Mater. **198-199**, 246 (1999)
33. L. Piraux, J. M. George, J. F. Despres, C. Leroy, E. Ferain, R. Legras, K. Ounadjela, and A. Fert: Appl. Phys. Lett. **65**, 2484 (1994)
34. A. Blondel, J. P. Meier, B. Doudin, And J.-Ph. Ansermet: Appl. Phys. Lett. **65**, 3019 (1994)
35. P.R. Evans, G. Yi, and W. Schwarzacher: Appl. Phys. Lett. **76**, 481 (2000)
36. W. Schwarzacher, O.I. Kasyutich, P.R. Evans, M.G. Darbyshire, Ge Yi, V. M. Fedosyuk, F. Rousseaux, E. Cambril, and D. Decanini: J. Magn. Magn. Mater. **198-199**, 185 (1999)
37. S. Dubois C. Marchal, J. M. Beuken, L. Piraux, J. L. Duvail, A. Fert, J. M. George, and J. L. Maurice: Appl. Phys. Lett. **70**, 396 (1997)
38. M. A. M. Gijs, S. K. J. Lenczowski, and J. B. Giesbers: Phys. Rev Lett. **70**, 3343 (1993)
39. S. F. Lee, W. P. Pratt, R. Loloee, P. A. Schroeder, and J. Bass: Phys. Rev. **B 46**, 548 (1992)
40. T. Valet, and A. Fert: Phys. Rev. **B 48**, 7099 (1993)
41. C. A. Ross: Annu. Rev. Mater. Sci. **24**, 159 (1994)
42. W. Schwarzacher, and D.S. Lashmore: IEEE Trans. Magn. **32**, 3133 (1996)
43. A. Fert, and L. Piraux: J. Magn. Magn. Mater. **200**, 338(1999)
44. A.J. Bard, and L.R. Faulkner: *Electrochemical Methods,* (John Wiley&Sons, New York 1980)
45. J. G. Wright: In: *Epitaxial Growth Part A, Material Science Series*, Edited J. W. Mattehews, (Academic Press, New York 1975)
46. L. M. Goldman, C. A. Ross, W. Ohashi, D. Wu, and F. Spaepen: Appl. Phys. Lett. **55**, 2182 (1989)
47. M. Alper: Electrodeposited Magnetic Superlattices. Ph. D. Thesis, University of Bristol, U.K. (1995)
48. M. Alper, W. Schwarzacher, S.J. Lane: J. Electrochem. Soc. **144**, 2346 (1997)
49. Y. Fujii: In: *Metallic Superlattices, Studies in Physical and Theoretical Chemistry,* edited by T. Shinjo and T. Takada, Vol. **49**, (Elsevier, Amsterdam 1987)
50. E. E. Fullerton, I. K. Schuller, H. Vanderstraeten, and Y. Bruynseraede: Phys. Rev. **B 45**, 9292 (1992)
51. T.P. Moffat: J. Electrochem. Soc. **142**, 3767 (1995)
52. L. J. Van der Pauw: Philips Res. Rep. **13**,1 (1958)
53. T. R. McGuire, and R. I. Potter: IEEE Trans. Magn., **Mag-11(4)**, 1018 (1975)
54. S. S. P. Parkin, R. F. Marks, R. F. C. Farrow, G. R. Harp, Q. H. Lam, and R. J. Savoy: Phys. Rev. **B 46**, 9262 (1992)
55. M. Alper, K. Attenborough, V. Baryshev, R. Hart,, D.S. Lashmore, and W. Schwarzacher: J. Appl. Phys. **75**, 6543 (1993)

8 Magnetic Anisotropies of FeO and NiMn Films

B. Aktaş[1], M. Özdemir[2], R. Yilgin[1], Y. Öner[3], T. Sato[4], and T. Ando[5]

[1] Department of Physics, Gebze Institute of Technology, 41400, Gebze, Kocaeli, Turkey
[2] Department of Physics, Marmara University, Goztepe, Istanbul, Turkey
[3] Department of Physics, Istanbul Technical University, Maslak, Istanbul, Turkey
[4] Department of Instrument. Eng., Fac. Sci. and Techno., Keio. Univ, Yokohama, Kanagava 223, Japan
[5] Department of Material Science, Fac. Sci. and Techno., Keio. Univ, Yokohama, Kanagava 223, Japan

Abstract. Epitaxial Fe_3O_4 films on MgO(100) substrate and polycrystalline reentrant $Ni_{77}Mn_{23}$ films of various thickness (50, 100, 300, 600, 1000, 3700 Å) on quartz substrate have been studied by Spin Wave Resonance (SWR) as a function of temperature. The SWR spectra have unusual uni-directional character upon cooling the sample in an external field H_c. All films show spin-glass properties in the temperature range below freezing temperature T_f (about 120 K for NiMn and 70K for Fe_3O_4). The number of peaks increases with film thickness, indicating long spin wave excitation across the film thickness.

The angular dependent SWR spectra were successfully analyzed by usual classical SWR theory. Using a field induced unidirectional exchange surface and bulk anisotropy together with usual magneto-crystalline anisotropy energy, we have simulated the experimental spectra and deduced magnetic parameters. The effective exchange stiffness parameter (D) remains practically the same for all films above T_f, while its value surprisingly increases below T_f, showing a strong correlation with spin-glass properties. The bulk anisotropy was found to strongly enhance with decreasing temperature and film thickness, while linearly temperature dependent surface anisotropy shows very weak thickness dependence. The easy-plane surface anisotropy on both substrate and free surfaces has been attributed to an intrinsic property for the films rather than accidental surface condition.

8.1 Introduction

The surface effects on magnetic properties of a material are becoming the subject of intense current research as the trend to higher magnetic recording densities creates a need for smaller magnetic particles having larger surface-to-volume ratios. If the surface magnetic properties differ from those of the bulk, they can dominate the overall magnetic behaviour of small particles or very thin films. Vassiliou et al. [1] showed that the magnetic anisotropy for these particles with diameter of about 30 Å is two orders of magnitude larger than that of bulk ferrite (γ-Fe_3O_4) crystals.

The ferrites have been used as magnetic recording media for a long time [2,3]. Scientists spent great efforts to explain their magnetic properties [4–8]. There are two kinds of sites for Fe^{++} and Fe^{+++} ions having different magnetic moments at room temperature. Site A has the tetrahedral oxygen configuration while site B has the octahedral configuration. A remarkable feature of the spinels is that all exchange integrals J_{AA}, J_{BB}, and J_{AB} are negative and favour *antiparallel* alignment of the spins connected by the interaction (see ref. [6,7]). However, since the interactions between non identical pairs (A-B) are much stronger than those between identical pairs (A-A or B-B), the resulting alignment is ferrimagnetic, that is, the A spins are parallel to each other and the B spins are parallel to each other so that A spins may be antiparallel to B spins (even the ground state correspond to a spiral configuration [8] at lower temperatures). As the temperature decreases the Fe_3O_4 shows a phase transition from cubic to orthorhombic crystalline structure and its magnetization is reduced [9]. Since the exchange interactions (super exchange) are mediated by oxygen ions a small distortion in the octahedra or tetrahedra may have considerable effects on the magnetic properties of ferrites.

Recently researchers have been applying various surface treatment techniques [10–14] to enhance and control the magnetic behavior of magnetic media. It is obvious that the surface anisotropy can be studied more easily on single crystal thin films than on fine particles. They hop that, the information gathered from thin films might be applied to fine particles in order to get a reasonable model for small magnetic particles.

On the other hand $Ni_{1-x}Mn_x$ alloys is one of the most interesting systems which exhibits unusual and rich magnetic properties [15–22] (i.e., displaced hysteresis loop upon cooling in an external field, the abrupt rise of the non-equilibrium magnetization with increasing temperature below T_f where strong magnetic irreversibility sets in, the frequency dependence of T_f , electrical resistivity, and so on). Most obvious and common behavior of $Ni_{1-x}Mn_x$ alloy is the sequence of magnetic phase transitions from paramagnetic at higher temperature, ferromagnetic-like at intermediate temperatures and spin glass-like properties with decreasing temperature. The neighboring spins are not aligned perfectly parallel to each other. These behaviors are believed to originate from the competing interactions between ferromagnetic Ni-Ni, Ni-Mn and antiferromagnetic Mn-Mn pairs.

Like many other researchers, we have also made some magnetic and electrical measurements on the system with various compositions in either bulk or thin film forms [23–29]. The magnetic properties of the system in thin film form are significantly different from that of bulk samples of the same alloy. The most common outcome of these experiments is that, the spinglass characteristics such as unidirectional anisotropy (H_a), and freezing temperature (T_f) vary with film thickness [30]. The general trend is the same for the films regardless of their composition rate and growth procedures. It has been shown that the magnetic parameters are highly affected from the surface anisotropy of the films [31,34]. The exchange ansotropy in this sytem can easily be manuplated by constituent ratios and the temperature. Thus the knowledge to be obtained for the field induced exchange ansiotropy of this system could be used to understand the other magnetic systems which are commonly applied in technology.

The ferromagnetic resonance (FMR) has proven to be one of the most sensitive and useful techniques [35–44] to study the surface anisotropy as well as other magnetic properties of magnetic thin films. With this motivation we used FMR technique to study the magnetic properties of both polycrystalline NiMn and epitaxial Fe_3O_4 thin films on MgO(100).

8.2 Experimental Procedures

8.2.1 Sample Preparation and Characterization

a) $Ni_{77}Mn_{23}$Films on Quartz Substrate. The thin films of $Ni_{77}Mn_{23}$ have been grown on polycrystalline quartz substrates by using electron beam gun evaporation method. The high purity Mn and Ni constituents have been put in two separate baskets symmetrically located under the substrate. The sample thickness and composition were monitored simultaneously during the growth and controlled by using quartz monitors. The composition was determined by EMPA (Electron-probe micro analyzer). The base vacuum pressure was 5×10^{-8} mbar and there was no control on the substrate temperature. Hence $Ni_{77}Mn_{23}$ films with various thicknesses of 50, 100, 300, 1000 and 3700 Å have been prepared on sufficiently pure quartz substrates.

b) Fe_3O_4 Films on MgO(100). Fe_3O_4 thin films were deposited on MgO(100) substrates using reactive dc sputtering technique [45] and characterized by x-ray diffraction, vibrating sample magnetometer, SQUID magnetometer, torque magnetometer and Mössbauer spectroscopy. x-ray diffraction studies of the 400 and 800 reflections indicated single crystal Fe_3O_4 films under a tensile stress. This single crystalline structure is consistent with torque measurements of the in-plane anisotropy of the films with thickness of 2000-3000 Å. The cubic anisotropy constant (K_1) obtained from torque measurement at high fields shows satisfactory agreement with the value for bulk Fe_3O_4 over a wide temperature range. However, these films exhibit several magnetic anomalies [45]. The magnetization shows a rapid increase with the field up to about 50 mT, and then it continues to increase linearly and very slowly even up to 7 T which is maximum experimental field value. For example the value of magnetization for some samples is only 85 % of the values for bulk Fe_3O_4 at 2 T. Also, Mössbauer spectroscopy shows that a very small fraction of the moments are nearly randomly distributed in three dimensions and not in-plane as dictated by film shape anisotropy. Lastly, magnetization measured in low applied fields gives a very small value for cubic anisotropy constant, K_1, compared to that for bulk.

8.2.2 FMR Measurements

FMR experiments were carried out at X-band (≈9.4 GHz) by using commercial Varian spectrometers. An ac modulation magnetic field of 100 kHz was applied

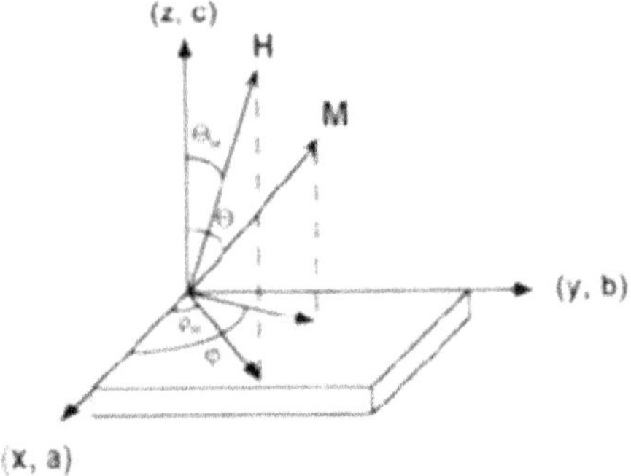

Fig. 8.1. Relative orientations of the applied magnetic field vector, H, equilibrium magnetization vector, M, and experimental references axes system which coincide the crystalline abc cubic axes system.

parallel to the dc field in order to record field-derivative curves of microwave resonant absorption. The angular dependencies of the field derivative of the FMR spectra were studied by using a goniometer to rotate the sample about vertical direction in the horizontal external magnetic field of the electromagnet. The sample temperature between 4-300 K was controlled by an Oxford continuous helium gas flow type cryostat. The precision of the temperature is better than 1 degree at lower temperatures and within 5 degrees at around room temperature.

8.3 Experimental Results

8.3.1 Fe_3O_4 Films on MgO(100)

When the applied dc magnetic field direction is close enough to that of the film normal, FMR spectra consist of two well-resolved peaks at room temperature. However, as the field is rotated away from the film normal these two lines come close to each other and overlap giving a relatively broader single absorption line. Fig. 8.2 shows in plane angular variation of room temperature resonance fields for Fe_3O_4 thin film of 3012 Å in thickness. The angle of the applied magnetic field is measured from one of the $\langle 100 \rangle$ direction. As seen from this figure, Fe_3O_4 film on MgO (100) exhibits a four fold symmetry with a period of 90 degrees in the film plane, suggesting a cubic crystalline structure for the film. However the maximum amplitude of the resonance field variation with the angle is about 3.5 mT. This value is at least one order of magnitude smaller than that one can theoretically predict from the cubic anisotropy parameter $2K_1/M$ (46 mT [46] for the bulk sample of the same compound) as will be explained in theoretical section of this paper. On the other hand, from dc magnetization we know that the FMR resonance field value is small to reach exact saturation of the magnetization (see Fig. 3 in ref. [45]). Also the torque measurement gives very small value for K_1 at lower field whereas one obtains a value close to that of the bulk sample at 2 T (see Fig. 7 in ref. [45]). This indicates that, although structurally single

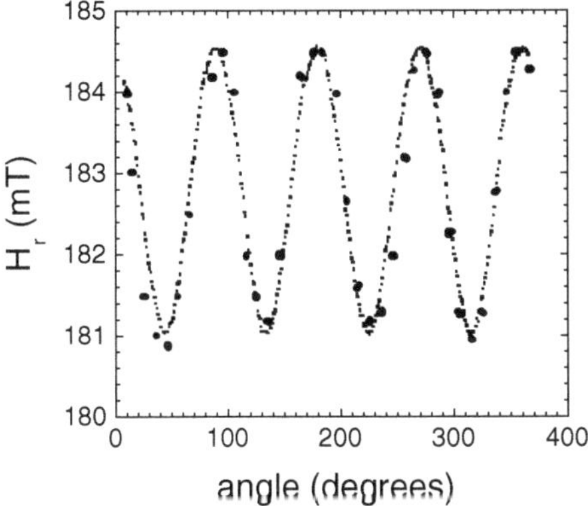

Fig. 8.2. The resonance field values of Fe_3O_4 at 300 K as a function of angle of applied magnetic field, H, in the sample plane. The filled circles are the experimental values while the continuous line represents to the theoretical values obtained from the general resonance equation (Eq.(8.2)) by using the fitted anisotropy parameters.

crystalline [45], at lower field the film is magnetically disordered due to some strong and randomly distributed microscopic anisotropy fields.

The temperature evolution of the FMR spectra is shown in Figs. 8.3 and 8.4. The samples were cooled down to 4.2 K in the remanent field (5 mT) of the electromagnet and then the spectra were recorded by sweeping the external field either parallel (n-ZFC case) or antiparallel (r-ZFC case) to the remanent field in the film plane. As shown in Fig. 8.3 there is a large hysteresis effect on the magnetic state for parallel geometry (PG, dc filed lies in sample palne) at low temperatures. The resonance field for n-ZFC case starts even at negative applied field region which is much lower than what we would expect for such a system having relatively small demagnetizing field due to the geometry ($4\pi M \simeq 0.5$ T). This indicates a strong anisotropy field induced along the cooling field direction. This field induced uni-directional anisotropy is coupled to the lattice so rigidly that it tries to align the magnetization parallel to the cooling field direction. For r-ZFC case the magnetization is directed in opposite direction of small magnetic field due to strong uni-directional anisotropy field induced along the cooling field (remanent field of the electromagnet) direction. Therefore the resonance (continues lines) starts again nearly at zero field and traces the negative field part of the resonance line for n-ZFC case. As the applied field is increased further the magnetization vector rotates towards the applied field direction and a second resonance absorption takes place at about -0.2 T in accord with the dispersion relation (Eq.(8.2)). In fact this resonance absorption corresponds to the main resonance peak for r-ZFC case. One can obtain the rigid component of the uni-directional anisotropy field H_A ($=\Delta H/2$) from the separation, (ΔH), between the main peaks for r-ZFC and n-ZFC cases.

Low field losses in soft ferromagnetic samples can usually give rise to a large increase in rf absorption due to domain resonance in multi-domain samples. How-

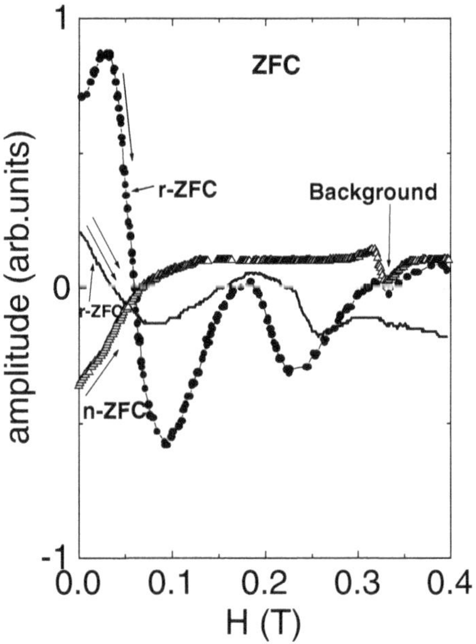

Fig. 8.3. FMR spectra of a Fe_3O_4 thin film. The open triangles represent the spectrum which was recorded by applying the measurement field parallel to the remnant field of the electromagnet, after cooling the sample down to 4.2 K (n-FC case). The dashed line corresponds to the spectrum which was plotted after the sample was rotated by 180 degrees relative to remnant field direction (r-FC case). Finally, the sample was rotated its original direction, applying a field of 0.4 T the sample is magnetized along the remanent field direction, the field was decreased down to zero value, the sample was rotated by 180 degrees again and then the spectrum (continuous line) was recorded a second time for r-ZFC case.

ever, even at low applied field there is a strong unidirectional induced anisotropy field as evidenced from a large amount of shift of the resonance field for r-FC case relative to that for n-FC case. Therefore the resonance absorption appearing at lower external field takes place at strong effective (internal) field. In other words, these peaks do not belong to the domain resonance. Low field losses for usual ferromagnetic samples should be the same for either direction of the external field (parallel or anti-parallel to the cooling field). So this uni-directional character is a clear sign for the field induced exchange uni-directional anisotropy. There is another field-induced anisotropy for bulk ferrite. But this anisotropy originates from a transition from cubic to orthorhombic structure. The easy axis (c-axis) becomes parallel to the external field during the cooling. This anisotropy field is bi-directional, that is, positive or negative direction of the c-axis is the same. However in our case the induced anisotropy includes both axial (bi-directional) and uni-directional components induced along the cooling field direction. As can be seen in Fig.8.3, the resonance field values and the line-shape are effected by successive sweeps of the field as well.

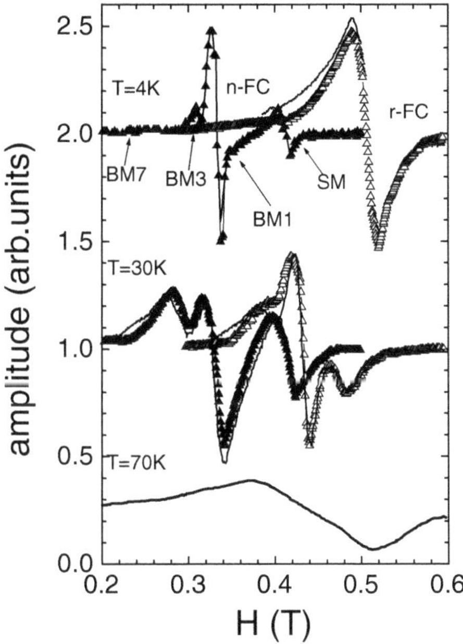

Fig. 8.4. FMR spectra of a Fe_3O_4 film for perpendicular field cooling case at some selected temperatures as indicated above each spectrum. The filled triangle represents the spectrum recorded by applying the external field parallel to the cooling field direction (n-FC) while the open triangles correspond to r-FC case in which the sample was rotated by 180 degrees relative to the cooling field in zero field after n-FC spectra. The continuous lines are the corresponding theoretical curves which were calculated by using both surface and bulk directional anisotropy in addition to usual crystalline anisotropy as described in the text.It should be noted that the spectrum is substantially changed about 70 K, suggesting a phase transition.

The most substantial effects on the spectra are observed when the cooling field (0.4 T) is perpendicular to the film plane (perpendicular geometry, PG). Fig. 8.4 shows the spectra taken by sweeping the applied field oriented parallel and/or antiparallel to the cooling field (which is perpendicular to the film plane) at some selected temperatures. At 4 K four relatively narrow and well resolved four resonance peaks (heavy triangles) were observed for n-FC case. However these peaks overlap and become a much broader single peak (open triangles) for r-FC case. As is well known, the multi-peak resonance line generally arises from the pinning of the spins at the surface of the ferromagnetic films and the relative intensities of the higher order bulk spin wave modes progressively decrease with mode number except for the surface mode. The simulated spin wave resonance spectra obtained from the model in the theoretical section are in a good agreement with the experimental spectra for perpendicular geometry. Also the angular behaviour of the experimental spectra shows a satisfactory agreement with the spin wave theory. Therefore this asymmetric behaviour of the FMR spectra with respect to the cooling field direction is an obvious sign for unidirectional surface anisotropy induced upon cooling the sample in an external field. The distance between the main resonance peaks (i.e. the most intense ones) for n-FC and r-FC cases is about 0.2 T and corresponds to an exchange unidirectional bulk anisotropy field H_A of about 0.1 T,. As the temperature increases, the single resonance peak for r-FC cases splits progressively and takes the same structure as that for n-FC cases in every respect. Accordingly, the separation of a particular resonance line for r-FC from the corresponding mode for the n-FC case progressively decreases and vanishes at about 70 K. Moreover the resonance peaks broaden with increasing temperature in this temperature regime.

Fig. 8.5 illustrates the angular dependencies of both the experimental (symbols) and the theoretical resonance values of various modes for FC cases for perpendicular geometry at 4 K. The angle was measured from the cooling field direction (perpendicular to the film) and the external field was rotated toward one of the cubic axis in the sample plane. The experimental resonance values were taken as an intersection of the resonance line and the base line. As the applied field is rotated away from the film normal, the line-width is increased and takes maximum value when the dc field lies in the sample plane. This implies a relatively strong random anisotropy field in the sample plane.

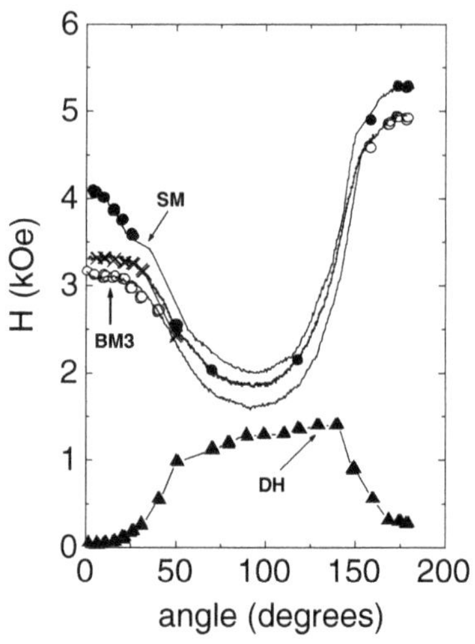

Fig. 8.5. Resonance field values for various FMR modes and peak-to-peak line width as a function of angle of the external field relative to cooling field direction (film normal) in the bc-plane. Continuous line are fitted values for corresponding FMR modes which they were calculated by using the Eq.(8.12) with the values for magnetic parameters given in Table 8.1.

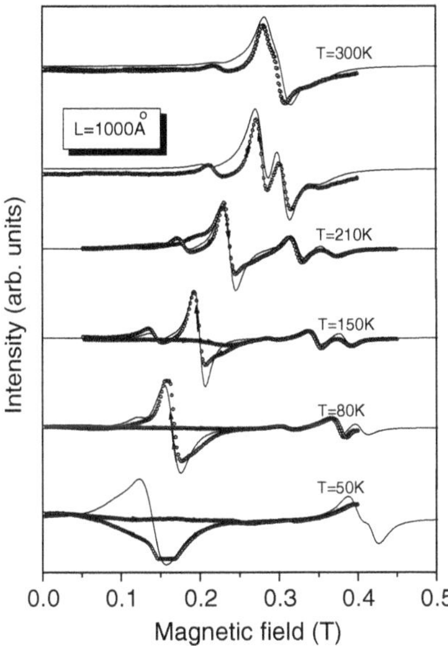

Fig. 8.6. The temperature dependence of SWR spectra recorded by applying the external field at right angles to the plane of 1000 Å thick film. The open circles represent the experimental data while continuous lines denote corresponding calculated spectra by using the model described in the text. The experimental absorption lines were traced for both increasing and decreasing direction of external field as marked by the arrows, to illustrate significant hysteresis effect even at relatively strong field like a few kG.

8.3.2 $Ni_{77}Mn_{23}$ Films

FMR spectra from the thin films of $Ni_{77}Mn_{23}$ alloy with 2700 Å, 1000 Å, 600 Å, 300 Å, 100 Å and 50 Å in thickness were recorded as a function of the angle between external magnetic field and film normal at the temperature range 4-300 K. The angular dependent multi component FMR spectra were attributed to the classical spin wave resonance (SWR) excitation. The number of the modes increases parallel to the film thickness in accordance with the SWR theory. Some examples of SWR spectra are given in Fig. 8.6 for perpendicular geometry of 1000 Å thick NiMn film at several temperatures. The qualitative behavior of the spectra is quite similar to that for the films with different thickness while the absolute values of measured quantities are still different. The distorted and quite broad resonance line seems to contain many overlapping (slightly resolved) resonance absorptions at 300 K. As the temperature is decreased this resonance line is split apart into two groups each with two peaks. The distance of the center of the gravity of the groups monotonically increases with decreasing temperature. The magnitude of this distance is very close to that of 600 Å 300 Å thick films. Also the peak at lower field intensifies and an additional weaker peak on the lower field side of this peak appears. As the temperature is further decreased, the two peaks at the high field side of the spectra weaken while their distance to the main peak increases and the minor peak at lowest field becomes more and more noticeable. The relative intensities of double peaks at higher fields decrease at lower temperatures. Hysteresis effects manifest themselves at 210 K for the sample and intensify as the temperature is further decreased. It should be noted that the low field resonance peaks recorded with increasing direction of the field are totally diminished while a considerable absorption appears for the case of decreasing direction of the field. These peculiarities are expected to originate from the exchange anisotropy of reentrant NiMn. The temperature dependence of the resonance spectra of the films with different thicknesses quite similar to that of thicker samples. However, the number of observable peaks is reduced to two for thinner films.

8.4 Theoretical Model

The theoretical resonance fields in Fig. 8.2 are calculated by using the following expression

$$E = -M.H + K_1(\alpha_1^2\alpha_2^2 + \alpha_1^2\alpha_3^2 + \alpha_2^2\alpha_3^2) + K_{eff}^2\alpha_3^2 \tag{8.1}$$

for the magnetic energy density, E, for a cubic single crystal film in an external field. Here the first term represents Zeeman energy density, the second term accounts for cubic magneto-crystalline energy density with anisotropy parameter, K_1, and , α's are the direction cosines of M, with respect to the cubic axes. $K_{eff} = K_u - 2\pi M^2)$ represents the effective uniaxial anisotropy parameter including both the shape anisotropy $(2\pi M^2)$ and any possible perpendicular anisotropy K_u, like magneto-elastic coupling due to lattice mismach [45] etc. For a general direction of the magnetic field, H, the basic dispersion relation for FMR is given [47] by

$$\begin{aligned}(\omega_o\gamma)^2 &= [\mathrm{Dk_n^2} + 1\mathrm{M_s}\sin^2\theta(\partial^2\mathrm{E}/\partial\varphi^2)][\mathrm{Dk_n^2} + 1\mathrm{M_s}(\partial^2\mathrm{E}/\partial\theta^2)] - \\ &- [1\mathrm{M_o}\sin\theta(\partial^2\mathrm{E}/\partial\theta\partial\varphi)]^2\end{aligned} \tag{8.2}$$

where γ is the gyro-magnetic ratio, ω is the microwave frequency, θ and φ are the usual spherical polar angles for M, D is the exchange stiffness parameter and k_n is the spin wave vector for nth mode. If the external dc field is applied in the film plane (θ_H=90°) and makes an angle φ_H with experimental x-axis (one of the cubic axes) as seen in Fig. 8.1, then, using the energy expression (1) in (2), one can calculate the resonance condition (when MH $\gg$K$_1$, K$_u$) for uniform mode (k=0). Using this expression we obtained best fit (continues line in Fig. 8.2) to the experimental values (filled circles) with the parameters $2K_1/M$=2.2 mT, $2K_1/M+4\pi M+2K_u/M$=380 mT, (ω/γ)=315 mT and g=2.14. The cubic anisotropy parameter, 2K$_1/M$, is more than one order of magnitude smaller than a value of 46 mT [46] for the bulk sample of the same compound. As mentioned above the resonance field value in FMR experiment is too small to reach exact saturation of the magnetization. That is, at lower field the film is magnetically disordered (unsaturated) although it is structurally single crystalline. Therefore each individual spin sees a different anisotropy field according to its own orientation in such a crystal. In order to explain the under-saturation effects on the measured anisotropy, we assumed a Gaussian angular distribution for the spin's orientation relative to the applied magnetic field. Therefore the number of spins directed within a solid angle Ω and $\Omega + \mathrm{d}\Omega$ around vector **H** is

$$n d\Omega = C \exp(-\frac{\beta}{\beta_o})^2 \sin\theta d\theta d\varphi \tag{8.3}$$

where C is a normalization constant so as to giving the total number of spins N in the unit volume, β is the angle of the spins with respect to the applied field H, β_o is a field-dependent parameter representing the mean angular distribution width. In expression (3), θ and φ are the spherical polar angles of the spins which are measured relative to Cartesian crystallographic axes with z (or c) directed along the film normal. Applying the external field at the angle φ_H in the sample plane the relation among the β, the spherical polar angles $\theta, \varphi, \varphi_H$ and becomes

$$\cos\beta = \sin\theta \cos(\varphi_H - \varphi) \tag{8.4}$$

If the moment of each spin is μ, then the average magnetization along the applied field **H** is simply written as

$$M_H = \int_{\Omega} \mu C \exp(-\frac{\beta}{\beta_o})^2 \cos\beta \sin\theta d\theta d\varphi \tag{8.5}$$

The z-component of the torque, T_z , arising from crystalline anisotropy on the spins oriented at the solid angle Ω is

$$T_z = \left(\frac{1}{\sin\theta} \left(\frac{dE}{d\varphi} \right) \right) \tag{8.6}$$

where

$$E = \frac{1}{N} \{ K_1(\alpha_1^2\alpha_2^2 + \alpha_1^2\alpha_3^2 + \alpha_2^2\alpha_3^2) + K_{eff}^2\alpha_3^2 \} \tag{8.7}$$

is the crystalline anisotropy energy of each individual spin having a direction cosines α_1, α_2, α_3 with respect to the cubic crystallographic axes a, b and c respectively. The z-component of the average torque acting on an individual spin by the crystalline anisotropy is obtained by using the energy expression (8.7) in the torque equation (8.6) as

$$T_z = \frac{1}{2} \frac{K_1 C}{N} \int_{\Omega} \exp(\frac{\beta}{\beta_o})^2 \sin^4\theta \sin(4\theta) d\theta d\varphi \tag{8.8}$$

This is the expression that has been used to calculate the torque corresponding to dc measurements for parallel geometry. The value for β is inserted from Eq.(8.5).

For perfect alignment of the spins along the applied field the M has its saturation values, M_s, and the torque expression reduces to

$$T_z = \frac{1}{2} K_1 \sin(4\varphi) \tag{8.9}$$

which would be obtained from well-known cubic anisotropy energy

$$E = K_1(\alpha_1^2\alpha_2^2 + \alpha_1^2\alpha_3^2 + \alpha_2^2\alpha_3^2) \tag{8.10}$$

Using this cubic energy in the dispersion relation Eq.(8.2) one can get magneto-crystalline anisotropy field, $H_k = 2K_1/M = 46\,\mathrm{mT}$ just as for the bulk samples.

The low temperature FMR spectra were analyzed using the following magnetic energy density for single crystalline ferrite film which is cubic above Verwey transition temperature

$$E = -M.H + M.(H_{AD} + H_{AR}) + (K_\alpha \alpha_a^2 + K_\alpha \alpha_a^4 + K_b \alpha_b^2 + K_b \alpha_b^4 + K_c \alpha_c^2 + K_c \alpha_c^4) \tag{8.11}$$

Here the first term is the Zeeman energy, the second term represents the unidirectional exchange anisotropy energy including both rigid (H_{AR} along the cooling field) and dynamic component(H_{AD} ,that is, an "isotropic anisotropy field[48–50] which is always directed along the direction of the applied field). The third term in Eq.(8.11) represents the magneto-crystalline anisotropy for a rhombic Fe_3O_4 crystal at low temperatures with easy the axis c along the direction of the cooling field which was taken along one of the cubic axis at higher temperature for perpendicular FC case. a and b-axes are rotated by 45 degrees relative to the other two cubic axes of the film [51]. However the rotation takes place either direction in equal probability on the various regions of the film plane(i.e. positive or negative 45 degrees' rotation). Taking average over a and b axes, the magneto-crystalline anisotropy term can be approximated to an uniaxial form having c direction as symmetry axis. Hence, Eq.(8.11) can be replaced by

$$E = -M.H + M.(H_{AD} + H_{AR}) + K_3\alpha_3^2 + K_3\alpha_3^4) \tag{8.12}$$

Here the term includes both the first order effective uniaxial magnetic anisotropy and the shape (demagnetization) anisotropy energy. For cooling in perpendicular field, both the uniaxial and unidirectional axes are induced along the film normal. Considering directional character of the FMR modes, we employed a surface anisotropy energy to determine the spin wave vector K as:

$$E_s = K_d \cos\theta + K_u \cos^2\theta \tag{8.13}$$

This includes both unidirectional (first term) and uniaxial surface anisotropy energy. Also the following expression [50]

$$K(T) = K(0)\left(1 - \frac{T}{T_{sg}}\right) \tag{8.14}$$

can be modelled for temperature dependence of both the surface and the bulk uni-directional exchange anisotropy field. Here K(0) is zero temperature value of either surface or bulk anisotropy and T_{sg} is spin glass transition temperature.

For policrystalline NiMn films the avaraged crystalline anisotropy naturally vanishes but uniaxial and unidirectional exchange anisotropy should be retained in addition to the Zeeman and shape anisotropy energy.

8.5 Theoretical Analysis

8.5.1 Fe_3O_4 Films on MgO(100)

We have developed a computer algorithm based on the theory to reconstruct the SWR spectra. Since the amplitudes of the spectra strictly depend on the experimental set up, an adjustable parameter for amplitude factor has been used in the calculation of the theoretical spectra. The magnetic parameters in the free energy have been determined by fitting the theoretical spectra to the experimental ones for general direction of the external field.

For room temperature analysis, using various values for β_o in Eqs.(8.5),(8.6), (8.8) and (8.10), we calculated the magnetization and the torque on M, as a function of angle, φ_H of the applied field **H** in the plane of the film. Fig. 8.7 exhibits the calculated results corresponding to the effective anisotropy field which would be obtained from both the FMR and the dc torque measurements. As can be seen from this figure, the effective anisotropy field progresively increases up to its expected value ($2K_1/M$=46 mT) as the magnetization rises to its saturation values, M_s, which corresponds to a perfect alignment of the spins along H. However the randomness of the spins affects the apparent (averaged over the spins) anisotropy field dramatically. Only about 10 % decrease in magnetization due to random distribution causes nearly 80 % decrease in the apparent (macroscopic) anisotropy field. Therefore when the applied field is not strong enough to saturate the magnetization, both FMR and dc torque measurements techniques inevitably give smaller values to the meassured anisotropy field.

The low temperature theoretical FMR spectra are calculated using the anisotropy energy given by Eqs.(8.12) and (8.13) in the calculation procedure in ref. [41]. The results are plotted in Fig. 8.4 together with the corresponding experimental ones for perpendicular geometry at some selected temperatures. As shown in this figure, simulated FMR spectra reproduce the experimental ones quite well. Some deviations, especially in line width, from experimental spectra come from some field training effects which we could not include in our model. Also the angular variations of the calculated resonance field values for various spin wave modes are shown in Fig. 8.5 together with corresponding experimental results at 4.2 K. Relatively small deviation from the theory can be tolerated by considering large line broadening with the angle, which makes it difficult to determine the resonance field more accurately. The fitted parameters are given in Table 8.1.

The g-value for Fe based compounds is generally 2.12. The dc magnetization was measured on the same sample [45]. From this analysis we have found that

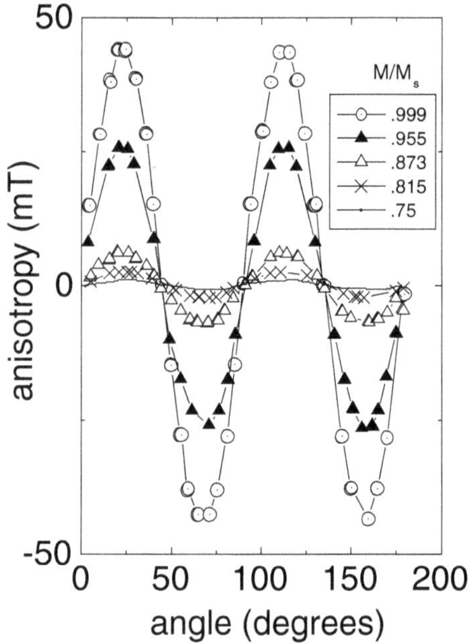

Fig. 8.7. The calculated effective anisotropy field by using the bulk anisotropy value for cubic anisotropy parameter, 2K1/M (46 mT) with a Gaussian distribution for spin orientation for various values of the distribution width (or average magnetization). It should be noted that the model give almost the bulk value for perfect alignment of the spins corresponding saturation magnetization Ms.

Table 8.1. Magnetic parameters deduced from the simulated spectra at different temperatures of a Fe_3O_4 film

Parameters	4K	30K	65K
$M(10^{-3}T)$	416	414	423
$H_{AR}(10^{-4}T)$	850	525	50
$H_{AD}(10^{-4}T)$	300	250	150
$K_3(10^2 J/m^3)$	-324	-330	-79
$K_3^{/}(10^2 J/m^3)$	-76	-78	-79
$K_{sa}(10^{-4} J/m^2)$	2	3	-3.5
$D(10^{-10} J/mT)$	4	5	6.6
$\Delta H^n(10^{-3}T)$	9	30	50
$\Delta H^r(10^{-3}T)$	25	30	50

the hard direction for surface spins is the film normal resulting in easy plane surface anisotropy over a wide temperature range 4-300 K. From the angular variation of the FMR spectra at higher temperature, the peak at the high field side was identified as a surface mode which appears for perpendicular direction

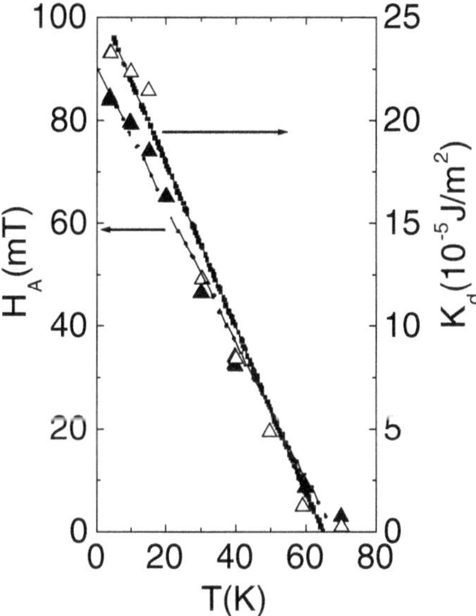

Fig. 8.8. The rigid components of both the unidirectional bulk (heavy triangles) and surface (open triangles) anisotropies as a function of temperature. The dotted and continuos line are respectively the calculated ones by assuming a linear temperature dependence given in the text.

of field. The drastic change in this surface mode is due to surface induced directional anisotropy. The temperature dependence of both the bulk and the surface directional anisotropies is shown in Fig. 8.8, as deduced by using the simulated spectra. The dotted and the continuous lines are respectively the corresponding fitted curves which were deduced from Eq.(8.16). There is a small deviation from the experimental results at lower temperature region. But the fitted values of the parameter, T_{sg}, are almost the same for both surface and bulk uni-directional anisotropys. This means that the physical origin of the bulk and surface exchange anisotropies is the same and the unidirectional anisotropy in the bulk reflects itself on the surface but the strength of this anisotropy at the surface varies due to lower symmetry at the surface [52].

Regarding the magnetic energy parameters in Table 8.1, the value of the magnetization below 70 K is smaller than bulk value at room temperature. This is a natural result of the Verwey transition of Fe_3O_4 from cubic to orthorhombic structure at lower temperatures. The rigid component, H_{AR} , of the unidirectional anisotropy at 4 K is relatively large compared to most of the spin-glass materials in the literature. That is the system is magnetically highly disordered. Value of H_{AR} for field cooling case decreases linearly and more rapidly than the dynamic component, H_{AD}, with increasing temperature up to 70 K. There is a strong correlation between rigid and dynamic component of the uni-directional anisotropy.

The bulk anisotropy parameters are practically constant below the Verwey transition temperature, T_v. Both uniaxial (K_3 which also include shape anisotropy, $2\pi M^2$) and bi-axial ($K_3^{/}$) components of the magneto-crystalline bulk anisotropy parameters in Eq.(8.12) are negative for the field cooled case. The easy direction for magnetization is perpendicular to the film plane. These results are consistent with well-known properties of the bulk Fe_3O_4 ferrite. As explained above, when a Fe_3O_4 single crystal is cooled down below T_v in the presence of an external field, the cubic Fe_3O_4 transforms into an orthorhombic crystalline structure in such a way that the easy axis (c-axis) becomes parallel to the external field. It should be remembered that the magnetic parameters in Table 8.1 were deduced from the spectra for perpendicular field cooling. Therefore the easy axis was induced along the direction perpendicular to the film plane.

The axial component (K_{sa}) of the surface anisotropy in Table 8.1 is smaller than the uni-directional component (K_d) in Fig. 8.8 at 4 K. However, K_{sa} increases while K_d decreases as the temperature rises. Perhaps the rigid component of the exchange anisotropy gains rotational character with the magnetization at relatively higher temperatures. The effect of this rotation manifests itself in the increase of K_{sa}. On the other hand the effective exchange interaction parameter, D, progressively decreases as the directional anisotropy grows at lower temperatures. Naturally, as a structural transition the Verwey transition should affect the exchange interaction parameter D.

Lastly, the peak-to-peak line widths ΔH^n and ΔH^r for n-FC and r-FC cases respectively increase with the increasing temperature. However the increase is much smaller for the n-FC case than for the r-FC case at 4 K. Similar behavior has been observed in literature [53] for reentrant NiMn alloys. This effect has been attributed to the elastic rotation of internal uni-directional anisotropy which was induced along the initial magnetization during the cooling of the sample. At lower temperature the anisotropy becomes much more rigid, resulting in a longer relaxation time (smaller line width). Increment in line width at higher temperatures in FC case should be originated from rapid relaxation of both the bulk and the surface unidirectional anisotropies.

8.5.2 $Ni_{77}Mn_{23}$ Films

Some examples from the calculated spectra (continuous lines) obtained by using the above model are shown in Fig. 8.6. For the sake of clarity not all the theoretical spectra corresponding to the experimental ones are plotted. As can be seen from Fig. 8.6 there is quite satisfactory agreement between the theoretical and the experimental data at relatively higher temperatures. The model gives almost all the main futures of the experimental spectra (number of peaks, relative intensities, relative distance and line-width of these peaks). However, one can easily note that there is disagreement between the experimental and theoretical peaks corresponding to the main mode at 50 K. This disagreement can be attributed to the well-known hysteresis effects and under saturation of magnetization due to frustration of the spins of the NiMn alloys of this composition at low temperatures. However, we have not included these hysteresis effects in our theoretical model. This frustration effect seems to manifest itself in the resonance line-shape.

The partially resolved two weak peaks at highest fields have been identified as surface modes which are induced at two opposite surface at intermediate temperatures. The relative distance of these resonance modes to the strongest one (which is identified to be as first order bulk mode) are increased with the increasing pining parameters, which in turn increased with surface anisotropy parameters. When the surface pinning parameters, Pi, increase then the spin waves corresponding to the surface modes are more and more localized on the surface region, giving weaker absorptions due to their relatively small contribution to high frequency magnetic susceptibility. This is consistent with the behavior of the experimental resonance lines. Conversely, as the temperature increases the shift from the main line decreases and the intensity of the surface modes increases. This implies that the pinning decreases with increasing temperatures as seen in the figure.

The minor peak at lowest field has been identified to be third order, showing odd symmetry with respect to the center point of the film bulk mode excited across the opposite surfaces. The distance of this mode to the first order bulk mode (main absorption peak) is seen to be proportional to the exchange parameter, providing the best way to determine its value from the experimental data.

The temperature variation of the exchange interaction parameter D is given in Fig. 8.9a. The deduced values at each temperature are connected by continuous and/or dotted lines as guide to the eye. Almost identical values are found for the three films with thickness 300 Å, 600 Å and 1000 Å. For a precise and independent determination of this parameter one needs at least two bulk modes in the experimental spectra. But the bulk modes are shifted to lower and lower fields (even to the negative field side) as the film becomes thinner and thinner. That is, in very thin films (50-100 Å) the wavelength of higher order spin wave modes is too short (spin wave energy is too high) to be excited by microwave photons. This makes exchange stiffness and surface anisotropy impossible to be

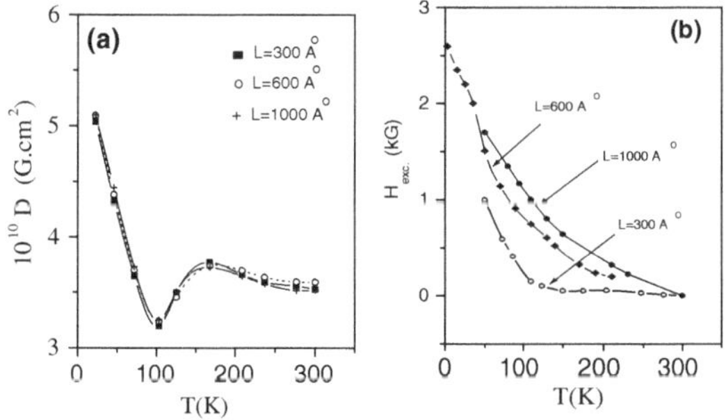

Fig. 8.9. Temperature variations of effective exchange stiffness parameters (a) and effective exchange anisotropy (b), deduced from the SWR spectra of three films with different thickness, by means of the model described in the text.

deduced independently. Therefore we have not attempted to determine D by using the data for the films thinner than 300 Å. The problem with the thickest film (3700 Å) arises from the partially overlapping multi-peak spectrum. Moreover, hysteric behavior further complicates the theoretical analysis of temperature dependent data. Nevertheless, the overall value for parameter D, which is roughly estimated from these spectra (not plotted in the figure), has been found to fall in the same range as for these three films. As seen from the Fig. 8.9a, the exchange parameter D first slowly increases down to 150 K (T_c), passes a relatively broad local maximum, decreases rapidly down to 110 K (T_f), and then passing through a minimum starts again to rise quite sharply with decreasing temperature. Similar behavior was also observed for other reentrant systems in the literature[34,35,40]. The first slow increase at higher temperature is expected as the system enters into the ferromagnetic regime from paramagnetic phase. The decrease after this first increase should originate from antiferromagnetic contribution of Mn atoms. The last rapid increase at lowest temperature can be attributed to the frustration of the spins. It should be noted that the effective spin wave stiffness (exchange interaction parameter D) is determined from energy of the macroscopic spin waves due to the small change of the relative orientation of the neighboring spins. As the spins are more and more frozen in the spin-glass regime, they effectively become more and more rigid requiring larger energy for excitation of the spin waves. That is, although the magnetization reduces in spin glass regime, the effective exchange stiffness parameter can increase because of freezing of disordered magnetic configuration due to continuous evolution of antiferromagnetic interactions between Mn-Mn pairs below the Nèel temperature.

Fig. 8.9b displays the exchange field deduced from the experimental data of the three films with intermediate thickness as a function of temperature. We have not plotted the results from the thickest and thinnest films because of the higher uncertainties due to the reasons mentioned above. Obviously any inaccuracy in the value of D would affect the precision in the value of the Hexc. As seen from the figure, the unidirectional anisotropy increases with decreasing temperature for all films. Although there seems to be some exponential temperature dependence, we have not attempted to fit the data to any theoretical expression. The exchange anisotropy field is larger for the 1000 Å thick film. That is, as the films become thinner the exchange anisotropy energy tends to diminish, giving more freedom for the spins to contribute to the overall macroscopic magnetization consistently with the measured magnetization data obtained either from dc magnetization or from SWR measurements. The resonance field for perpendicular geometry of 100 Å thick film is larger than that for thicker films, due to the larger demagnetizing field. Basically the exchange anisotropy originates from the competing interactions between ferromagnetic Ni-Ni, Ni-Mn and antiferromagnetic Mn-Mn pairs. Naturally these interactions are strictly affected by the distances of corresponding atoms depending on the crystalline structure. Since the environment of any magnetic pair at the surface is totally different from that in the volume, the exchange anisotropy induced at the surface can dominate over

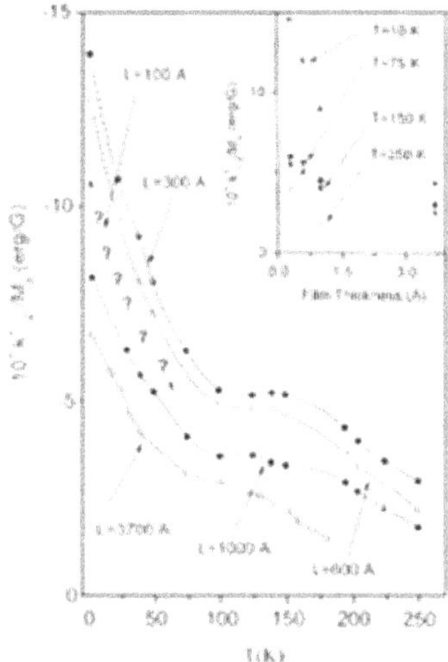

Fig. 8.10. Temperature variation of effective surface pinning parameter deduced from the SWR spectra of $Ni_{77}Mn_{23}$ films with different thickness indicated by arrows in the figure. The ? mark denotes the values of the surface pinning parameter, which obtained by taking zero values for the bulk anisotropy energy for 100 Å thick film. The inset shows thickness dependence of this parameter at some selected temperatures.

its value in the bulk, giving relatively larger effective bulk anisotropy values for thinner films.

The normalized surface pinning parameters are plotted against temperature in Fig. 8.10. In this case we have added the data from the thicket (3700 Å) and the thinner (100 Å) films. For these latter films we have used the same values for D that were the same for the films with intermediate thickness. The procedure used to obtain the magnetic parameters for these extreme films will be given blow. The ? mark in the figure implies high uncertainty of the data for 100 Å thick film. The general trend of the temperature behavior of this parameter seems to be the same for all the films while its absolute value is thickness dependent. A comparison of the Figs. 8.9 and 8.10 show a very close correlation among the effective exchange parameter, unidirectional bulk anisotropy and the surface anisotropy parameters. The absolute values of the surface anisotropy parameters first increase down to 150 K, pass a very broad maximum-like plateau until 110 K, and then start to increase rapidly like the exchange parameters D of the same samples. The values of both the exchange anisotropy field and exchange stiffness parameters qualitatively follow each other as a function of temperature. There seems to be weak thickness dependence for the surface anisotropy parameter [41]. The effective value of the easy plane surface anisotropy is somewhat larger for thinner films. This makes the surface anisotropy as a dominant term on the behavior of the magnetization and gives a larger magnetization measured at lower fields. We have also plotted the thickness dependence of the surface anisotropy in the inset for some selected temperature. As seen from this inset, as the temperature rises the slope of the line tends to diminish in accordance with the observation in literature [41].

8.6 Conclusion

8.6.1 Fe_3O_4 Films

The epitaxial Fe_3O_4 film on MgO(100) shows some differences in magnetic properties as compared to those in bulk single crystal Fe_3O_4. The FMR spectra of Fe_3O_4 thin films include well-resolved surface induced modes in addition to the bulk FMR modes. Moreover, both surface and bulk modes have uni-directional character at low temperatures, upon cooling the sample in an external field. This is a clear sign for an exchange anisotropy. The exchange anisotropy value at the surface differs from that of the bulk of the same film. For thin film, a strong exchange anisotropy field is induced upon cooling the sample in the presence of an external magnetic field. As well -known, the exchange anisotropy generally originates from a competing ferromagnetic and anti-ferromagnetic interactions in a magnetic system. However, all kinds of interactions in the bulk Fe_3O_4 magnetite are anti-ferromagnetic as explained above. Possible exchange interactions mediated by oxygen ions (super exchange) between A-A, A-B, and B-B pairs are different in magnitude. Normally the A-B interaction is much larger than the A-A and B-B interactions. As a result Fe_3O_4 has ferromagnetic order (to be more precise a spiral spin configuration) which gives rise to a relatively sharp hysteresis loop for magnetization. It should be noted that there have already been competing interactions between A-B and B-B (or A-A) sites to orient the spins in site B (or A) in opposite direction of any other nearest neighboring spin. None of the spins likes another spin to be oriented along its own direction. Such kind of interaction naturally causes a frustration for spin alignments but in the stoichiometric Fe_3O_4 or Fe_2O_3 magnetite they result in a well ordered spiral spin configuration. Therefore one might expect that a small change of the crystalline parameters of epitaxial Fe_3O_4 on MgO (100) due to a little misfit at the interface, modifies the relative exchange interactions among the different pairs. This in turn might cause a disorder in the spin alignments and a big exchange anisotropy field in Fe_3O_4 films. But Krebs et al. [54] showed that a single crystal Fe_3O_4 grown on MgO by Molecular Beam Epitaxy (MBE) behaves essentially like a bulk sample. So the exchange anisotropy seems to originate from some other effects in our samples.

On the other hand Aragon et. al. [55] studied non-stoichiometric $Fe_{3(1-\delta)}O_4$ ($0< \delta <0$) magnetite and found that a slight deviation from exact composition rate has profound effects on physical properties. For instance, Verwey transition, remnant magnetization, cubic magnetic anisotropy constant K_1 and exchange parameter(J_{AB}, dominant interaction between A and B sites) are reduced with increasing non-stoichiometry. Even some Fe_3O_4 samples from our different runs give different magnetic cubic and exchange anisotropy values depending on a very small change of preparation conditions (partial oxygen pressure and temperature). Comparing the values for the exchange parameter J_{AB} from this work(The value for D in Table 8.1 corresponds to a value of 30 K for J_{AB}) and from Aragon's study for various Fe concentrations one can conclude that our sample is very close to the exact Fe_3O_4 composition. A fast decrease in resonance field at the temperatures below 90 K implies a decrease in Verwey transition in our case. We also could make some Fe_3O_4 films giving very small exchange anisotropies and magneto-crystalline cubic anisotropy which are very close to those of bulk sample even at relatively small applied field. The results will be published in a separate paper. All findings above suggest a very small Fe deficiency(i.e. a tendency from Fe_3O_4 to Fe_2O_3 composition) to be responsible for these unexpected magnetic behaviors.

Finally unidirectional surface anisotropy field was phenomenologically introduced by Puszkarski [44] for surface spin pinning. However, the observed angular behavior of most of the FMR spectra in the literature [42] show generally a higher symmetry than that of a unidirectional surface field. As far as we are aware ours is the first [45] clear observation of unidirectional bulk and surface anisotropy in FMR lines of Fe_3O_4 thin films.

It should also be noted that recently Avgin [56] et.al. calculated spin wave stiffness for Ca doped YIG where uncompensated Ca causes a strong random ferromagnetic interaction between Fe ions occupying different sublattices. Similar interaction picture due to a slight Fe deficiency may also occur in our system. We will do further investigation along that direction.

8.6.2 NiMn Films

The strong correlation of all the parameters can be understood as follows: As explained in a previous paper [34], these films exhibit some anomalous anisotropies arising from deposition conditions. The films were deposited by e-beam evaporation of pure Ni and Mn powders contained in separate baskets located at different positions under the substrate. Each beam impinges on the film with a different angle depending on the relative basket location. It can be expected that, at very early stage of growth, each incident beam tends to tilt the grains of the NiMn alloy in the film along its own direction. Since the atomic percentage of Ni is much higher than that of Mn, the resultant granular structure may be grown along the Ni beam direction. In addition to this, the Ni and Mn beams tend to deposit on different sides of the grains, that is, the side facing Ni may be Ni rich, while the other side, facing the Mn beam, is Mn rich compared to the nominal value of 23 % at the Mn sites in the alloy. Therefore, it is quite possible that the concentration in the granular structure slightly varies from the Mn to the Ni side in the plane of the film. This may be assumed as an additional effect that gives local growth induced (geometric or oblique) anisotropy for the spins.

On the other hand, it should be remembered that the surface anisotropy in NiMn films varies considerably with the temperature. Normally, a magneto crystalline surface anisotropy is not expected to exhibit such strong temperature dependence. That is, the origin of this anisotropy can not be attributed to the usual magneto-crystalline interaction. Considering the rapid enhancement of all the above parameters just below the same characteristic spin-glass temperature, all of these effects may be expected to originate from the exchange anisotropy field (D-M anisotropy), which is easy plane in nature and induced at the surface of each individual granular. The contribution to the average energy per spin from surface anisotropy is much larger compared to that from bulk anisotropy energy. As the surface to volume ratio of these granular increases for thinner films, this local surface pinning effect becomes dominant for the overall behavior of the magnetic properties as well as for the resonance line-shapes.

The authors would like to thank Dr. David Margulies for his kindness to supply the characterized samples. This work has been supported by Scientific and Technical Research Council of Turkey and Islamic Development Bank.

References

1. J.K.Vassiliou, V.Mehrotra, M.W.Russell, E.P.Giannelis, R.D.McMichael, R.D.Shull, R.F.Ziolo, J. Appl. Phys. 73(10), 5109(1993)
2. F.Jorgensen, The complete Handbook of Magnetic Recording (TAB Books Inc. 1980), pp.191
3. A.E.Berkowitz, IEEE Trans. Magn. MAG-22, 466(1986)
4. L.Neel, Ann. Phys. 3, 137(1948)
5. E.W.Gorter, J.A.Schulkes, Phys. Rev. 90, 487(1953)
6. Y.Yafet, C.Kittel, Phys. Rev. 87, 290(1952)
7. C Kittel, An Introduction to Solid State Physics, Seventh Ed., Page: 459, 1996
8. T.A.Kaplan, K.Dwight, D.Lyons and N.Menyuk, J. Appl. Phys. 32, (1961) 13S
9. E.J.W.Verwey, Nature 144, 327(1939)
10. F.J.Parker, A.E.Berkowitz, IEEE Trans. Magn. MA G-22, 466(1986)
11. F.Itoh and M.Satou, Jpn. J. Appl. Phys. 14, 2091(1975)
12. F.Itoh, M.Satou and Y.Yamazaki, IEEE Trans. Magn. MAG-13, 1385(1977)
13. F.E.Spada, A.E.Berkowitz and N.J.Prokey, J. Appl. Phys., 69(8), 4475(1991)
14. A.E.Berkowitz, F.E.Parker, E.L.Hall, and G.Podolsky, IEEE Trans. Magn. MAG-24, 2871(1988)
15. J.S.Kouvel and C. D. Graham, J. Phys. Chem. Solids 11 (1959) 220;
16. W.Abdul-Razzaq and J.S.Kouvel, Phys. Rev. B, 35 (1987) 1764;
17. W.Abdul-Razzaq and J.S.Kouvel, and Kh. Zig., Phys. Rev. B, 35 (1987) 1768
18. N.V.Volkenshtein., N.I.Kourov and Yu.N.Tskioukin, Fiz. Met. Metalloved, 42 (1976) 213 (Phiz.Met.Metallogr. (USSR) 42 (1976) 190)
19. T.Sato, R.B.Goldfarb and C.E.Patton, Phys. Rev. B, 18 (1978) 3684
20. R.B.Goldfarb and C.E.Patton, Phys. Rev. B, 24 (1981) 1360
21. S.Senoussi and Y.Öner, J. Appl. Phys. 55 (1984) 1472
22. S.Senoussi and Y.Öner, Phys. Rev. B, 28 (1983) 455
23. B.Aktaş, Y.Öner, and E.A.Harris, Phys. Rev. B, 39 (1989) 528;
24. Y.Öner, B.Aktaş, F.Apaydın and E.A.Harris, Phys. Rev. B, 37 (1988) 5866
25. B.Aktaş, Y.Öner, and H.Z.Durusoy, J. Magn. Magn. Mater., 119 (1993) 339
26. Y.Öner, B.Aktaş, F.Apaydın and C.Kaptanoglu, J. Magn. Magn. Mater., 109 (1992) 223
27. Y.Öner, B.Aktaş, Phys. Rev. B, 42 (1990) 2425
28. H. Z.Durusoy and Y.Öner, Phys. Rev. B, 42 (1990) 6831
29. Y.Öner, A Kılıç, H.Çelik, Physica B, 215 (1995) 205

30. T.Sato, N.Nogeyama, Y.Torri and T.Ando, J.Magn.Magn.Mater.,90&91 (1990) 349
31. M.Özdemir, B.Aktaş, Y.Öner, T.Sato, T.Ando, J.Magn.Magn.Mater.,164 (1996) 53
32. M.Özdemir, B.Aktaş, Y.Öner, T.Sato, T.Ando, J.Phys.Condens. Matter., 9 (1997) 6433
33. T.G. Knorr and R.W.Hoffman, Phys. Rev. 113, 4 (1959);
34. O.Kohmoto and C.Alexander, Jr., J.Magn.Magn.Mater.,116 (1992) 405
35. B.Heinrich, K.B.Urquhant, A.S.Arrott, J.F.Cochran, K.Myrtle, and S.T.Purcell, Phys. Rev. Lett. 59, 1756(1987)
36. G.A.Prinz, G.T Rado, and J.J.Krebs, J. Appl. Phys. 53, 2087(1982)
37. Z.Zhang, P.E.Wigen, and S.S.P.Parkin, J. Appl. Phys. 69, 5649(1991)
38. C.Kittel, Phys.Rev. 110, 1295(1958)
39. A.Layadi, J.O.Artman, B.O.Hall, R.A.Hoffman, C.L.Jensen, D.J.Chakrabarti and D.A.Saunders, J. Appl. Phys. 64, 5760(1988)
40. P.E.Wigen, Thin Solid Films, 114, 135(1984)
41. B.Aktas, and M.Özdemir, Physica B. 193, 125(1994)
42. B.Aktas, Sol. Stat. Comm. 87, 1067(1993)
43. B.Aktas, Y.Öner, H.Z.Durusoy, J.Magn.Magn Mater. 119, 339(1993)
44. H.Puszkarski, 1979 Progr. Surf. 9, 191 and references therein
45. D.T.Margulies, F.T.Parker, F.E.Spada, R.S.Goldman, J.Li, R.Sinclair and A.E.Berkowitz, Phys.Rev.B, 53, 9175(1996)
46. B.A.Calhoun, Phys. Rev. 94, 1577(1954)
47. L.J. Maksymowicz and D.Sendorek, J. Magn. Magn. Mater. 37, 177(1983)
48. M.A.Manheimer, S.M.Bhagat, and D.J.Wolf, J. Appl. Phys. 51, 347(1985)
49. P.Jamet, J.C.Dumais, J.Seiden, and K.Knor, J. Magn. Magn. Mater. 15, 197(1980)
50. D.Youm, and S.Schultz, Phys. Rev. B 34, 7958(1986)
51. H.J.Williams, R.M.Bozorth, and M.Goertz, Phys.Rev. 91, 1107(1953)
52. J.F.Cochran, B.Heinrich, and A.S.Arrott, Phys. Rev. B 34, 7788(1988)
53. B.Aktaş, M.Özdemir, R.Yilgin, Y.Öner, T.Sato, T.Ando, Physica B 305 (2001) 298
54. J.J.Krebs, D.M.Lind, and S.D.Berry, J. Appl. Phys. 73, 6457(1993)
55. R.Aragon, Phys. Rev. B 46, 5328-5334(1992)
56. I. Avgin and O.L.Huber, J. Appl. Phys. 75, 5518(1994).

9 Low Temperature Magnetic Properties of Nanocrystalline Iron

A. Hernando[1], P. Crespo[1], M.S. Flores[1], Z. Sefrioui[2], J.L. Menéndez[2], A. Cebollada[2], and F. Briones[2]

[1] Instituto de Magnetismo Aplicado, U.C.M.-RENFE, P.O. Box 155, 28230 Madrid, Spain
[2] Instituto de Microelectrónica de Madrid-IMM (CNM-CSIC), Isaac Newton 8 (PTM) Tres Cantos, Madrid 28760, Spain

Abstract. Nanocrystalline iron has been obtained by sputtering on different substrates.Resistivity measurements point out an anomalous behaviour of the thermal dependence of the resistivity at low temperatures which dissapears when a high magnetic field is applied. Strong changes in the spin dependent scattering in islanded Fe (110) thin films are due to a low temperature spin freezing of the island boundary magnetic regions, producing a suppression of the exchange coupling between islands. A consequence of the magnetic decoupling is the random arrangement of the individual magnetization, determined by the magnetocrystalline anisotropy of each island, which results in a spin-dependent induced increase of resistivity below freezing temperature. After application of a magnetic field, magnetic ordering recovers regardless of temperature, obtaining a typical metallic behavior for the temperature dependence of the resistivity. Ratio of low temperature resistivities, measured with and without magnetic field, yields magnetoresistance values of up to 5.5 % for 16.5 % nm in-plane island sizes. The following two conclusions are obtained, i) nanocrystalline Fe can be considered a low temperature GMR-like system and ii) Fe grain boundaries are not ferromagnetic at low temperature, but behave as reentrant spin-glass.

9.1 Introduction

There exists a great interest in the development of nanostructured magnetic materials due to novel behaviors observed. When the size of the magnetic entities is reduced to the nanoscale range, the structural correlation length becomes of the same order as the exchange correlation length, thus giving rise to new phenomena. In particular, the relative number of atoms located at the grain boundaries increases drastically. Since magnetic properties are closely linked to short range order that fluctuates at the boundaries, differences in the magnetic behavior of grain and grain boundaries could be expected. In this work, the evolution of the magnetic properties of nanocrystalline Fe grown by sputtering with different in-plane and out of plane dimension is shown. We present the low temperature magnetic behavior of this system by measurements of transverse Kerr effect, the thermal dependence of the magnetization and electrical resistance. These measurements evidence the existence of a low temperature magnetic disordered state at the grain boundaries.

Nanostructured magnetic materials have attracted a lot of interest since their macroscopic magnetic properties are very different from those of the bulk. These materials are formed by an assembly of magnetic grains, called nanocrystals, embedded in a magnetic or non-magnetic matrix. Nanocrystals are generally interconnected by either the grain boundaries or a different matrix. As the size of the crystallites is reduced, the number of atoms located at the grain boundary becomes comparable to the number of atoms on regular lattice sites. A simple calculation reveals that when the mean grain size D is around 10 nm, the fraction of atoms at the interface between the crystallites can be as high as 30 %. Finite size and interface effects are claimed to be the main responsible for the novel physical properties exhibited by bulk nanocrystalline materials.

Let us focus on ferromagnetic nanocrystals. Generally we mean by ferromagnetic nanocrystalline sample any sample composed of ferromagnetic nanocrystals embedded in a matrix. The relevant aspect of magnetic nanocrystals is related to the coincidence of the grain size scale (nanometers) with the critical magnetic length. One important characteristic of nanocrystalline ferromagnetic samples is the orientation fluctuation of the local magnetization easy axes. The correlation length of the anisotropy axes is typically of the order of D. The second important characteristic is the degree of coupling, exchange or magnetostatic, between adjacent crystallites carried out through the atoms of the intergranular region which in general exhibit different magnetic properties from the atoms place inside the nanocrystals. The difference between exchange and magnetostatic coupling is remarkable. Exchange is more intense and localized, whereas dipole-dipole interactions are smaller but long ranged. Exchange from ferromagnetic toward paramagnetic metals can penetrate a few interatomic distances decaying exponentially. But its boundary value at the ferromagnetic side reaches effective fields of 1000 T. Exchange interactions also govern the coupling through insulators driving the spin dependence of electronic tunnelling. On the other hand, the stray fields carrying magnetostatic coupling are of the order of a few teslas but extend over the overall crystallite size. Both types of interaction lead to ordered ground states at low temperatures. The ground state is in general difficult to predict since many types of interaction are present.

From the point of view of their practical uses, coercivity is by far the most relevant parameter characterizing the behaviour of nanostructured magnetic materials. The coercivity is firstly related to the intrinsic properties of the main phase forming the material. The central intrinsic property influencing coercivity is the magnetic anisotropy giving the magnitude of the maximum energy barrier to be overcome during the magnetization reversal. Other intrinsic quantities influencing the field at which the global magnetization reverses are the saturation magnetization, M_s and the order temperature.

The influence of the nanostructure on the macroscopic anisotropy of a ferromagnetic solid is a beautiful example of the dependence of properties on the ratio between structural lengths and characteristic magnetic lengths. In a ferromagnetic homogeneous specimen the following characteristic lengths are normally defined:

(1) Exchange length, L_{ex}, measures the particle size below which the magnetostatic energy associated with an uniform magnetization is smaller than the exchange cost required to demagnetize the particle, therefore the material is expected to be a single domain when is smaller than Lex. When the particle size is larger than L_{ex} the decomposition in domains becomes energetically favourable. L_{ex} is defined as:

$$L_{ex} = (2A/\mu_o M_s)^{\frac{1}{2}}$$

where A is the exchange constant, typically of the order of $10^{-11}\,\mathrm{Jm}^{-1}$ in 3d ferromagnetic metals.

(2) Exchange correlation length or wall thickness, L, is defined as

$$L = (A/k)^{\frac{1}{2}}$$

where k is the anisotropy constant. L measures the distance at which a local fluctuation of the spin alignment propagates through the material.

It is important to remark that the characteristic lengths defined above are in the range of nanometers. Thus, the magnetic behaviours of the nanocrystals are actually expected to be strongly dependent of the ratio D/L. Particles in the size range of nanometers should exhibit peculiar and anomalous thermal dependence due to the change induced by temperature in the ratio D/L [1].

It is somehow amazing that Fe, that has long been considered the archetype of a ferromagnetic material and has been carefully scrutinized for decades, exhibits new and singular magnetic behaviour when it is in the form of nanostructures. In this sense, pure iron nanostructures are perhaps the most fundamental case of physical processes described above: the appearance of unique magnetic properties due to the different physical behaviour of interface atoms with respect to those in the bulk.

In recent years many experimental observations have pointed out the anomalous magnetic behaviour of the atoms located at the particle surface compared with that of the inner atoms. The experimental results have shown that there exist different degrees of coupling between inner and surface atoms and that the surface spin structure is different from the inner particle magnetic structure.

It has been recently shown that nanocrytalline iron, obtained by the ball milling technique, exhibits a decrease of ac susceptibility, zero field cooled magnetization and thermoremanence below a specific temperature. This observation was interpreted by the authors as a consequence of the low-temperature freezing of the spin at the grain boundaries into a non collinear order [2]. The observed magnetic behaviour suggested us that a noticeable influence of this effect on the electronic transport properties should be observable at his temperature range. It will be shown here that disordered magnetism at boundaries separating crystal regions with different strain constitute an alternative way to induce GMR.

In this article we present experimental evidence of low temperature spin freezing in a different ferromagnetic system, epitaxial iron thin films, and the consequent appearance of a minimum in the thermal dependence of the resistivity at T_f, the presence of GMR below T_f, and a decrease of zero-field-cooled magnetization. This new observation points out that the origin of these phenomena is common for quite different iron based magnetic systems and related to the speromagnetic behaviour of lattice regions at which competitive exchange interactions take place.

9.2 Experimental Details

Experimental results were obtained for a two dimensional Fe system formed by exchange coupled nanocrystals with controlled anisotropy. Two set of samples have been prepared.

First, epitaxial (001) Fe thin films formed of nanometer size islands were grown by sputtering. Fe (110) islands were deposited on Al_2O_3 (0001) substrates in a UHV triode sputtering system at 700°C with typical deposition rates of 1 nm/minute. In all the cases, a 2.5 nm Pt capping layer was grown at room temperature to protect from oxidation [3]. X Ray Diffraction (XRD) and Transmission Electron Microscopy (TEM) measurements indicate the (110) textured growth of bcc Fe and three in plane equivalent azimuthally orientations ([110] Fe//Al_2O_3) are expected from the symmetry of the substrate.

Second, epitaxial (001) Fe thin films have been obtained by sputtering on (001) MgO at different substrate temperatures, from RT to 700°C. Film thickness was 20 nm and to avoid oxidation in air they were capped in situ with a thin (1,5 nm) layer of TiN. All samples show clear biaxial in-plane magnetic anisotropy along the [100] and [010] crystalline directions [4]. Atomic Force Microscopy (AFM) has been performed to caracterise the morphology of the system. Resistivity was measured by using standard four terminal method.

The island size and morphology for different Fe deposition times was studied with AFM. As an example, Fig. 9.1 shows the atomic force microscopy images for Fe grown on Al_2O_3 substrates for different deposition times. Island size ranges from 11 nm in-plane diameter and 2 nm heights up to 100 nm in plane and 23 nm heights. Both average islands diameter and height do increase for larger deposition times. The small islands show a homogeneous size distribution. We observe a reduction in the number of islands per unit area with increasing deposition time due to coalescence of small islands to form bigger ones. AFM observations indicate that the island tops are separated from each other, but unfortunately the AFM tip cannot reach the island bases due to the proximity between adjacent islands. Fig. 9.1(d) shows a TEM planar view for the sample with 3 minutes deposition time. Moire patterns are observed in some of the islands, indicating the crystalline character of the structure. As shown in Fig. 9.2, the presence of three possible equivalent in-plane crystalline orientations of the Fe islands on the substrate implies that the orientation of the in-plane (100) Fe magnetic easy axis can only jump between adjacent islands at discrete values of 60°. As mentioned previously, samples grown on MgO substrates show biaxial in-plane magnetic anisotropy along the [100] and [010] crystalline direction.

The hysteresis loops have been obtained through MOKE effect. This is a very useful technique to characterise the magnetization of magnetic thin films, since

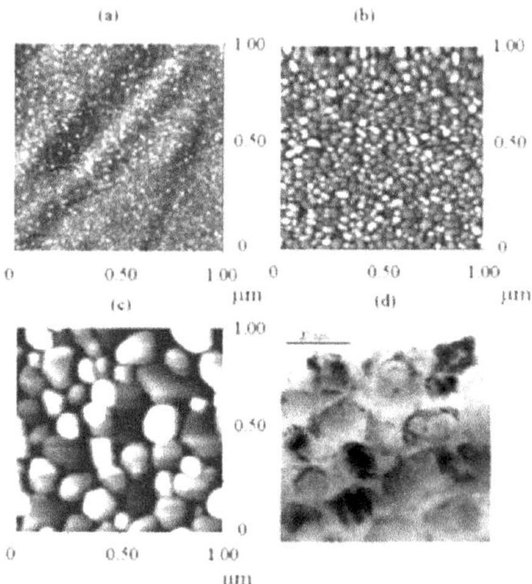

Fig. 9.1. (a-c) AFM images for three Fe structures grown on Al_2O_3 with different deposition times: (a) 20 seconds, (b) 5 minutes, (c) 20 minutes, (d) TEM planar view of a 3 minutes deposited sample.

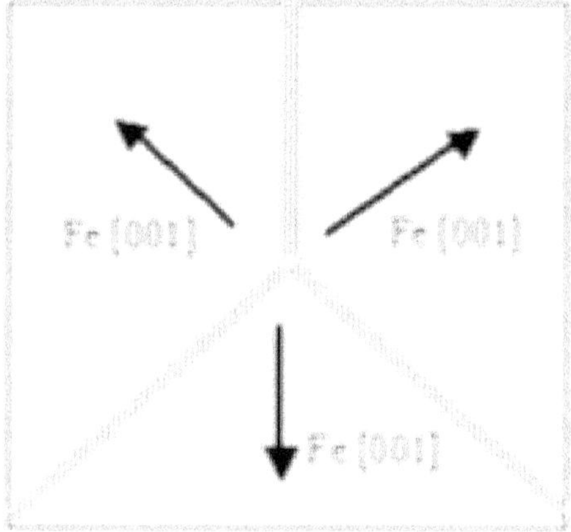

Fig. 9.2. Sketch of the three equivalent epitaxial orientations of Fe (110)deposited on Al_2O_3

the Kerr effect probes a quantity which is proportional to the magnetization on the surface. Magneto optic effects arise within ferromagnetic materials due to the optical anisotropy present within the materials. The source of the optical anisotropy is the magnetisation M within surface domains. The anisotropy alters the state of linearly polarised light reflecting off ferromagnetic materials, or there is a change in the intensity of the light itself.

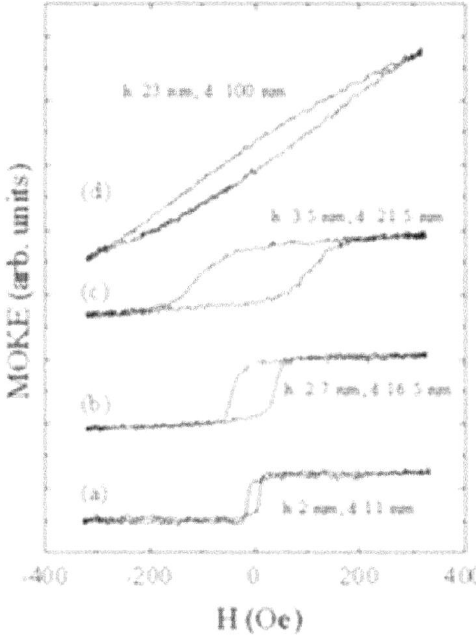

Fig. 9.3. Kerr hysteresis loops of deposited Fe on Al_2O_3 for a series of samples with (a) 2, (b) 2.7, (c) 3.5 (d) 23 nm nominal thickness

9.3 Results and Discussion

Regarding the magnetic behavior, in Fig. 9.3, we show the hysteresis loops taken with Kerr effect measurements for a series of samples with different deposition times and therefore different island size and contact area between islands. The first important point to remark is the lack of superparamagnetism, even though for the smallest islands the typical size would correspond to the superparamagnetic regime. Note that the maximum volume for the onset of superparamagnetism in bcc Fe at room temperature is around 10^3 nm^3. This is presumably an indication of the presence of contact between the islands, which leads to an exchange interaction through the in-contact islands boundaries. Thus we assume that the islands are in physical contact and this contact must be carry through the islands boundaries.

The second important effect is shown in Fig. 9.4, where the continuous increase of the coercive field with islands size, with values ranging from 11 Oe for the smallest islands up to 170 Oe for the big ones, is illustrated. It is remarkable the hardening observed in the sample with the highest amount of Fe (sample d), for which the coercive field cannot be determined because the loop obtained with the maximum field achievable is a minor loop. Since magnetic domain wall size in Fe is around 15 nm the islands with dimension smaller than 15 nm are

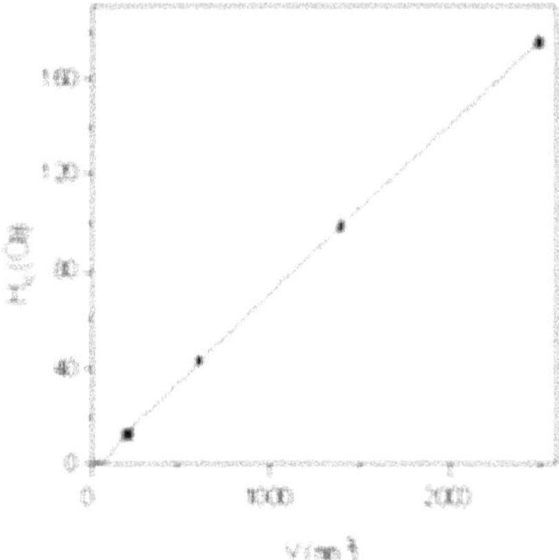

Fig. 9.4. Coercive field, H_c, vs. average island volume.

expected to be single domain. If this were the case, and the islands were magnetically decoupled, the coercivity would decrease with increasing the grain size, just with the size dependence corresponding to the evolution of single domain to multidomain structure. However, the experimentally observed dependence of coercivity with island volume is the opposite. Therefore, the observed increase of coercivity with increasing island size also seems to indicate that the islands are magnetically coupled through the island interface [5].

According to a two phase generalized random anisotropy model, RAM [6,7], the exchange correlation length through the intergrain boundary, L*, is of the order of $(\gamma A/k)^{1/2}$, where A is the exchange coupling coefficient, γ accounts for the possible modification of the interface structure interactions and k is the experimental anisotropy constant of bcc Fe [8]. As has been previously shown, for L*>d a self-consistent procedure [7] applied to the two dimensional case, 2D, yields $\langle k\rangle = k^2 v^{2/3}/(\gamma A)$. Note that L*>d implies that $\gamma > k\ d^2/A$, therefore for the case of Fe with smaller islands (11 nm) it is required that become that $\gamma > 0.6$. For γ values comprising between $0.6 < \gamma < 1$ there will be coupling between islands. In any ferromagnetic sample the coercivity can be expressed with generality as $\alpha = \langle k\rangle/\mu o Ms$ [1] where a is a parameter, $0 < \alpha < 2$, that depends on the type of magnetization process and on initial domain structure and Ms is the spontaneous magnetization. Hence, for samples with islands smaller in lateral size than 15 nm the coercivity, H_c, is expected to behave as

$$H_c = (\alpha/\gamma)(k^2\nu^{2/3}/A)(1/\mu_o Ms) = (\alpha/\gamma)1, 1.1020d^2 \qquad (9.1)$$

where d corresponds to the islands average in-plane diameter. This expression accounts at least qualitatively for the two more important characteristics of coercivity: i) As a result of inter-island coupling, γA, the samples are neither superparamagnetic nor single domain. ii) Even though from (1) H_c is proportional to $d^2\gamma$, we experimentally observe a linear dependence with d^3.

To further understand the magnetic behavior in these structures, field cooling (FC) and zero field cooling (ZFC) measurements have been performed on selected samples, as shown in Fig. 9.5. The experimental data have been obtained by either cooling the sample from room temperature to 5 K under zero field conditions (ZFC curve) or by cooling the sample with a constant applied field (FC curve). Both curves are then obtained by measuring the magnetization with a constant field of 1000 Oe while the sample is heated to room temperature. The most striking result is the reduction in M measured in the ZFC curve at temperatures below ca. 40 K. This reduction is detected in all the samples studied, being clearer for smaller Fe islands. In all cases the deviation from the normal ferromagnetic behavior occurs at temperatures around 40 K, becoming broader the larger the islands. This indicates that a spin freezing phenomena is taking place below a freezing temperature, T_f. A broadening in this feature can be due to an increase in the island size dispersion with amount of deposited Fe. The dispersion is a growth induced phenomenon and therefore it is difficult to obtain a narrow island size distribution for every specific nominal thickness. In addition, the thermal dependence of the magnetization is not the one expected in superparamagnetic systems, it seems rather due to a demagnetization process dominated by spin-wave excitations. The experimental data, for temperatures

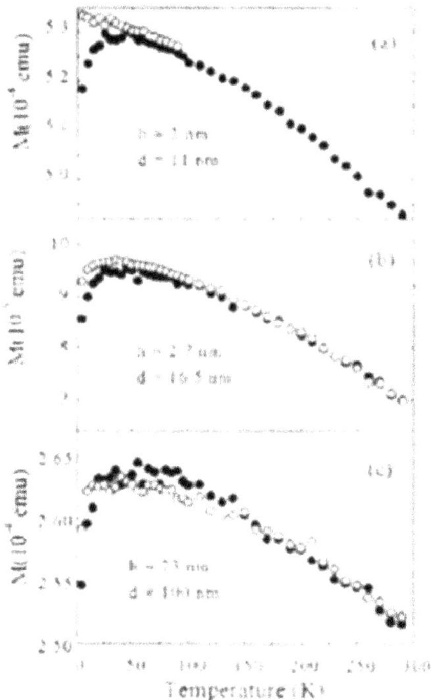

Fig. 9.5. Field cooling (·) and zero-field cooling (°) magnetization as a function of temperaure for the samples (a), (b) and (d) of Fig. 9.2. Applied magnetic field H=1000 Oe

above 40 K, can be fitted by a Bloch law, confirming a ferromagnetic spin configuration. Therefore, the magnetization curves suggest that a transition from a low-temperature disordered frozen state to a ferromagnetic state occurs for temperatures below T_f, which is around 40 K for these samples [9].

It should be noted that a similar decrease of ZFC magnetization, real susceptibility and thermoremanence has been previously observed and reported in nanocrystalline Fe obtained by ball milling [2]. The origin of such anomalous low temperature behavior was also suggested to be a spin freezing of the grain boundary. By applying a high enough magnetic field a ferromagnetic configuration of the magnetization is obtained. Therefore, it seems reasonable to expect changes in the electrical resisivity of nanocrystalline Fe under the presence of a magnetic field, since the above results indicate that it is possible to switch between a random distribution of the magnetization to a ferromagnetic alignment. Our samples, obtained by sputtering, are indeed very suitable to test the existence of spin freezing by measuring the magnetic component of the electrical resistivity, $\rho(T)$. As is known the influence of magnetic disorder in the spin dependent scattering shows its clear manifestation in the well-known giant magnetoresistance (GMR) effect [10-11]. This phenomenon is due to the change in resistivity of systems in which the magnetization directions of individual magnetic entities can be aligned with respect to each other upon application of a magnetic field. A key parameter is the spin diffusion length which determines an upper limit for the extension of the magnetic disorder to give rise to magnetoresistive effects. In other words, granular ferromagnetic thin films formed of nanometer size islands in physical contact are very adequate systems to study transport properties. Due to the reduced island height for all the fabricated structures, it is very reasonable to assume that every individual island is a single crystal [9]. The experimental thermal dependence of resistivity under ZFC and FC conditions is outlined below.

Fig. 9.6 shows the temperature dependence of electrical resistivity measured at zero magnetic field and in a magnetic field of 0.5 T, for the sample with in-plane dimensions of 16.5 nm. It is clear that the resistivity exhibits a non metallic behaviour below 150 K, when measured under zero field conditions. This anomaly disappears under the action of a magnetic field of 0.5 T. Therefore, below 150 K, the system exhibits negative GMR which increases with further cooling down up to a value of 5.5 % at 10 K.

This anomaly has been observed in both sets of samples provided the island size is small enough. Fig. 9.7 shows the thermal dependence of $\rho(T)$ for samples grown on MgO with different grain sizes, l_f, which have been deduced from the XRD peaks half-width using the Scherrer formula.

High temperature resistivity curves are essentially linear and parallel. The linear increase of residual residual resistivity with the inverse of l_f points out the outstanding role of boundaries in determining the electronic mean free path. However, the most relevant feature is the observed minimum of the resistivity, more noticeable as l_f is reduced, that appears a temperature ranging from 50 to 30 K for samples with l_f varying from 5 to 7 nm, respectively. As observed previ-

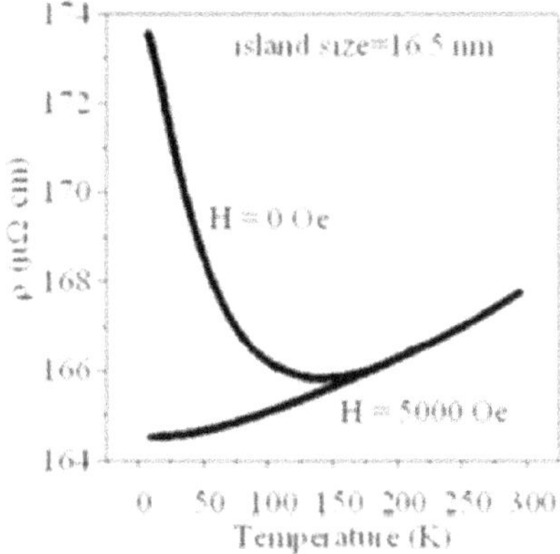

Fig. 9.6. Electrical resistivity at zero magnetic field and in magnetic field of 0.5 T for a samples with in-plane dimension of 16.5 nm.

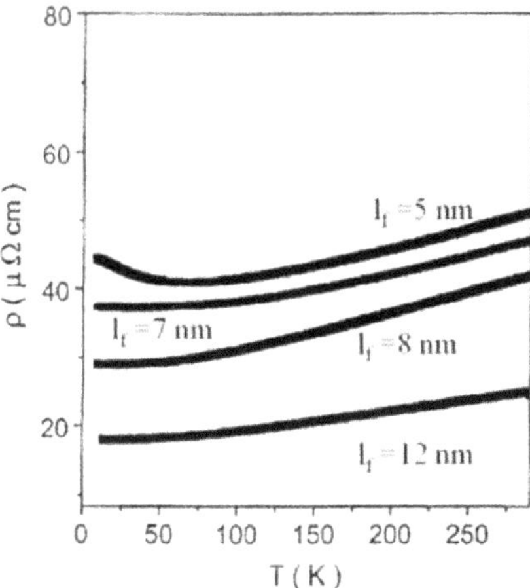

Fig. 9.7. Electrical resistivity at zero magnetic field for samples with lf = 5, 7, 8 and 12 nm deposited on MgO.

ously the resistivity minimum dissapears under the action of a magnetic field of 0.5 T as shown in Fig. 9.8. The presence of GMR indicates that below T_f the iron epitaxial system behaves as an uncoupled magnetic system. Moreover, the combination of magnetic transport measurements indicates that this ferromagnetic system evolves towards a magnetic structure not completely collinear.

The magnetic origin of this novel phenomenon, is pointed out by the ZFC-FC magnetization measurements depicted in Fig. 9.5 that clearly shows the correlation between the upturn of the resistivity at low temperatures and the suppression of the magnetization in these systems. The occurrence of the upturn

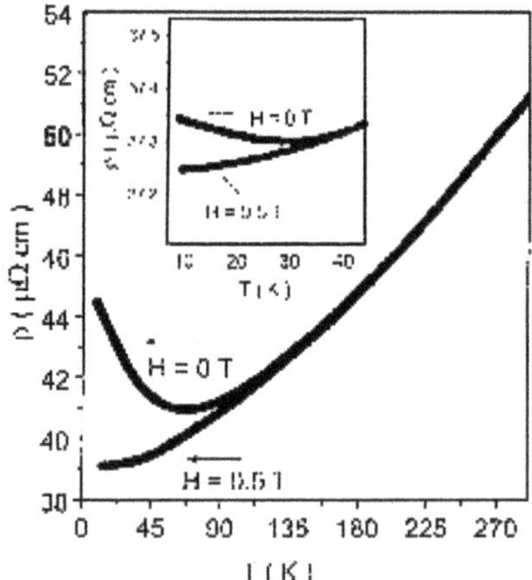

Fig. 9.8. Electrical resistivity at zero magnetic field and in magnetic field of 0.5 T for a samples with lf= 5nm. Inset: electrical resistivity at zero magnetic field and in magnetic field of 0.5 T for samples with lf=7 nm.

of the resistivity together with the suppression of the magnetization allows us to conclude that below a certain temperature, T_f the magnetization becomes inhomogeneous with a typical correlation length smaller than the electron mean free path. As Fe is ferromagnetic at low temperatures, the decrease of magnetization observed below T_f can only be a consequence of the magnetic behavior of grain boundaries. The magnetization of boundaries is an important parameter governing the intergrain exchange coupling. For temperatures above T_f the spins of the boundary are magnetized along the direction of the adjacent grain by contact to their molecular field. The whole system is ferromagnetic and the thermal dependence of the resistivity follows a typical metallic behavior. Below T_f the boundary spins start to freeze toward the spin disordered ground state reducing the exchange coupling between islands. This reduction leads to a tendency of the grain magnetization to orient along the easy axis, randomly distributed in direction, hence giving rise to a progressive decrease of the overall magnetization. If the distance between different oriented easy axes, that is the grain size, is lower than the electron mean free path, the spin dissalignement should give rise to an increase of the electrical resistivity via an increase of the spin dependent scattering.

9.4 Conclusions

We have presented a new granular magnetic system in thin film form that exhibits properties that strongly depend on the intergrain interaction. The behavior observed and reported in this work can be considered as a strong evidence of the non ferromagnetic character of the Fe grain boundaries at low temperature.

The decrease of magnetization at low temperatures, similar to that previously observed in powders of ball milled iron, indicates that it is a characteristic of nanocrystalline Fe, independent of the production method and structure of the sample. The existence of a minimum for the resistivity below a certain temperature which depends on Fe island size and the large magnetoresistance associated with this effect confirms the spin glass like behaviour of grain boundaries in nanocrystalline Fe.

References

1. A. Hernando, J. Phys., Condens., Mater. 11, 9455 (1999)
2. F. Bonetti, L. Del Bianco, D. Fiorani, D. Rinaldi, R. Caciuffo and A. Hernando, Phys. Rev. Lett. 83, 2829 (1999).
3. J.L.Menéndez, G.Armelles, C.Quintana and A.Cebollada. Surface Science (accepted).
4. J.L.Menéndez, G.Armelles, A.Cebollada, D. Weller and A. Delin, Phys. Rev. B 62, 10498 (2000).
5. G. Herzer, IEEE Trans. Magn. 25, 3327 (1989), Mater. Sci. Eng. A133, 1 (1991).
6. R.Alben et.al., Journal of Applied Physics, vol. 49, p. 1653, 1978. E.M.Chudnovsky et.al., Physical Review B, vol. 33, p. 251, 1986.
7. A. Hernando, M. Vázquez, T. Kulik, C. Prados, Phys. Rev. B 51, 3581 (1995).
8. B.D. Cullity, Introduction to Magnetic Materials, Ed. Addison Wesley, 1972
9. Z.Sefrioui, J.L.Menéndez, E. Navarro, A.Cebollada, F.Briones, P. Crespo and A. Hernando, Phys. Rev. B 64, (2001), in press.
10. P. Levy, S. Zhang, and A. Fert, Phys. Rev. Lett. 65, 1643 (1990). S. S. P. Parkin, N. More, and K. P. Roche, Phys. Rev. Lett. 64, 2304 (1990)
11. G. Binasch, P. Grünberg, F. Saurenbach, and W. Zinn, Phys. Rev. B. 39, 4828 (1989)

10 Dynamics of Magnetization Reversal in Models of Magnetic Nanoparticles and Ultrathin Films

Per Arne Rikvold[1,2], Gregory Brown[2,3], Steven J. Mitchell[1,2], and M.A. Novotny[4]

[1] Center for Materials Research and Technology and Department of Physics, Florida State University, Tallahassee, FL 32306-4351, USA
[2] School of Computational Science and Information Technology, Florida State University, Tallahassee, FL 32306-4120, USA
[3] Center for Computational Sciences, Oak Ridge National Laboratory, P.O. Box 2008 Mail Stop 6114, Oak Ridge, TN 37831-6114, USA
[4] Department of Physics and Astronomy, Mississippi State University, Mississippi State, MS 39762, USA

Abstract. We discuss numerical and theoretical results for models of magnetization switching in nanoparticles and ultrathin films. The models and computational methods include kinetic Ising and classical Heisenberg models of highly anisotropic magnets which are simulated by dynamic Monte Carlo methods, and micromagnetics models of continuum-spin systems that are studied by finite-temperature Langevin simulations. The theoretical analysis builds on the fact that a magnetic particle or film that is magnetized in a direction antiparallel to the applied field is in a metastable state. Nucleation theory is therefore used to analyze magnetization reversal as the decay of this metastable phase to equilibrium. We present numerical results on magnetization reversal in models of nanoparticles and films, and on hysteresis in magnets driven by oscillating external fields.

10.1 Introduction

In recent years, the interest in nanostructured magnetic materials has soared for a variety of reasons. For one, it is only quite recently that it has become possible to synthesize and measure nanometer-sized magnetic particles in small, ordered arrays, often by techniques that involve modern, atomic-resolution microscopies, such as scanning-tunneling microscopy (STM), atomic force microscopy (AFM), or magnetic force microscopy (MFM) [1–5]. AFM and MFM pictures of an array of nanometer-sized iron pillars, fabricated by STM-assisted chemical vapor deposition [6], are shown in Fig. 10.1. The techniques are currently becoming precise enough to even allow investigation of individual nanoparticles. At the same time, computers have had a profound influence in two different ways. The need for ever higher data-recording densities has driven the size of particles used in recording media down into the nanometer range [7–9], while the rapidly increasing power of computers has made it feasible to perform simulations of the dynamic properties of realistic model systems of sizes comparable to experimental ones [10–12].

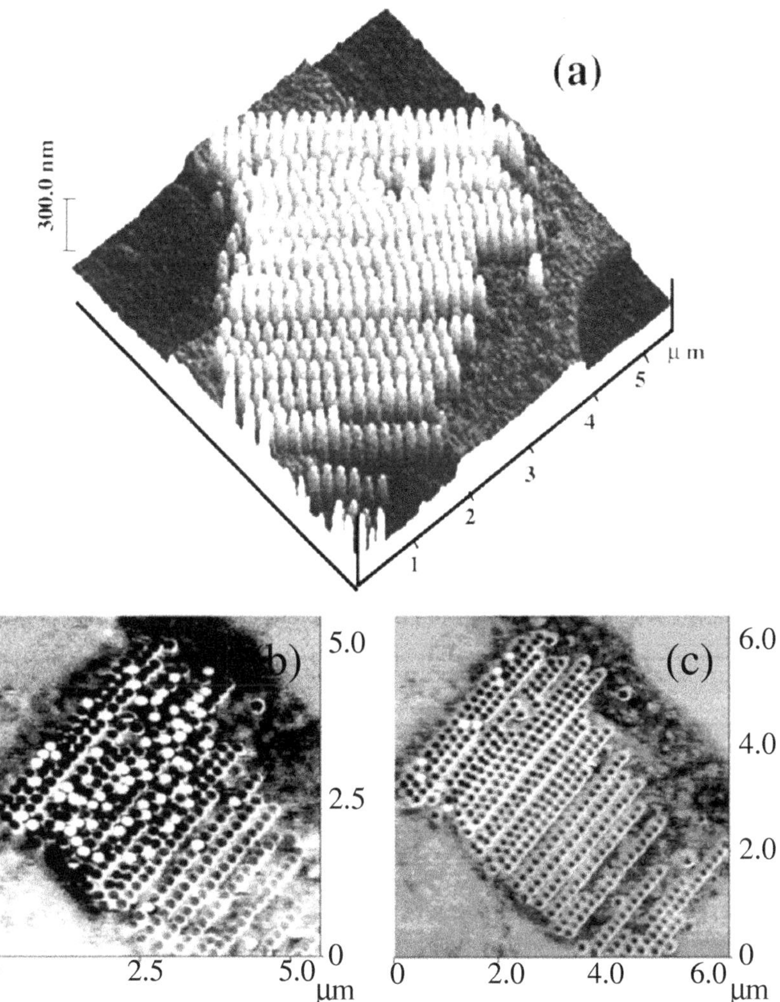

Fig. 10.1. Array of nanoscopic iron pillars of dimensions approximately $40 \times 40 \times 200\,\mathrm{nm}^3$, grown by STM-assisted chemical vapor deposition. (**a**) AFM image of the array of pillars, as grown on top of a μm-size Hall-effect magnetometer. (**b**) MFM image of the array after thermal randomization in near-zero applied field. The magnetic field from each pillar is imaged and seen to point along the major axis of the pillars, either up (white) or down (black). (**c**) MFM image of the array in an applied field of 200 G. Almost all the magnets are aligned with the field. Image data courtesy of D.D. Awschalom and J. Shi. (**a**) and (**b**) after [6].

Modern magnetic recording technologies involve particles that are near the *superparamagnetic limit*. In this limit, the energy barrier separating the two energetically degenerate magnetic orientations is small enough that thermal fluctuations frequently lead to spontaneous switching of the orientation. As a result, the magnetic coercivity decreases with decreasing particle size for particles below

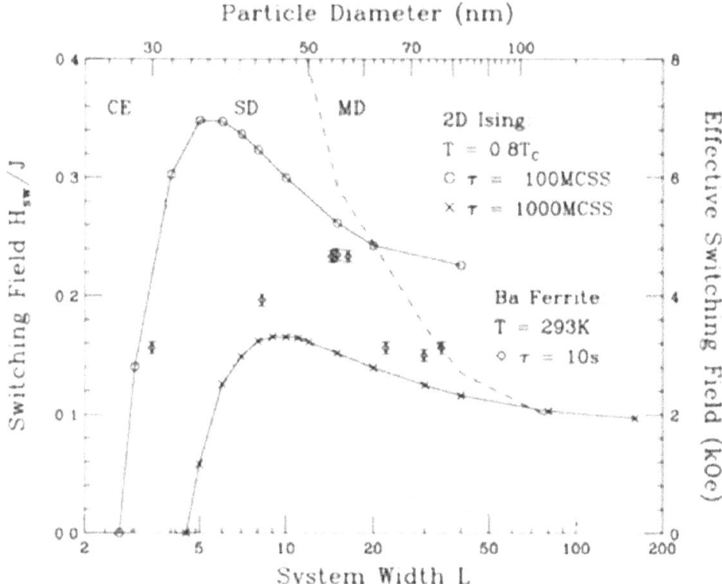

Fig. 10.2. Effective switching field (analogous to coercivity) versus particle size for single-domain ferromagnetic barium ferrite particles (◇ with error bars, right vertical and top horizontal axes, experimental results) and for two-dimensional $L \times L$ Ising systems (data points connected by solid lines, left vertical and bottom horizontal axes). The barium ferrite data were digitized from Fig. 5 of [15]. The Ising data are Monte Carlo simulations from [16] for waiting times τ=100 Monte Carlo Steps per Site (MCSS) (○) and 1000 MCSS (×) at T=0.8T_c, where T_c is the exact Ising critical temperature. After [17]

the superparamagnetic size limit. The limit appears as a maximum in a curve showing switching field or coercivity versus particle size, such as in Fig. 10.2. Since the random magnetization reversals in particles below the superparamagnetic limit degrade recorded information, the engineering challenge has been to keep the energy barrier in the individual particles high enough to make spontaneous switching infrequent while keeping the material magnetically soft enough to facilitate recording. As the volumes of the magnetic particles have shrunk to reach recording densities on the order of 100 Gb/in^2 or more [8,13], materials with higher coercivities due to strong crystalline anisotropies have been employed [14]. In order to enhance engineering practices, it is essential to extend the physical understanding of the superparamagnetic limit past the theories of uniformly mangetized particles to include magnetization reversal dynamics that proceed through localized regions of reversed magnetization that subsequently spread throughout the magnetic element.

The remainder of this article is organized as follows. In Sec.10.2 we summarize some aspects of the theory of magnetization switching in anisotropic magnets, including effects of anisotropy (Sec.10.2.1), nucleation theory (Sec.10.2.2), and model systems (Sec.10.2.3). In Sec.10.3 we give some new results of finite-temperature micromagnetic simulations of magnetic nanoparticles. In Sec.10.4 we discuss hysteresis in nanoparticles and ultrathin films, in particular the frequency dependence of hysteresis loops (Sec.10.4.1) and a dynamic order-disorder phase transition (Sec.10.4.2). A brief summary and conclusions are given in Sec.10.5.

10.2 Theory of Magnetization Switching in Anisotropic Magnets

10.2.1 Effects of Magnetic Anisotropy

The most common description of magnetization switching is the mean field, uniform-rotation theory of Néel [18] and Brown [19,20]. One assumes uniform rotation of all localized moments in the particle to avoid an energy barrier due to exchange interactions of strength J. The remaining barrier, Δ, is caused by magnetic anisotropy – a combination of crystal-field and magnetostatic effects. The equilibrium thickness of a wall separating oppositely magnetized domains is $\ell_{\mathrm{w}} \propto \sqrt{J/\Delta}$. For particles smaller than ℓ_{w} with small anisotropy, the uniform-rotation picture is reasonable. If the anisotropy is largely magnetostatic, the competition between exchange interactions and the demagnetizing field favors domains of opposite magnetization in particles larger than ℓ_{w}. The domains control switching through the field-driven motion of preexisting domain walls [21–23]. However, if the anisotropy is largely crystalline, there exists a range of single-domain particle sizes that are larger than ℓ_{w} but smaller than the size at which the particle becomes multidomain (often the case in ultrathin films [24]).

In anisotropic nanomagnets the state of uniform magnetization opposite to the applied field constitutes a *metastable phase*. This nonequilibrium phase decays by thermally assisted nucleation and subsequent growth of localized regions, inside which the magnetization is parallel with the field [16]. These growing regions are referred to as *droplets* to distinguish them from equilibrium domains. This mechanism yields results very similar to recent experiments on single-domain nanoscale ferromagnets [25].

10.2.2 Application of Nucleation Theory to Magnetization Reversal

Here we present a short summary of homogeneous nucleation theory as it applies to uniaxial magnets. This theory covers situations in which the switching events are nucleated by thermal fluctuations, without the influence of defects. Further details are available in [10,16,17,26–29].

The central problems in nucleation theory are to identify the fluctuations that lead to the decay of the metastable phase and to obtain their free-energy cost relative to the metastable phase. For anisotropic systems dominated by short-range interactions, these fluctuations are compact droplets of radius R. The free energy of the droplet has two competing terms: a positive surface term $\propto \sigma(T)R^{d-1}$ and a negative bulk term $\propto |H|R^d$ where d is the spatial dimension,

$\sigma(T)$ is the surface tension of the droplet wall, and H is the applied magnetic field along the easy axis. Their competition yields a critical droplet radius, $R_c(H,T) \propto \sigma(T)/|H|$. Droplets with $R < R_c$ most likely decay, whereas droplets with $R > R_c$ most likely grow to complete the switching process. The free-energy cost of the critical droplet ($R = R_c$) is $\Delta F(H,T) \propto \sigma(T)^d/|H|^{d-1}$. Nucleation of critical droplets at nonzero temperature T is a stochastic process with nucleation rate per unit volume given by an Arrhenius relation:

$$I(H,T) \propto |H|^K \exp\left[-\beta \Delta F(H,T)\right] \equiv |H|^K \exp\left[-\beta \Xi(T)/|H|^{d-1}\right] , \quad (10.1)$$

where $\beta = 1/k_\mathrm{B}T$ (k_B is Boltzmann's constant), $\Xi(T)$ is the H-independent part of ΔF, and the prefactor exponent K is known for many models from field-theoretical arguments [17,30–32]. The particles are of finite size L, and the dominant reversal mechanism depends on H, T, and L.

In the weakest applied fields, the particles are in the "Coexistence" (CE) regime, with the average metastable lifetime $\tau_\mathrm{CE}(H,T,L) \sim \exp\left[2\beta\sigma(T)L^{d-1}\right]$. (This result is nearly independent of the boundary conditions [27].) The regime corresponds to $R_\mathrm{c} > L$, and the associated L-dependent crossover field is called the Thermodynamic Spinodal (ThSp) [17,28,29]. Estimating its value by assuming $R_\mathrm{c}(H,T,L) \approx L$, one finds $H_\mathrm{ThSp}(T,L) \sim L^{-1}$. The L-dependence of H_ThSp is given by the dotted curve in Fig. 10.2.

For $|H| > H_\mathrm{ThSp}$ (but not too large), the lifetime is determined by the inverse of the total nucleation rate,

$$\tau_\mathrm{SD}(H,T,L) \approx \left(L^d I(H,T)\right)^{-1} \propto L^{-d}|H|^K \exp\left[\beta \Xi(T)|H|^{d-1}\right] , \quad (10.2)$$

which is inversely proportional to the particle volume, L^d (see Fig. 10.3). The subscript SD stands for Single Droplet and indicates that in this regime the switching is normally completed by the first droplet to reach R_c. In both of the stochastic reversal regimes(CE and SD) the probability that switching has not taken place within a time t after the field reversal, $P_\mathrm{not}(t)$, takes the form $P_\mathrm{not}(t) = \exp(-t/\tau)$.

A second crossover, called the Dynamic Spinodal (DSp) [17,28,29], is a consequence of the finite velocity, $v \approx \nu|H|$, of the surface of a growing supercritical droplet [23]. A reasonable criterion to locate the DSp is that the average time between nucleation events, which is τ_SD, should equal the time it takes a droplet to grow to a size comparable to L. This yields the asymptotic relation $H_\mathrm{DSp}(T,L) \sim [\ln(L)]^{1/(d-1)}$. The L-dependence of H_DSp is given by the dashed curve in Fig. 10.2. For $|H| > H_\mathrm{DSp}$, the metastable phase decays through many droplets which nucleate and grow independently in different parts of the system. In this Multidroplet (MD) regime [17,28,29], the classical Kolmogorov-Johnson-Mehl-Avrami (KJMA) theory of metastable decay in large systems [37–39] gives the lifetime

$$\tau_\mathrm{MD}(H,T) \propto \left[I(H,T)(\nu|H|)^d/(d+1)\ln 2\right]^{-1/(d+1)} , \quad (10.3)$$

independent of L (see Fig. 10.3). In the MD regime $P_\mathrm{not}(t) \approx \mathrm{erfc}\left((t-\tau)/\Delta\right)$ [16], where the width Δ of the switching-time distribution depends on H, T, and L.

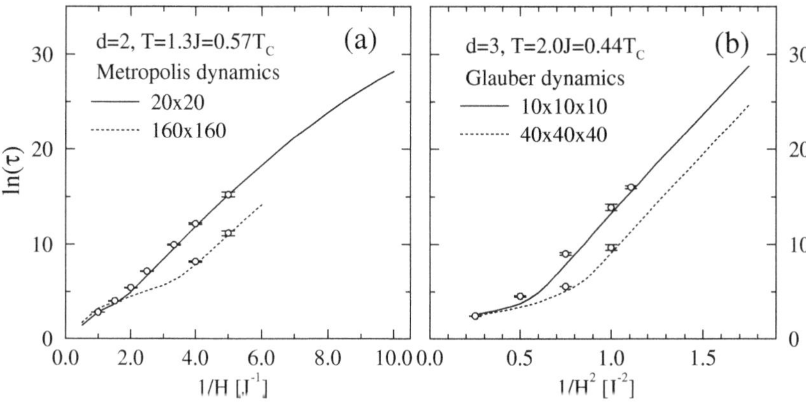

Fig. 10.3. Lifetimes for two-dimensional L^2 Ising systems with $L = 20$ and 160 at $T = 0.57T_c$ (**a**), and three-dimensional L^3 Ising systems with $L = 10$ and 40 at $T = 0.44T_c$ (**b**). The data points are direct Monte Carlo simulation results, while the lines are extrapolations with the Projected Dynamics (PD) accelerated dynamics algorithm [33–35], based on the smallest system at the weakest field. The sharp changes in slope correspond to the DSp, with the deterministic regime to the left and the stochastic regime to the right. The ratio of the slopes of the curve in the single-droplet and multidroplet regimes is $(d + 1)$, in agreement with (10.2) and (10.3). After [34]

For very strong fields nucleation theory becomes irrelevant to the switching behavior. A reasonable way to estimate the crossover to this Strong-Field (SF) regime is to require that the critical radius should be on the order of the lattice constant a. Specifically, requiring $R_c = a/2$, we get the crossover field called the mean-field spinodal (MFSp) $H_{\rm MFSp}(T) \approx 2\sigma(T)/m_{\rm eq}(T)$, where $m_{\rm eq}(T)$ is the zero-field equilibrium magnetization. The "metastable phase diagram" in Fig. 10.4 shows $H_{\rm MFSp}(T)$, as well as $H_{\rm DSp}(T)$ for two-dimensional Ising systems of widely varying sizes. Note the logarithmically slow convergence to zero of $H_{\rm DSp}$ with increasing system size. As a result, even macroscopic metastable systems may be "small" in the sense that they decay via the single-droplet mechanism.

The switching field, $H_{\rm sw}(t_{\rm w}, T, L)$, is the field required to observe a specified average waiting time, $t_{\rm w}$. It is found by solving τ in the relevant region (CE, SD, or MD) for H with $\tau = t_{\rm w}$. The resulting L dependence of $H_{\rm sw}$ is a steep increase with L in the CE regime, peaking near the ThSp, followed by a decrease in the SD regime towards a plateau in the MD regime [16,26]. This behavior is illustrated in Fig. 10.2. Note that the maximum in $H_{\rm sw}$ (related to the maximum coercivity) occurs *even in the absence of dipole-dipole interactions*. For other boundary conditions and in systems with dipole interactions the $H_{\rm sw}$ versus L curve can even have more that one maximum [27]. Maximizing the coercive field is important in magnetic recording applications.

10.2.3 Statistical-Mechanical Model Systems

Simplified statistical-mechanical models are often amenable to analytic solutions, with no free fitting parameters, that agree well with the numerical results.

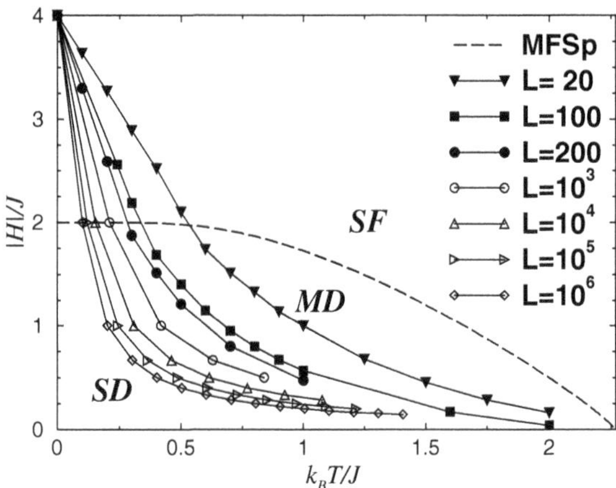

Fig. 10.4. Metastable phase diagram for the two-dimensional Ising model at temperatures up to T_c. The dashed curve represents $H_{\mathrm{MFSp}}(T)$. Data points connected by solid line segments represent $H_{\mathrm{DSp}}(T)$ for several values of L between 20 and 10^6. The filled data points are the results of Monte Carlo simulations, while the empty data points represent a low-temperature approximation [40] for large systems

Despite their lack of realism, they are therefore important as testing grounds for theoretical descriptions of different switching mechanisms. In more realistic models, that correspond to larger numbers of actual materials, analytic results are difficult to come by, but the physical insights gained from the simpler models can be readily applied. Here we introduce three such models in order of increasing complexity: the kinetic Ising model, the classical Heisenberg model, and finite-temperature Langevin micromagnetic models.

The Ising Model. The simplest microscopic model of a ferromagnet is the nearest-neighbor Ising model, in which discrete spins, $s_i = \pm 1$, are placed on the sites (labeled i) of a two- or three-dimensional lattice. The spins interact with their neighbors with a strength J, so that the model is described by the Hamiltonian

$$\mathcal{H} = -J \sum_{\langle i,j \rangle} s_i s_j - H \sum_i s_i \; . \tag{10.4}$$

The model can easily be generalized to longer-range interactions, different lattice geometries, etc. Despite its apparent simplicity, it has many of the attributes of more complicated systems, while many of its properties are exactly known. It is therefore a very commonly studied model.

The Ising Hamiltonian, (10.4), is not a true quantum-mechanical Hamiltonian, and the Ising model therefore does not have an intrinsic dynamic. To simulate thermal fluctuations one uses Monte Carlo simulation of a local stochastic dynamic which does not conserve the order parameter. An often-used example is the Metropolis [41] dynamic with the spin-flip probability

$$W_{\mathrm{M}}(\beta \Delta E) = \min\left[1, \exp\left(-\beta \Delta E\right)\right] \; , \tag{10.5}$$

where ΔE is the energy change that would ensue if the flip were to occur. Another popular choice is the Glauber dynamic [42], defined by

$$W_{\mathrm{G}}(\beta \Delta E) = \frac{\exp(-\beta \Delta E)}{1 + \exp(-\beta \Delta E)} . \tag{10.6}$$

The basic time scale of the Monte Carlo simulation is not known from first principles, but it is expected to be on the order of a typical inverse phonon frequency, 10^{-9}–10^{-13} s. In dynamics such as these, where each potential flip is accepted or rejected randomly, flips can become very rare when rejection rates are high. To perform simulations on the very long time scales necessary to observe metastable decay, one needs to use rejection-free Monte Carlo algorithms [43–46] and other advanced algorithms [33–35].

Analogous stochastic time evolutions can also be imposed on models whose spins have continuous degrees of freedom. Here we briefly discuss one such model, the anisotropic Heisenberg model.

The Anisotropic Heisenberg Model. Like the Ising model, the Heisenberg model consists of spins located at discrete points on a lattice. However, unlike the spins in the Ising model, which equal ± 1, Heisenberg spins are n-dimensional vectors of unit length. When $n = 2$, this model is usually referred to as the XY or plane-rotor model. The Hamiltonian for the nearest-neighbor Heisenberg model with only interaction anisotropy is

$$\mathcal{H} = -\sum_{\langle i,j \rangle} \sum_{n} \left[J_n s_{n,i} s_{n,j} \right] - \boldsymbol{H} \cdot \sum_{i} \boldsymbol{s}_i , \tag{10.7}$$

where $s_{n,i}$ is the n-th component of the i-th spin vector $\boldsymbol{s}_i$, J_n are coupling constants, and $\boldsymbol{H}$ is the external magnetic field. For the example of this model presented here, $n = 3$, $J_x = J_y = 1$, $J_z = 2$, $\boldsymbol{H} = H_z \hat{z}$, and the lattice is a two-dimensional $L \times L$ square lattice.

There are many stochastic dynamics for the Heisenberg model which yield identical equilibrium results, but have different relaxation dynamics. The dynamic assumed here consists of selecting a spin vector at random, then choosing a new orientation for that spin, uniformly distributed over the unit sphere, and then accepting or rejecting the new configuration based on (10.6). Other dynamics exist which make only small changes to spin orientations, but these are not discussed here.

The simulation begins with all $\boldsymbol{s}_i = -\hat{\boldsymbol{z}}$ and $H_z > 0$. This metastable phase then decays in a manner consistent with homogeneous nucleation and growth. However, unlike the Ising model, the continuous degrees of freedom add additional complications, such as effective long-range interactions between droplets. We have not yet attempted to quantify these differences. Figure 10.5 shows lifetimes and configuration snapshots in the single-droplet, multidroplet, and strong-field regimes for an anisotropic Heisenberg model at a temperature below criticality. The field dependence of the lifetime is seen to be very similar to that of the two-dimensional Ising model, shown in Fig. 10.3(**a**).

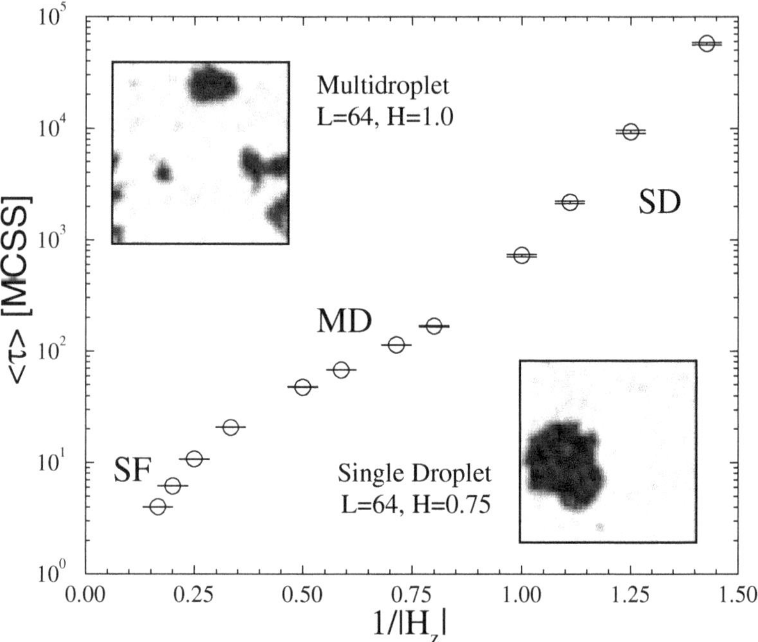

Fig. 10.5. Switching behavior for an anisotropic $n = 3$ Heisenberg model on a two-dimensional $L \times L$ square lattice. The parameters are $J_x = J_y = 1$, $J_z = 2$, and $T = 1$, which is below the critical temperature for this model. The circles are for a system with $L = 16$, and the single-droplet (SD), multidroplet (MD), and strong-field (SF) regimes are labeled. The insets show typical system configurations during the switching process for systems with $L = 64$. Note that the dynamic spinodal (DSp) depends on L, and thus the SD and MD regimes appear at different fields for $L = 16$ and $L = 64$. The grayscale for the insets shows s_z with lighter shades indicating metastable spins and darker shades indicating more stable spins

Finite-Temperature Langevin Micromagnetics. More realistic representations of nanoscale magnetic systems can be obtained by micromagnetic modeling. In this method the "spins" are coarse-grained magnetization vectors $\boldsymbol{M}(\boldsymbol{r}_i)$; each represents the magnetization within a cell centered at position $\boldsymbol{r}_i$. In this low-temperature model [47], the vectors have a fixed magnitude M_s corresponding to the bulk saturation magnetization density. The time evolution of each spin is governed by the damped precessional motion given by the Landau-Lifshitz-Gilbert (LLG) equation [48,49]

$$\frac{\mathrm{d}\boldsymbol{M}(\boldsymbol{r}_i)}{\mathrm{d}t} = \frac{\gamma_0}{1+\alpha^2}\boldsymbol{M}(\boldsymbol{r}_i) \times \left(\boldsymbol{H}(\boldsymbol{r}_i) - \frac{\alpha}{M_s}\boldsymbol{M}(\boldsymbol{r}_i) \times \boldsymbol{H}(\boldsymbol{r}_i)\right) , \qquad (10.8)$$

where the electron gyromagnetic ratio is $\gamma_0 = 1.76 \times 10^7$ Hz/Oe [49], and α is a phenomenological damping parameter. The local field at the i-th spin, $\boldsymbol{H}(\boldsymbol{r}_i)$, is generally different at each location. This field is a linear superposition of fields, one for each type of interaction in the system. Typical examples include fields from external sources, exchange, crystalline anisotropy, and dipole-dipole interactions. Thermal fluctuations may also contribute a term: a stochastic field $\boldsymbol{H}_\mathrm{n}(\boldsymbol{r}_i)$ that is assumed to fluctuate independently for each spin [20]. The fluctuations are assumed Gaussian, each with zero first moment and with the second moments given by the fluctuation-dissipation relation [20]

$$\langle H_{\mathrm{n}\mu}(\boldsymbol{r}_i, t) H_{\mathrm{n}\mu'}(\boldsymbol{r}'_i, t')\rangle = \frac{2\alpha k_\mathrm{B} T}{\gamma_0 M_s V} \delta\left(t - t'\right) \delta_{\mu,\mu'} \delta_{i,i'} , \tag{10.9}$$

where $H_{\mathrm{n}\mu}$ indicates one of the Cartesian components of $\boldsymbol{H}_\mathrm{n}$. Here $V = (\Delta r)^3$ is the discretization volume of the cell, $\delta_{\mu,\mu'}$ is the Kronecker delta representing the orthogonality of the Cartesian coordinates, and $\delta(t - t')$ is the Dirac delta function.

While this stochastic term necessitates careful treatment of the numerical integration in time of this stochastic differential (Langevin) equation, the most computationally intensive part of the calculation involves the dipole-dipole term. For systems with more than a few hundred model spins, it is necessary to use a sophisticated algorithm such as the fast multipole method (FMM) [50,51]. An extensive discussion of the issues involved in finite-temperature simulations of micromagnetics is presented in [12]. The growth of a droplet during the switching of an iron nanopillar at T=20 K is shown in Fig. 10.6. Nucleation is observed to occur at the ends of the pillars [12].

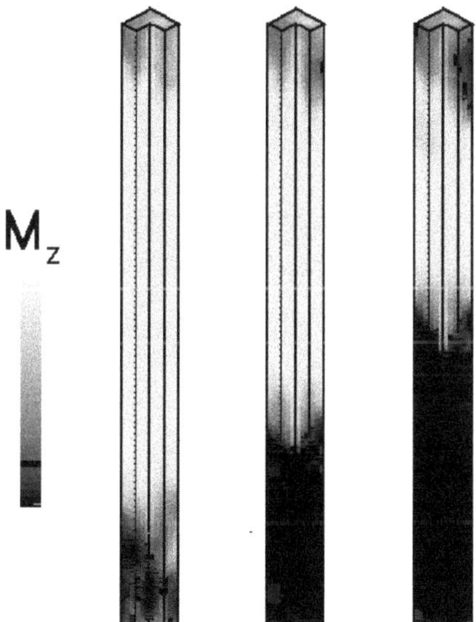

Fig. 10.6. Magnetization along the pillar long axis, M_z, at three times during the switching process. Light shades represent the metastable orientation and dark shades the equilibrium orientation. The micromagnetic simulation shown in this figure models an individual pillar using a 7×7×49 lattice at T=20 K. The pillar is shown in a one-quarter cut-away view

10.3 Finite-Temperature Micromagnetics Results for Nanoparticles

Micromagnetic simulations have been applied to the study of iron pillars modeled after those shown in Fig. 10.1. Unless otherwise noted, the model iron pillars discussed here are 5.2 nm × 5.2 nm × 88.4 nm. The cross-sectional dimensions are small enough, about two exchange lengths, that the only significant inhomogeneities in the magnetization are those in the z direction, *i.e.* along the long axis of the pillar [52,53]. In light of this, the pillars have been modeled as a linear system of magnetic cubes with side 5.2 nm. This model, discussed previously in [12,54], includes thermal fluctuations, exchange, and dipole-dipole interactions.

The results for $P_{\mathrm{not}}(t)$, with T=100 K are shown in Fig. 10.7 for applied fields of H=1000 Oe and 800 Oe. Here switching is defined to occur when the z-component of the total magnetization, M_z, passes through zero. The form of $P_{\mathrm{not}}(t)$ is not exponential, which can be explained by the fact that nucleation of the reversed droplets is easier at the ends of the pillars than in the middle. Assuming that the nucleation rate for droplets at the end of the pillars is constant, I, and that the earliest time switching can occur because of the finite velocity of droplet growth is t_0, it can be shown that the probability of not switching is [12]

$$P_{\mathrm{not}}(t) = \begin{cases} 1 & t < t_0 \\ e^{-2I(t-t_0)}\left[1 + 2I(t-t_0)\right] & t_0 \le t < 2t_0 \\ e^{-2I(t-t_0)}\left[1 + 2It_0\right] & 2t_0 \le t \end{cases} . \tag{10.10}$$

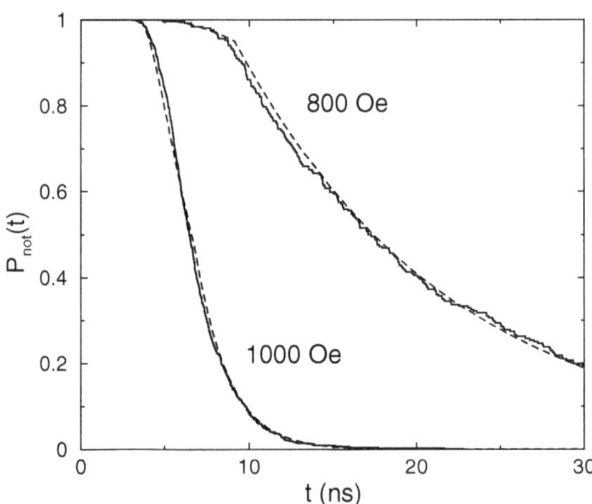

Fig. 10.7. Probability of not switching before time t, $P_{\mathrm{not}}(t)$, for micromagnetic simulations at T=100 K with applied fields of 1000 Oe and 800 Oe as labeled. The solid curves are simulation data for 624 and 252 switches, respectively, for pillars modeled as a one-dimensional chain of spins. The dashed curves are fits to the theoretical model, (10.10)

The parameters are fitted by matching the first and second moments of the simulation results to those of the theoretical forms. As long as the applied field is relatively weak, the agreement between the theory and the simulations is quite good. Switching at times $t_0 \leq t < 2t_0$ is possible only when nucleation occurs at both ends.

Since the nucleation occurs at the ends, the dependence of the switching on the size of the system is different from that seen in isotropic models. Results for the parameters t_0 (squares) and $1/I$ (circles) at T=100 K and H=1000 Oe are shown in Fig. 10.8 for pillars of different lengths, *i.e.* composed of different numbers of cubes 5.2 nm on a side. The nucleation rate is nearly constant, indicating that the size of the energy barrier does not depend on the pillar length. The growth time, indicated by t_0, however, increases as the droplets have to grow farther to switch the magnetization. The nearly linear increase with pillar length indicates that the interface velocity is not significantly affected by the demagnetizing field associated with the high aspect ratio of the pillars.

Finally, changes in the switching mode as the field is changed are shown in Fig. 10.9. Here the mean switching time, $\langle t_{\text{sw}} \rangle$, and standard deviation, σ_t, are shown versus applied field for the 88.4 nm long pillars at T = 100 K. The mean, $\langle t_{\text{sw}} \rangle$, and standard deviation, σ_t, of the switching time t_{sw} versus inverse applied field for pillars of the same type as considered in Fig. 10.7. At weak fields the mean and standard deviation are nearly equal, where the exponential tail at $t > 2t_0$ dominates (10.10). As the applied field is increased, the barrier to nucleation decreases, and the exponential behavior becomes less dominant.

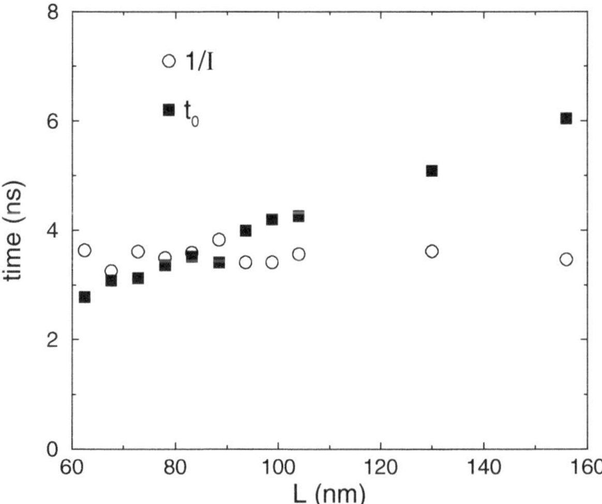

Fig. 10.8. Inverse nucleation rate, I^{-1}, and earliest switching time, t_0, from (10.10), for different pillar lengths. The rate of nucleation at the ends of the pillars depends only weakly on the pillar length, and therefore likewise the energy barrier. The earliest switching time increases since the droplets have to grow for a longer time to switch longer pillars

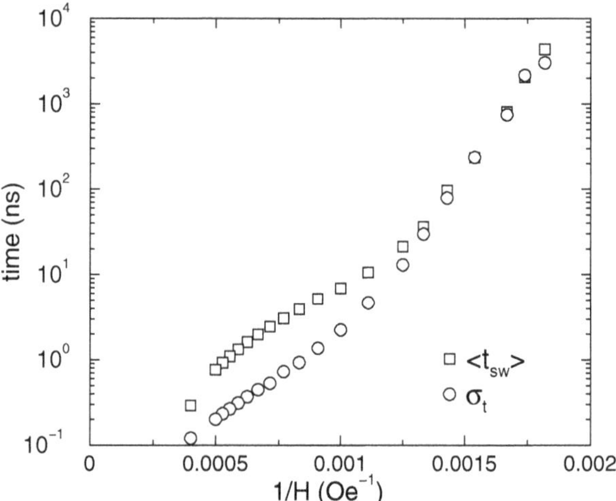

Fig. 10.9. For very weak fields, $\langle t_{\mathrm{sw}} \rangle$ and σ_t are equal, indicating an exponential probability density for the switching times in a single-droplet mode. At higher fields σ_t decreases faster as the switching process becomes nearly deterministic. This is associated with the multidroplet switching mode

Eventually, (10.10) breaks down as the multidroplet reversal mechanism becomes important. The multidroplet nature of the reversal has been verified by direct observation of the switching, which shows droplets nucleating away from the ends at $H = 1000\,\mathrm{Oe}$.

10.4 Hysteresis

Hysteresis is common in many nonlinear systems driven by an oscillating external force, including nanostructured magnets in an oscillating field. It occurs when the dynamics of the system is too sluggish to keep pace with the force. The term was coined by Ewing in the context of magnetoelasticity [55] from the Greek word *husterein* ($\acute{\upsilon}\sigma\tau\epsilon\rho\acute{\epsilon}\omega$) which means “to be behind.”

10.4.1 Hysteresis-Loop Areas

Among the earliest aspects of hysteresis to receive sustained interest is the hysteresis-loop area. In the magnetic context of this article, the hysteresis loop is a plot of magnetization versus applied field, and its area is given by the integral $A = -\oint m(H)\mathrm{d}H$. A typical hysteresis loop for a small system with thermal noise is shown in Fig. 10.10. The particular importance of the loop area is that it corresponds to the energy dissipation per period of the applied field. It is thus relevant to the performance of most electrical and electronic equipment. Recent experiments on ultrathin Fe and Co films with Ising-like anisotropy have considered the frequency dependence of the hysteresis-loop areas [57–60]. The results

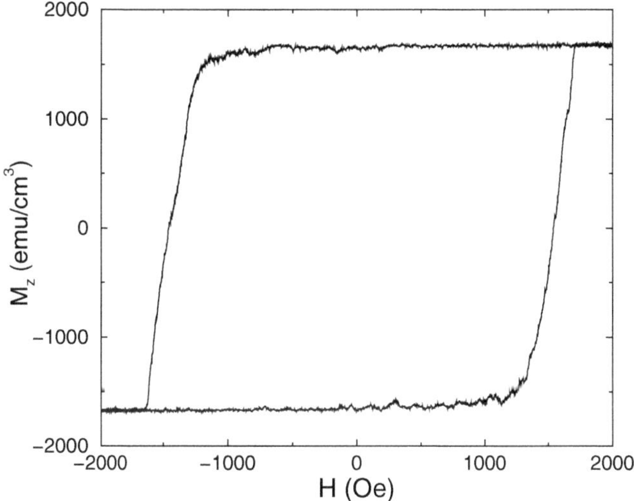

Fig. 10.10. Hysteresis loop for one model Fe pillar of the same type considered in Fig. 10.7 at $T = 100$ K for an oscillating field with a period of 75 ns

of these studies were interpreted in terms of power laws, but with exponents that vary widely between experiments. The experimental situation thus may appear somewhat unclear.

A resolution is provided by the nucleation-and-growth picture of magnetization switching presented here. We assume that the system is driven by a sinusoidally oscillating field, $H(t) = H_0 \sin(\omega t)$. Since the nucleation rate, which has the dimension of a frequency per unit volume, is proportional to $\exp\left[-\beta \Xi(T)/|H|^{d-1}\right]$ by (10.1), one would expect that the field at which the magnetization changes sign should depend on the frequency as $-\left(\ln \omega\right)^{-1/(d-1)}$. The loop area is approximately proportional to the switching field multiplied by the saturation magnetization (see Fig. 10.10). Thus, one would expect the loop area to show this logarithmic frequency dependence in the asymptotic low-frequency limit. Analytic calculations have confirmed this asymptotic result. However, for higher, but still low, frequencies they show a very slow crossover to the asymptotic behavior, which is confirmed by Monte Carlo simulations. Such a slow crossover could easily be mistaken for a power law, even when observed over several decades in frequency [56,61,62]. This result was shown to hold, both when the magnetization reversal occurs via the single-droplet mechanism [62] and the multidroplet mechanism [56], even though the details are different. The behavior is illustrated in the top part of Fig. 10.11. Recently we have also found analogous behavior in micromagnetics simulations of nanometer-sized iron pillars, see the bottom part of Fig. 10.11. In these figures the frequency is given in terms of the dimensionless frequency $1/R = \omega\tau(H_0, T)/2\pi$.

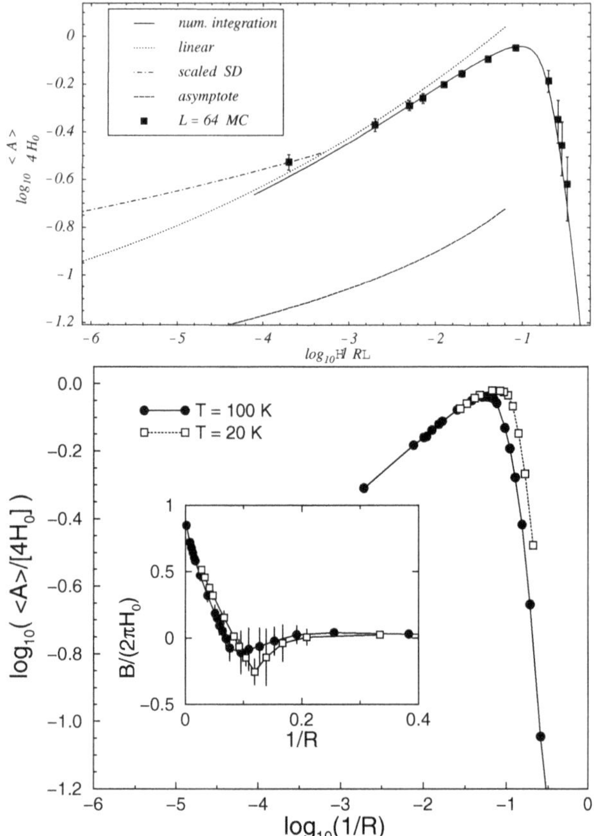

Fig. 10.11. The hysteresis-loop area, A, versus the dimensionless frequency, $1/R$. *(Top)* Two-dimensional Ising model of an ultrathin film. Data points: Monte Carlo simulations for L=64, T=0.8 T_c, and H_0=0.3 J. For these parameters the magnetization switching occurs via the multidroplet mechanism, except for the lowest frequencies. Solid curve: numerical integration for sinusoidally varying field. Dotted curve: numerical integration for linearly varying field. Dot-dashed curve: numerical integration assuming magnetization reversal via the single-droplet mechanism. Dashed curve: low-frequency asymptotic solution. Power-law fits would yield very different effective exponents for fits centered at different frequencies. After [56]. *(Bottom)* Micromagnetic model of iron pillars with length 88.4 nm and square cross-section 5.2 nm × 5.2 nm for T=100 K (circles) and T=20 K (squares) with H_0=2000 Oe. The lines are guides to the eye. The inset shows the correlation between the magnetization and field versus frequency. The lowest-frequency zero-crossing, indicating a resonance condition, occurs at roughly the same frequency as the maximum in A

10.4.2 Dynamic Phase Transition

Different phenomena occur in hysteretic systems as the driving frequency is increased. Eventually the field will vary too quickly for the system to have time to switch during a single period. Thus, low frequencies lead to symmetric hysteresis loops, such as in Fig. 10.10, while high frequencies produce loops in which the magnetization oscillates about one or the other of its zero-field equilibrium values. For small systems or weak field amplitudes, such that the magnetization reversal occurs via the single-droplet mechanism, this results in stochastic resonance [62].

For large systems or stronger fields, such that the magnetization switching occurs via the multidroplet mechanism, the transition from symmetric to asymmetric hysteresis loops becomes a genuine critical phenomenon at a sharply defined critical frequency. The transition is essentially due to a competition between two time scales: the metastable lifetime, $\tau(H_0, T)$, and the frequency of the applied field, $\omega/2\pi$. As a result, the critical value of the reduced frequency $1/R$ is on the order of unity. This nonequilibrium phase transition was first observed in numerical solutions of mean-field equations of motion for ferromagnets in oscillating fields [63,64]. Subsequently it has been observed in numerous Monte Carlo simulations of kinetic Ising systems [56,65–75] and in further mean-field studies [68,70,71,73,76]. It may also have been experimentally observed in ultrathin films of Co on Cu(100) [58,77].

In this far-from-equilibrium phase transition the role of order parameter is played by the period-averaged magnetization, $Q = (\omega/2\pi) \oint m(t)\mathrm{d}t$. This quantity is shown in Fig. 10.12(**a**) versus the dimensionless period R for several system sizes. The order-parameter fluctuation strength, $X_L^Q = L^2 \left[\langle Q^2 \rangle - \langle |Q| \rangle^2\right]$, which corresponds to the susceptibility in an equilibrium system, is shown for several values of L in Fig. 10.12(**b**). Both the order parameter and its fluctuations depend on L in a way very similar to data from simulations of *equilibrium* phase transitions. And, indeed, formal finite-size scaling analysis of the Monte Carlo data [56,65,75], as well as analytical arguments [78,79], have shown that

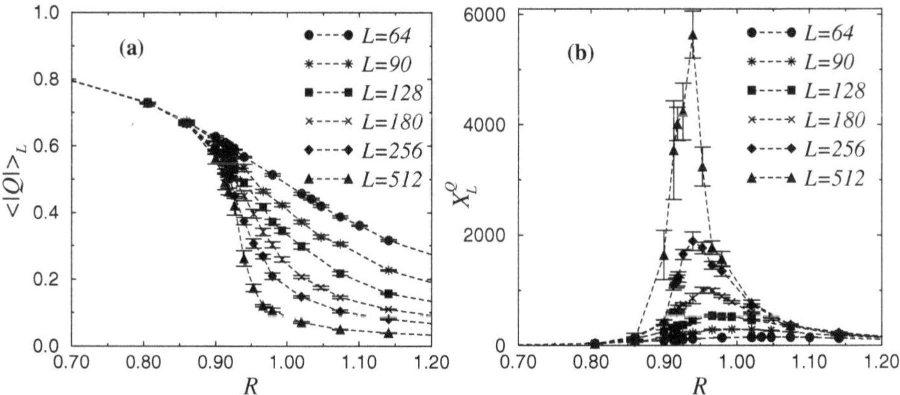

Fig. 10.12. The dynamic phase transition in a two-dimensional Ising system at T=0.8 T_c, driven by a square-wave oscillating field of amplitude $H_0 = 0.3\,J$. After [65]. (**a**) The dynamic order parameter, $\langle |Q| \rangle$, versus the dimensionless period R for several system sizes. (**b**) The order-parameter fluctuation strength, X_L^Q, versus R for several system sizes

this far-from-equilibrium phase transition belongs to the same universality class as the equilibrium phase transition in the Ising model in zero field. This is a quite remarkable result, as it extends the scope of an equilibrium universality class to a far-from equilibrium system.

10.5 Summary

In this article we have presented numerical and theoretical results on magnetization reversal and hysteresis in models of magnetic nanoparticles and ultrathin films. Models that were explicitly considered are kinetic Ising and classical Heisenberg models, which were studied by dynamic Monte Carlo simulations, and continuum-spin micromagnetics models, which were studied by finite-temperature Langevin-equation methods. The simulation results were interpreted within the context of nucleation theory, and it was shown how the reversal modes change from single-droplet to multidroplet upon increasing the strength of the applied field or the size of the system. Computer simulations of model systems such as those presented here enable one to study in detail the statistical properties of the reversal processes, as well as the time dependent internal magnetization structure. Such simulation results have now attained sufficient quality that they can fruitfully be compared with present and future experiments.

Acknowledgments

We are happy to acknowledge the collaborators in our studies of magnetization switching phenomena: M. Kolesik, G. Korniss, H.L. Richards, S.W. Sides, D.M. Townsley, and C.J. White. We also thank S. Wirth and S. von Molnár for useful conversations, and D.D. Awschalom and J. Shi for the image data on which Fig. 10.1 is based.

Supported in part by U.S. National Science Foundation Grant No. DMR-9871455, and by Florida State University through the Center for Materials Research and Technology and the School of Computational Science and Information Technology. Supercomputer time was provided by Florida State University and by the U.S. Department of Energy through the National Energy Research Scientific Computing Center (DOE-AC03-76SF00098).

References

1. J.L. Dormann, D. Fiorani, eds.: *Magnetic Properties of Fine Particles*, (North Holland, New York 1992)
2. M.F. Crommie, C.P. Lutz, D.M. Eigler: Science **262**, 218 (1993).
3. A.D. Kent, T.M. Shaw, S. von Molnár, D.D. Awschalom: Science **262**, 1250 (1993)
4. A.D. Kent, S. von Molnár, S. Gider, D. D. Awschalom: J. Appl. Phys. **76**, 6656 (1994)

5. W.W. Pai, J.D. Zhang, J.F. Wendelken, R.J. Warmack: J. Vacuum Science & Technology B **15**, 785 (1997)
6. S. Gider et al.: Appl. Phys. Lett. **69**, 3269 (1996)
7. H.N. Bertram: *Theory of Magnetic Recording* (Cambridge U. Press, Cambridge 1994)
8. S.Y. Chou, M.S. Wei, P.R. Krauss, P.B. Fischer: J. Appl. Phys. **76**, 6673 (1994)
9. M.A. Novotny, P.A. Rikvold: 'Magnetic Particles'. In: *Encyclopedia of Electrical and Electronics Engineering, Vol. 12*, ed. by J.G. Webster (Wiley, New York 1999) p. 64
10. P.A. Rikvold, M.A. Novotny, M. Kolesik, H.L. Richards: 'Nucleation Theory of Magnetization Reversal in Nanoscale Ferromagnets'. In: *Dynamical Phenomena in Unconventional Magnetic Systems*, ed. by A.T. Skjeltorp, D. Sherrington (Kluwer, Dordrecht 1998) p. 307, and references therein
11. U. Nowak: 'Thermally Activated Reversal in Magnetic Nanostructures'. In: *Annual Review of Computational Physics IX*, ed. by D. Stauffer (World Scientific, Singapore 2001) p. 105
12. G. Brown, M.A. Novotny, P.A. Rikvold: Phys. Rev. B **64**, in press (2001), e-print cond-mat/0101477
13. J.J.M. Ruigrok, R. Coehorn, S.R. Cumpson, H.W. Kestern: J. Appl. Phys. **87**, 5398 (2000)
14. D. Weller, M.F. Doerner: Ann. Rev. Mater. Sci. **30**, 611 (2000)
15. T. Chang, J.-G. Zhu, J.H. Judy: J. Appl. Phys. **73**, 6716 (1993).
16. H.L. Richards, S.W. Sides, M.A. Novotny, P.A. Rikvold: J. Magn. Magn. Mater. **150**, 37 (1995)
17. P.A. Rikvold, B.M. Gorman: 'Recent Results on the Decay of Metastable Phases'. In: *Annual Reviews of Computational Physics I*, ed. by D. Stauffer (World Scientific, Singapore 1994) p. 149, and references therein
18. L. Néel: Ann. Géophys. **5**, 99 (1949)
19. W.F. Brown: J. Appl. Phys. **30**, 130S (1959)
20. W.F. Brown: Phys. Rev. **130**, 1677 (1963)
21. M. Kolesik, M.A. Novotny, P.A. Rikvold: Phys. Rev. B **56**, 11790 (1997)
22. M. Kolesik, M.A. Novotny, P.A. Rikvold: Mater. Res. Soc. Conf. Proc. Ser. **492**, 313 (1998)
23. P.A. Rikvold, M. Kolesik: J. Stat. Phys. **100**, 377 (2000)
24. H.J. Hug et al.: J. Appl. Phys. **79**, 5609 (1996)
25. R.H. Koch et al.: Phys. Rev. Lett. **84**, 5419 (2000).
26. H.L. Richards, M.A. Novotny, P.A. Rikvold: Phys. Rev. B **54**, 4113 (1996)
27. H.L. Richards et al.: Phys. Rev. B **55**, 11521 (1997)
28. H. Tomita, S. Miyashita: Phys. Rev. B **46**, 8886 (1992)
29. P.A. Rikvold, H. Tomita, S. Miyashita, S.W. Sides: Phys. Rev. E **49**, 5080 (1994)
30. J.S. Langer: Ann. Phys. (N.Y.) **41**, 108 (1967)
31. J.S. Langer: Ann. Phys. (N.Y.) **54**, 258 (1969)
32. N.J. Günther, D.A. Nicole, D.J. Wallace: J. Phys. A **13**, 1755 (1980)
33. M. Kolesik, M.A. Novotny, P.A. Rikvold: Phys. Rev. Lett. **80**, 3384 (1998)
34. M. Kolesik, M.A. Novotny, P.A. Rikvold, D.M. Townsley: 'Projected Dynamics for Metastable Decay in Ising Models'. In: *Computer Simulation Studies in Condensed Matter Physics X*, ed. by D.P. Landau, K.K. Mon, H.-B. Schüttler (Springer, Berlin 1998), p. 246
35. M.A. Novotny, M. Kolesik, P.A. Rikvold: Computer Phys. Commun. **121-122**, 330 (1999)

36. R.H. Schonmann: Commun. Math. Phys. **161**, 1 (1994)
37. A.N. Kolmogorov: Bull. Acad. Sci. USSR, Phys. Ser. **1**, 355 (1937)
38. W.A. Johnson, R.F. Mehl: Trans. Am. Inst. Mining and Metallurgical Engineers **135**, 416 (1939)
39. M. Avrami: J. Chem. Phys. **7**, 1103 (1939); **8**, 212 (1940); **9**, 177 (1941)
40. J. Lee, M.A. Novotny, P.A. Rikvold: Phys. Rev. E **52**, 356 (1995)
41. N. Metropolis et al.: J. Chem. Phys. **21**, 1087 (1953)
42. R. J. Glauber: J. Math. Phys. **4**, 294 (1963)
43. M.A. Novotny: Phys. Rev. Lett. **74**, 1 (1995); erratum **75**, 1424 (1995)
44. A.B. Bortz, M.H. Kalos, J.L. Lebowitz: J. Comput. Phys. **17**, 10 (1975)
45. G.H. Gilmer: J. Crystal Growth **35**, 15 (1976)
46. M.A. Novotny: Computers in Physics **9**, 46 (1995)
47. D.A. Garanin: Phys. Rev. B **55**, 3050 (1997)
48. W.F. Brown: *Micromagnetics* (Wiley, New York 1963)
49. A. Aharoni: *Introduction to the Theory of Ferromagnetism* (Clarendon, Oxford 1996)
50. L. Greengard, V. Rohklin: J. Comput. Phys. **73**, 325 (1987)
51. L. Greengard: Science **265**, 909 (1994)
52. D. Hinzke, U. Nowak: Comput. Phys. Commun. **122**, 334 (1999)
53. U. Nowak, D. Hinzke: J. Appl. Phys. **85**, 4337 (1999)
54. E. D. Boerner, H. N. Bertram: IEEE Trans. Magn. **33**, 3052 (1997)
55. J. A. Ewing: Proc. R. Soc. London **33**, 21 (1881)
56. S.W. Sides, P.A. Rikvold, M.A. Novotny: Phys. Rev. E **59**, 2710 (1999)
57. Y.-L. He, G.-C. Wang: Phys. Rev. Lett. **70**, 2336 (1993)
58. Q. Jiang, H.-N. Yang, G.-C. Wang: Phys. Rev. B **52**, 14911 (1995)
59. J.H. Suen, J. L. Erskine: Phys. Rev. Lett. **78**, 3567 (1997)
60. J.H. Suen, M.H. Lee, G. Teeter, J. L. Erskine: Phys. Rev. B **59**, 4249 (1999)
61. S.W. Sides, P.A. Rikvold, M. A. Novotny: J. Appl. Phys. **83**, 6494 (1998)
62. S.W. Sides, P.A. Rikvold, M.A. Novotny: Phys. Rev. E **57**, 6512 (1998)
63. T. Tomé, M.J. de Oliveira: Phys. Rev. A **41**, 4251 (1990)
64. J.F.F. Mendes, E.J.S. Lage: J. Stat. Phys. **64**, 653 (1991)
65. G. Korniss, C.J. White, P.A. Rikvold, M.A. Novotny: Phys. Rev. E **63**, 016120 (2001)
66. W.S. Lo, R.A. Pelcovits: Phys. Rev. A **42**, 7471 (1990)
67. M. Acharyya, B.K. Chakrabarti: 'Ising System in Oscillating Field: Hysteretic Response'. In: *Annual Reviews of Computational Physics I*, ed. by D. Stauffer (World Scientific, Singapore 1994) p. 107
68. M. Acharyya, B. Chakrabarti: Phys. Rev. B **52**, 6550 (1995)
69. M. Acharyya: Phys. Rev. E **56**, 1234 (1997)
70. M. Acharyya: Phys. Rev. E **56**, 2407 (1997)
71. M. Acharyya: Phys. Rev. E **58**, 179 (1998)
72. B. Chakrabarti, M. Acharyya: Rev. Mod. Phys. **71**, 847 (1999)
73. G. M. Buendía, E. Machado: Phys. Rev. E **58**, 1260 (1998)
74. G. M. Buendía, E. Machado: Phys. Rev. B **61**, 14686 (2000)
75. S.W. Sides, P.A. Rikvold, M.A. Novotny: Phys. Rev. Lett. **81**, 834 (1998)
76. M. Zimmer: Phys. Rev. E **47**, 3950 (1993)
77. Q. Jiang, H.-N. Yang, G.-C. Wang: J. Appl. Phys. **79**, 5122 (1996)
78. G. Grinstein, C. Jayaprakash, Y. He: Phys. Rev. Lett. **55**, 2527 (1985)
79. H. Fujisaka, H. Tutu, P.A. Rikvold: Phys. Rev. E **63**, 036109 (2001); erratum **63**, 059903 (2001)

11 Dipolar Effects in Magnetic Nanostructures

Jean-Claude Serge Lévy[1] and Abdelwahab Ghazali[2]

[1] Laboratoire de Physique Théorique de la Matière Condensée, Université Paris 7, 75251 Paris Cedex 05, France
[2] Groupe de Physique des Solides, UMR 7588 CNRS, Universités Paris 6 et Paris 7, 75251 Paris Cedex 05, France

Abstract. The medium scale magnetic arrangement of nanostructures results from the competition of long ranged interactions such as dipolar ones with short ranged interactions such as exchange and anisotropy. A brief report of the experimental situation and of related numerical simulations is given with attention to the specific patterns such as vortices and pin-shaped domains which are observed in nanostructures. A new method of determining the magnetic ground state of a 2D nanostructure is reported. The indirect influence of the lattice symmetry on the magnetic symmetry by means of long ranged interactions is demonstrated to occur as observed experimentally. Different approximate solutions of the ground state equation are classified according to their level of approximation with evidence for topological defects such as vortices, spirals and labyrinths as metastable solutions in 2D samples. Vortices are shown to be stable in nanostructures. The relative stability of pin-shaped domains is demonstrated.

11.1 Introduction

The large scale magnetic organization of ultrathin films and magnetic dots is now starting to be known experimentally [1–5] with evidence for labyrinthine domains in the case of Ising-like spins with magnetization normal to the film (or to the flattened dimension) and for magnetic vortices in the case of Heisenberg-like spins with magnetization in the plane of the film. These structures are in a first approximation quite similar to those already observed in thicker films [6]. The main differences between magnetic structures observed in nanostructures and that observed in films and ultrathin films are the abundance in vortex-like and labyrinthine structures in nanostructured materials [5] as well as the small stripe sizes which are already known to be thickness dependent for thin films. Both this basic similarity and these differences require an explanation. It defines a first goal of this paper which is to obtain a simple formalism to compare the numerous nanostructures with 2D macroscopic materials.

Recent experimental progress in magnetic force microscopy and related magnetic surface tunneling microscopy techniques [7] gives the hope of observing atomic scale details in magnetic patterns of ultrathin films and dots in the very near future. On the other hand the knowledge of stable versus metastable magnetic arrangements in nanostructures will be of great interest for high density magnetic recording and reading, since the existence of metastable states can very efficiently slow down the change in magnetization when an external field is applied. A further related interest lies in the accurate knowledge of magnetization processes in such nanostructures since these processes are quite fundamental for recording and reading operations.

Because of these various features of practical interest, many numerical studies of temperature dependent patterns in ultrathin films with or without periodic boundary conditions have been performed recently by using Monte-Carlo techniques, or micromagnetism equations of evolution, in complement of experimental observations which are still lacking at a very atomic scale. The authors considered different values of magnetic anisotropy and exchange interactions as compared to dipolar interactions and the sum of their results give a picture of this still unobserved atomic scale of magnetic structures [5,8–11]. These results on magnetic nanostructures give evidence for the presence of both stable and metastable topological defects as well as for the presence of uniform domains. Among these defects, stripes, chevrons and labyrinthine structures, vortices, antivortices and saddle points are usual. Their observation is quite in accordance with the experimental results obtained on thicker samples [6].

The magnetization processes during the field variation have also been studied numerically [11] with evidence for soliton propagation. One further numerical result is the stability of very narrow domains in finite 2D samples. The stability of such linear pin-shaped domains in the presence of external fields has been observed recently [12,13] by means of optical sampling of conveniently delayed laser pulses as realized in an earlier experiment on garnet films by Suzuki et al. [14]. The understanding of the stability of such sharp domains is another challenge to be dealt with in this paper. In 2D spin systems, vortices were predicted theoretically a long time ago and recognized as topological defects [15]. In the present case the existence of vortices is due to the competition between long ranged dipolar interactions issuing from different sites. Vortices of both chiralities were already observed in many thin magnetic films by means of Foucault electron microscopy [16–18]. These defects are also involved in the details of the magnetization reversal dynamical processes in magnetic dots as suggested by preliminary observations of anomalous dynamics in dots made of various materials with different shapes and sizes [1–5,19–21]. The stabilization of vortices in dots is a quite established fact from the numerous micromagnetism calculations which were done about these experiments. Theoretically, because of the long range dipolar interactions, the macroscopic magnetic structures strongly depend on the sample size. Then the usual Fourier integral transform cannot be used, and a local treatment of the Hamiltonian is necessary. Here we are interested in an analytical treatment of two-dimensional dipolar interactions which can be valid to deal with both an infinite lattice and finite dots. So we present an approach of the two dimensional long ranged properties by using a continuous Taylor expansion of the dipolar field. This approach enables us to derive a local Landau-like Hamiltonian of the spin field and its derivatives of all orders [22]. This local Hamiltonian contains many lattice sums which are coefficients relative to the spin field derivatives. The first lattice sums are finite and just define the magnetic anisotropy which is induced by dipolar interactions acting on a lattice. Of course, this anisotropy depends on the lattice symmetry. The other nonzero lattice sums are increasing functions of the sample size. They would diverge for infinite samples, and thus their corresponding factors in the Hamilonian must be set equal to zero for an infinite sample, in order to obtain a finite energy per spin which is a minimal requirement for energy optimization. In this way a set of nonlinear equations on the spin field must be satisfied in the ground state and in metastable states. The first non trivial equation deduced from the part of the lowest order divergent lattice sums, which are calculated as integrals, involves in a nonlinear way the spin field itself and its second order derivatives. This basic equation admits a uniform spin configuration as a solution for an infinite 2D sample and topological defects such as vortices, spirals and saddle points as approximate solutions. The classification of such corrective terms according to their magnitudes enables us to derive a hierarchy of topological defects as metastable states, and to deduce from simple estimations how they can be stabilized in nanostructures, with some frequency of appearance and some spatial extension. The terms associated with higher order derivatives correspond to sharp spatial variations of the spin field. They can be neglected in a first approach. However they are important when discussing the stability of pin-shaped domains as they are observed experimentally and numerically in presence of external field, since the spin field varies sharply within these structures. Finally, this local description of the dipolar interaction can be used not only to deduce the static properties of ultrathin films and dots, but also to deduce their dynamic properties, as they can be observed in newly proposed local resonance experiments [23]. Obviously such a description requires an inertial term for spins as it was early developed in connection with spin wave resonance by means of the inductive effect [24]. This dynamic treatment has the advantage to predict magnetic topological defect resonance. The first part of this paper is devoted to a general presentation of this theoretical treatment. The special case of Heisenberg spins without uniaxial anisotropy is considered in the second part with attention paid to defects. A third part introduces the treatment of local resonance of topological defects. As a conclusion, a comparison with experimental and numerical results is given.

11.2 The Local Dipolar Interaction

With a Taylor expansion of the spin field [22–25], the spin at site j reads as a function of the local spin field at site i in the same plane:

$$\mathbf{S}_j = \sum_{p,q=0}^{\infty} \frac{x_{ij}^p y_{ij}^q}{p!q!} \left(\frac{\partial^{p+q}\mathbf{S}}{\partial x^p \partial y^q} \right)_i \tag{11.1}$$

where the vector $\mathbf{r}_{ij}$ of coordinates (x_{ij}, y_{ij}) joins the lattice sites i and j in the plane layer $z = 0$. This enables us to consider the non local version of the dipolar Hamiltonian between spins:

$$H = \sum_{i,j \neq i} \frac{\mathbf{S}_i.\mathbf{S}_j}{r_{ij}^3} - 3 \sum_{i,j \neq i} \frac{(\mathbf{S}_i.\mathbf{r}_{ij})(\mathbf{S}_j.\mathbf{r}_{ij})}{r_{ij}^5} \tag{11.2}$$

and to derive the local version of this Hamiltonian when introducing convenient lattice sums:

$$H = \sum_i H_i$$
$$H_i = \sum_{p,q} \frac{1}{p!q!} \mathbf{S}_i . \frac{\partial^{p+q} \mathbf{S}_i}{\partial x^p \partial y^q} I_{p,\ q} - 3 \sum_{p,q,\alpha,\beta} \frac{(\mathbf{S}_i)_\alpha}{p!q!} \frac{\partial^{p+q} (\mathbf{S}_i)_\beta}{\partial x^p \partial y^q} L_{p,q,\alpha,\beta} \quad (11.3)$$

This expression is deduced for an infinite lattice with translational invariance. Two kinds of lattice sums appear, namely the isotropic sum $I_{p,q}$ and the anisotropic sum $L_{p,q,\alpha,\beta}$ with the definition:

$$I_{p,q} = \sum_j{}' \frac{x_{ij}^p y_{ij}^q}{(x_{ij}^2 + y_{ij}^2)^{3/2}}$$
$$L_{p,q,\alpha,\beta} = \sum_j{}' \frac{(r_{ij})_\alpha (r_{ij})_\beta x_{ij}^p y_{ij}^q}{(x_{ij}^2 + y_{ij}^2)^{5/2}} \quad (11.4)$$

For a finite sample, these sums are site dependent. It must be noted that many lattice sums vanish because of symmetry rules in the general case of an infinite lattice.

The zeroth order lattice sums $I_{0,0}$ and $L_{0,0,\alpha,\beta}$ define the magnetic anisotropy induced from the lattice symmetry by the way of dipolar interactions. These sums can be calculated numerically and lead to distinct results according to the lattice symmetry [22].

For a simple square lattice, the resulting zero order local spin Hamiltonian $H_{sq,\alpha,\beta}$ exhibits a perfect fourfold symmetry. The involved sums were calculated numerically by previous authors who took advantage of their fast convergence [26]:

$$H_{sq,0,i} = I_{sq,0,0} \mathbf{S}_i^2 - 3 L_{sq,0,0,1,1} \left(S_{i,x}^2 + S_{i,y}^2 \right) \quad (11.5)$$

For a hexagonal lattice with the same site density, i.e. with a lattice parameter a' related to the square lattice parameter a by $a' = 2^{1/2} 3^{-1/4} a \simeq 1.075a$, the resulting zero order local spin Hamiltonian $H_{hex,0,i}$ exhibits a perfect hexagonal symmetry and thus does not follow the standard fourfold symmetry:

$$H_{hex,0,i} = I_{hex,0,0} S_{i,z}^2 + \left[I_{hex,0,0} - \frac{3}{2} \left(L_{hex,0,0,1,1} + L_{hex,0,0,2,2} \right) \right] \left(S_{i,x}^2 + S_{i,y}^2 \right)$$
$$- \frac{3}{2} \left(L_{hex,0,0,1,1} - L_{hex,0,0,2,2} \right) \left(S_{i,x}^2 - S_{i,y}^2 \right) \quad (11.6)$$

with

$$- \frac{3}{2} \left(L_{hex,0,0,1,1} - L_{hex,0,0,2,2} \right) \left(S_{i,x}^2 - S_{i,y}^2 \right) = \frac{1.723}{a^3} \left(S_{i,x}^2 - S_{i,y}^2 \right) \quad (11.6a)$$

Thus the lattice induced anisotropy has six easy magnetization directions in the lattice plane and these directions are normal to the six dense directions of the hexagonal lattice as observed in numerical simulations [11].

11.2.1 Other Sources of Anisotropy

In samples where a superimposed uniaxial anisotropy normal to the plane is strong enough to induce a quasi-Ising spin arrangement, within the domain walls, i.e. in regions where competing interactions balance themselves, the relative easy orientations are due to the dipolar interactions which actually drive the magnetization in such a way as spins are nearly in-plane [1–4] as they are in the case of in-plane Heisenberg spins. Such a behavior is confirmed by numerical simulations [11]. When the superimposed uniaxial anisotropy normal to the plane is weaker than before, as it can be simulated numerically [11] or as it occurs when the effective number of layers is varied, a gradual reorientation transition occurs [1]. In the limiting cases, when there is no anisotropy, there is an in-plane configuration and when anisotropy is very high the spin configuration is perpendicular to the plane. This reorientation transition has been largely observed and numerically studied.

11.2.2 Lattice Sums and Lattice Integrals

In order to consider lattice sums of all orders, a continuous treatment by means of integrals instead of sums is required. This treatment induces a perfect rotational invariance and integration occurs over concentric circular rings. Now, high order terms are considered. For these terms the relative weight of sites far from the center is considerably larger than it is for terms of low order. Within a ring of finite size these sites are quite numerous and rather dense. per unit length. Thus the assumed rotational invariance does not perturb strongly the result for lattice sums. Then radial and angular integrals appear in the analysis of the former isotropic sums $I_{p,q}$ with:

$$I_{p,q} = \sum_j{}' \frac{x_{ij}^p y_{ij}^q}{(x_{ij}^2 + y_{ij}^2)^{3/2}} = \frac{1}{a^2}\int_a^L r^{p+q-2} dr \int_0^{2\pi} \cos^p\theta \sin^q\theta d\theta = K_{p,q}N_{p,q} \tag{11.7}$$

where the radial integrals $K_{p,q} = K_{p+q}$ are defined by:

$$K_{p,q} = \int_a^L \frac{r^{p+q-2}}{a^2} dr = \frac{L^{p+q-1} - a^{p+q-1}}{a^2(p+q-1)} = K_{p+q} \tag{11.7a}$$

of which the size exponent is easily derived: $p+q-1 = n-1$. Note that there is a divergence of the radial sums K_n for $n > 1$ when the sample becomes large.

The angular integrals $N_{p,q}$ are defined by:

$$N_{p,q} = \int_0^{2\pi} \cos^p\theta \sin^q\theta d\theta \tag{11.8}$$

Asymptotic formulae for the non null values of the $N_{p,q}$'s are easily derived using the gamma function with the basic result:

$$N_{p,q} \simeq 2\left(\frac{2}{p}\right)^{(q+1)/2} \Gamma\left((q+1)/2\right); \quad N_{p,0} \simeq 2\sqrt{\frac{2\pi}{p}} \tag{11.9}$$

From this continuous treatment the anisotropic lattice sums $K_{p,q,\alpha,\beta}$ are also deduced from the same radial and angular integrals, with:

$$L_{p,q,\alpha,\beta} = K_{p+q} N_{p',q'} \tag{11.10}$$

where the indices α and β are both lower than three for non null values of these integrals since all spins belong to the same layer $z = 0$. The explicit link between the anisotropic sums $L_{p,q,\alpha,\beta}$ and the radial K_{p+q} and angular $N_{p,q}$ integrals is given by the following rules to define p' and q' from p and q:

$$\begin{aligned} &\alpha = 1,\ \beta = 1 \Rightarrow p' = p + 2,\ q' = q \\ &\alpha = 1,\ \beta = 2 \Rightarrow p' = p + 1,\ q' = q + 1 \\ &\alpha = 2,\ \beta = 1 \Rightarrow p' = p + 1,\ q' = q + 1 \\ &\alpha = 2,\ \beta = 2 \Rightarrow p' = p,\ q' = q + 2 \end{aligned} \tag{11.10a}$$

Because of symmetry rules, all the Hamiltonian terms which are non zero are factors of K_{2n} with $n \in N$. Thus an obvious classification of the Hamiltonian terms according to the index n occurs since the divergence strength of K_{2n} for an infinite sample would increase with n as seen in (11.7a).

Direct calculations in the continuous approximation lead us to write the first term $H_{0,0}$ of the local dipolar Hamiltonian as a special case:

$$H_{0,0} = \pi K_0 \left(-S_{i,x}^2 - S_{i,y}^2 + 2S_{i,z}^2\right) \tag{11.11}$$

In agreement with our previous accurate calculations on square and hexagonal lattices reported in [5] and [6], $H_{0,0}$ remains finite in the continuous approximation. Since the value of K_0 is positive, Eq.(11.11) just shows that the anisotropy induced by dipolar interaction favors an in-plane magnetization in this 2D-rotationally invariant lattice. The same situation is also expected to occur in magnetic ultrathin films from classical magnetostatic arguments [6].

The next term H_2 in the expansion of the site dipolar Hamiltonian H_i contains the factor K_2. From the previous expansion this term reads:

$$H_2 = \frac{\pi}{8} K_2 \begin{bmatrix} 4\mathbf{S}.\left(\frac{\partial^2 \mathbf{S}}{\partial x^2} + \frac{\partial^2 \mathbf{S}}{\partial y^2})\right) - 6S_x \frac{\partial^2 S_y}{\partial x \partial y} - 6S_y \frac{\partial^2 S_x}{\partial x \partial y} \\ -9S_x \frac{\partial^2 S_x}{\partial x^2} - 3S_x \frac{\partial^2 S_x}{\partial y^2} - 3S_y \frac{\partial^2 S_y}{\partial x^2} - 9S_y \frac{\partial^2 S_y}{\partial x^2} \end{bmatrix} \tag{11.12}$$

Since the factor K_2 would be infinite for an infinite sample, the ground state solution of an infinite sample as well as metastable states must satisfy the following nonlinear partial differential equation for the spin field in order to yield a finite site energy:

$$\begin{aligned} &4\mathbf{S}.\left(\frac{\partial^2 \mathbf{S}}{\partial x^2} + \frac{\partial^2 \mathbf{S}}{\partial y^2})\right) - 6S_x \frac{\partial^2 S_y}{\partial x \partial y} - 6S_y \frac{\partial^2 S_x}{\partial x \partial y} \\ &- 9S_x \frac{\partial^2 S_x}{\partial x^2} - 3S_x \frac{\partial^2 S_x}{\partial y^2} - 3S_y \frac{\partial^2 S_y}{\partial x^2} - 9S_y \frac{\partial^2 S_y}{\partial x^2} = 0 \end{aligned} \tag{11.13}$$

Of course for a finite sample as a nanostructure, (11.13) becomes an inequality defining a stable or metastable state:

$$
\begin{aligned}
&4\mathbf{S}.\left(\frac{\partial^2\mathbf{S}}{\partial x^2}+\frac{\partial^2\mathbf{S}}{\partial y^2})\right)-6S_x\frac{\partial^2 S_y}{\partial x\partial y}-6S_y\frac{\partial^2 S_x}{\partial x\partial y}\\
&-9S_x\frac{\partial^2 S_x}{\partial x^2}-3S_x\frac{\partial^2 S_x}{\partial y^2}-3S_y\frac{\partial^2 S_y}{\partial x^2}-9S_y\frac{\partial^2 S_y}{\partial x^2}\leq 0 \qquad (11.13a)
\end{aligned}
$$

In practice high order derivatives must be nearly equal to zero in order to avoid the unphysical case of functions varying significantly over unoccupied space because of the discrete character of the lattice. Thus the next equation or inequality which is of fourth order in spin derivatives as well as the following equations or inequalities must just be understood as corrective terms to the ground state as defined by (11.13) and are neglected in the main part of this approach.

Thus basically the ground state defined by (11.13) is nonlinear and has a fourfold symmetry. Then in the basic case considered here without any other source of anisotropy, spins lie in the plane as deduced from (11.11). This simple case enables us to reduce the spin field phase space as done in the next part which is devoted to Heisenberg spins without uniaxial anisotropy.

11.3 Stable and Metastable States for Vector (XY) Spins

With Heisenberg spins without uniaxial anisotropy the magnetization configuration is planar and can be just described by a polar angle θ with:

$$S_x=\cos\theta,\ S_y=\sin\theta,\ S_z=0$$

Eq.(11.13) now reads as a nonlinear partial differential equation for the polar angle θ as a function of the lattice coordinates:

$$
\begin{aligned}
&3\sin 2\theta\left(\frac{\partial^2\theta}{\partial x^2}-\frac{\partial^2\theta}{\partial y^2})\right)-6\cos 2\theta\frac{\partial^2\theta}{\partial x\partial y}+6\sin 2\theta\frac{\partial\theta}{\partial x}\frac{\partial\theta}{\partial y}\\
&+6\sin 2\theta\frac{\partial\theta}{\partial x}\frac{\partial\theta}{\partial y}+(2+3\cos 2\theta)\left(\frac{\partial\theta}{\partial x}\right)^2+(2-3\cos 2\theta)\left(\frac{\partial\theta}{\partial y}\right)^2=0 \quad (11.14)
\end{aligned}
$$

Then the rotational symmetry of Eq.(11.14) leads us to introduce polar coordinates for the plane ($x=r\cos\phi,\ y=r\sin\phi$). Equation (11.14) now reads with the usual notations for partial derivatives:

$$
\begin{aligned}
&\left[\theta_{r^2}-\frac{\theta_{\varphi^2}}{r^2}-\frac{\theta_r}{r}+2\frac{\theta_r\theta_\varphi}{r}\right]\sin[2(\theta-\varphi)]\\
&+\left[-2\frac{\theta_{r\varphi}}{r}+2\frac{\theta_\varphi}{r^2}+\theta_r^2-\frac{\theta_\varphi^2}{r^2}\right]\cos[2(\theta-\varphi)]+\frac{2}{3}\left[\theta_r^2+\frac{\theta_\varphi^2}{r^2}\right]=0 \ (11.15)
\end{aligned}
$$

So different test functions can be considered for solutions of this nonlinear equation.

11.3.1 Magnetic Domains

Eq.(11.15) admits constant values of θ as solutions, i.e. physical uniform magnetic domains of any orientation. Such solutions are observed both experimentally and numerically within domains between walls or in presence of an applied external field strong enough to saturate the sample [5,11]. In addition, these solutions also satisfy quite obviously the full set of partial differential equations which generalize Eq.(11.13) at all orders since each spin derivative is zero for these solutions.

11.3.2 Topological Defects

The presence of the argument $2(\theta - \phi)$ in Eq.(11.15) leads us to consider for test functions simple spin fields which correspond to topological defects with constant values of $\theta - \phi$.

Among these spin fields are the *vortex* models with the two possible chiralities: $\theta = \phi + \epsilon\pi/2$ either with *anticlockwise* vortex for $\epsilon = +1$ or with *clockwise* vortex for $\epsilon = -1$. The value of the left hand term of Eq. (11.15) for such functions is $-1/3r^2$. Thus these vortices are approximate solutions of the spin field equation far from the origin. In finite samples they are optimal because of inequality (11.13a) which means that up to corrective terms due to finite size effects, the presence of vortices decreases the magnetic energy and thus stabilize the magnetic configuration. In infinite samples the correction to the deviation from Eq.(11.15) can be balanced by a radial variation of the spin angle θ. Note that this left hand term of Eq.(11.15) is negative and of order L^{-2}; thus it leads to an extra energy $\propto L^{-2}K_2 \propto L^{-1}$ per site which tends to zero for large samples. This term can be easily balanced by a roughening of the vortex shape.

Different processes can stabilize these vortices in infinite samples. The site energy of those test functions could be lowered by the introduction of an effective vortex made of several adjacent domains each one with a uniform value of θ. Such non regular spin arrangements appear in numerical simulations [8,9,11]. Here the origin of the 2D space defines the vortex core. Since the origin choice is arbitrary, the presence of vortices of any chirality and of coupled vortices is expected to occur in the whole sample, as observed at least in numerical simulations related to an effective large sample [11].

Another pair of topological defects to be considered here as test functions for magnetization in Eq.(11.15) are a *source* ($\theta = \phi$) and an *anti-source* or *sink* ($\theta = \phi + \pi$) since they both give the same simple value to the argument $2(\theta - \phi)$. Sources and sinks are observed experimentally for rather thick samples [6]. Yet numerically for single layered films such defects are observed only between adjacent vortices of opposite chiralities [11]. These sources and sinks do not even appear in their full extension; they are present just over a small part of the sample. In fact, these magnetic sources lead to a value of the left term of Eq.(11.15) equal to $5/3r^2$. This positive extra term is similar to the term calculated for vortices but positive. The positive extra energy explains the infrequent observation in numerical simulation of sources in finite samples [11] and their non observation in nanostructures [1–5,19–21]. The experimental observation of sources and sinks in large and thicker samples [6] is in agreement with the possible radial variations which can balance the extra contribution. Another explanation could be the simultaneous presence of a perpendicular component of the magnetization which allows an extra correction because of this escape dimension. The fact that common topological defects such as vortices give rise to a negative extra energy in an approximate way explains their presence, especially in finite samples even if in an infinite lattice they would be unstable without radial variation.

The general class of simple topological defects to be considered are *logarithmic spirals*: $\theta = \phi + \alpha$ with $r = a\exp(cot\alpha)$. They include vortices and sinks. Then the left hand side of equation (11.15) is equal to $(3\cos 2\alpha + 2)/3r^2$ which can be negative or positive according to the value of α. Precisely if $65.91^o < \alpha < 114.09^o$ this left hand side is negative. In particular for values of α close enough to $\pm\pi/2$, such logarithmic spirals which are close to vortices are stable for finite samples. For other values of α, this part of equation is positive, thus such logarithmic spirals are metastable both for finite and for infinite samples. These remarks are in excellent agreement with the observation of logarithmic spirals as excited states, so called anger states [12], where these spirals are close to pure vortices.

Finally the weak extra energy which is due to topological defects makes them weakly excited states. Because of the effective boundary conditions in a real space such excited states can be stabilized. This explains why they are indeed observed both numerically and experimentally, especially when an alternative magnetic field with a suitable frequency is added which stimulates the appearance of vortices and spirals [12,13].

11.3.3 Generalization to Non XY Spins

When the sample is also submitted to a strong intrinsic uniaxial anisotropy which stabilizes a magnetization direction perpendicular to the film, domains with alternate up and down magnetization appear. Within the walls which lie between such adjacent domains, in-plane components of the spin field must be accounted for. The previous result on the approximate solutions of Eqs. (11.13-11.15) are still valid for these in-plane parts of the lattice plane which are located within the domain walls. Such an argument makes clear the strong topological similarity between magnetic patterns observed in magnetic films with strong uniaxial anisotropy and those observed in magnetic films with weak uniaxial anisotropy. In the first case, parallel stripes, chevry domains and whirled labyrinthine domains occur [27], while in the second case, uniform domains, successive 60 deg and vortices occur with a complete similarity. Locally, the longest side of an Ising domain defines the spin direction of the corresponding XY model in accordance with basic magnetostatic considerations of the parallelism between magnetization orientation and border line [6,28]. This duality between XY models and Ising models is effective because the striped nature of Ising domains in stripes, chevrons and labyrinthine arrangements ensures the non-degeneracy of the choice of its longest side and thus defines the local XY arrangement. This confirms the numerically observed states [11].

11.3.4 Beyond Second Order Derivatives

When the variation of the spin field is rapid, higher order spin field derivatives cannot be neglected any longer. Such configurations appear numerically [11] and experimentally [12,13] in presence of an external field for both Heisenberg and Ising spins. Then it is necessary to go further in the Hamiltonian expansion, i.e.:

$$
\begin{aligned}
\frac{64H_4}{\pi K_4} = & 2S_z\frac{\partial^4 S_z}{\partial x^4} - 3S_x\frac{\partial^4 S_x}{\partial x^4} + S_y\frac{\partial^4 S_y}{\partial x^4} + 2S_z\frac{\partial^4 S_z}{\partial y^4} + S_x\frac{\partial^4 S_x}{\partial y^4} - 3S_y\frac{\partial^4 S_y}{\partial y^4} \\
& + 4S_z\frac{\partial^4 S_z}{\partial x^2\partial y^2} - 2S_x\frac{\partial^4 S_x}{\partial x^2\partial y^2} - 2S_y\frac{\partial^4 S_y}{\partial x^2\partial y^2} - 4S_x\frac{\partial^4 S_y}{\partial x\partial y^3} \\
& - 4S_x\frac{\partial^4 S_y}{\partial x^3\partial y} - 4S_y\frac{\partial^4 S_x}{\partial x\partial y^3} - 4S_y\frac{\partial^4 S_x}{\partial x^3\partial y}
\end{aligned}
$$

where K_4 is positive. Then it is easy to consider a delta function-like variation of the spin in some direction to simulate a pin-shaped domain. The result for the right hand side of the above equation is positive or negative according to the direction of the domain. For finite samples, it means that pin-shaped domains are stable for some directions and unstable or metastable for others. This result is also in general agreement with observations [12,13], but the selected directions must be defined carefully.

Another case of spatial singularity occurs in chevry domains [27] and in labyrinthine domains [29] where the elbows between successive stripes are aligned over rather large distances. In this case too the fourth order term of the Hamiltonian must be considered.

11.4 Dynamics: Magnetic Resonance of Topological Defects

Experimentally it becomes now possible to investigate magnetic resonance in the narrow region surrounding the cantilever of a surface tunneling microscope [23]. Practically this means that specific magnetic resonance associated with magnetic topological defects is accessible. As generally noticed about the competition between dipolar interactions, exchange and anisotropy interactions, there are scale invariance properties [11] which are also associated with topological defects. Thus topological defects of various sizes are obtained. The next point to notice is that each topological defect, namely vortex, saddle point, source, or spiral, must have its own set of resonance modes and frequencies.

Then for a given defect such as a vortex in a given sample, the set of resonance modes includes several modes such as the classical breathing or shrinking mode which corresponds to a mere uniform extension or diminution of the vortex size. Among these modes, numerous splitting modes correspond to the vortex splitting into parts of given symmetries. Other resonance modes characterize whole motions of the defect, in one or two dimensions, around an equilibrium position. More complex modes can couple motion and internal rearrangement. Of course the details of the magnetic field orientation around the cantilever are decisive for selecting the mode or the modes which are resonantly activated.

Without giving a detailed analysis of this resonance, it is easy to derive the principles of such a calculation. When working within the harmonic approximation, which means a moderate intensity of the resonance field, the variation of the total contribution of the term H_2 during the mode display gives a quadratic contribution in the mode coordinate. The inertial part, also quadratic in the time derivative of the mode coordinate comes from induction and dipolar interaction, as classical in the study of Bloch wall oscillations [24]. These calculations are easily done in samples with a single topological defect, but they are more difficult to handle when different topological defects occur simultaneously in the sample.

11.5 Comparison with Experimental and Numerical Results

The first element of comparison is reported in Fig. 11.1 where a low temperature spin configuration of a disk of 10192 spins deduced from Monte-Carlo simulation is shown in the case of a rather weak dipolar effect with $D/Ja^3 = 0.1$ without any anisotropy. Several vortices or logarithmic spirals appear in this configuration. Far from the vortex centers, large domains with uniform magnetization are noted. There is also a special defect which is rather similar to a part of a source. It occurs between vortices and is quite less extended than vortices. At the border, the magnetization is nearly everywhere parallel to the border. All these details are in agreement with the previous discussion.

In Fig. 11.2 the dipolar effect is stronger than in Fig. 11.1 with $D/Ja^3 = 1$. More vortices or spirals appear as well as a lot of domains with uniform magnetization and walls. This is in agreement with the remark on space renormalization:

Fig. 11.1. Triangular lattice with vector spins at low temperature: $k_B T/J = 0.01$, $D/Ja^3 = 0.1$

Fig. 11.2. Same as Fig. 11.1 but with $D/Ja^3 = 1$

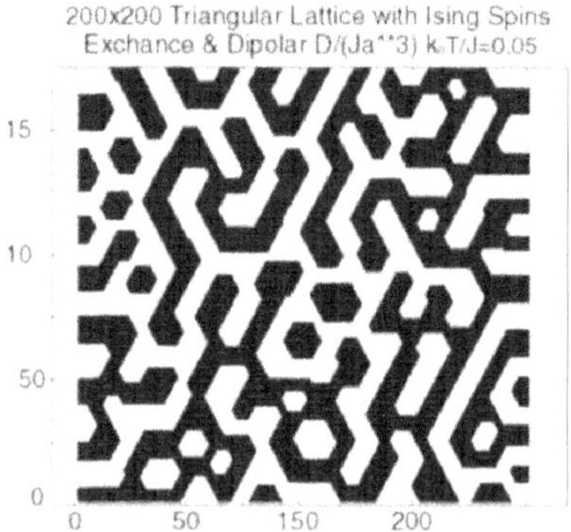

Fig. 11.3. Magnetic structure at low temperature: $D/Ja^3 = 0.3$, $k_BT/J = 0.05$

a large dipolar effect is equivalent to a large size, thus in a larger disk more vortices are observed [11]. A special feature to be noticed is the absence of any crystalline organization of vortices. More exactly a random distribution of vortices of both chiralities is quite generally observed from the pair distribution function in these calculations which use free boundary conditions. This means that the interaction between vortices is quite weak.

The case of a rectangular block of 40000 Ising spins is reported in Figs. 11.3-11.6 from Monte-Carlo simulations performed at different temperatures when the dipolar effect is such as the stripe width allows to consider stripe arrangements within the sample. Here the calculations are done with $D/Ja^3 = 0.3$. Mazes with nearly circular arrangements are observed at all temperatures lower than Curie temperature as expected from the previous arguments. At low temperature walls are linear and well defined, while at high temperature walls are bent with indentations of various sizes and a nearly fractal roughness. This temperature effect is not directly related to the previous considerations; however it is connected with the existence of abundant low energy excited states. The appearance of more and more loops and curly shapes when the temperature is

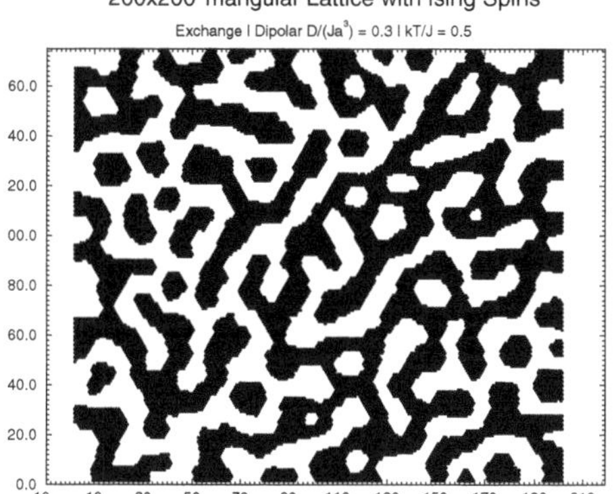

Fig. 11.4. Same as Fig. 11.3 with $k_BT/J = 0.5$

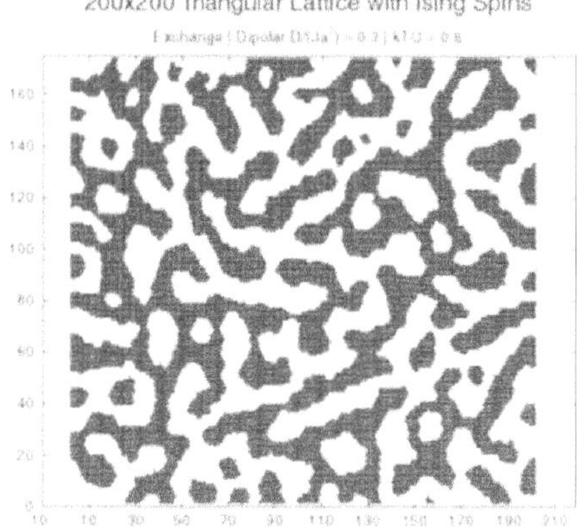

Fig. 11.5. Same as Fig. 11.3 with $k_BT/J = 0.6$

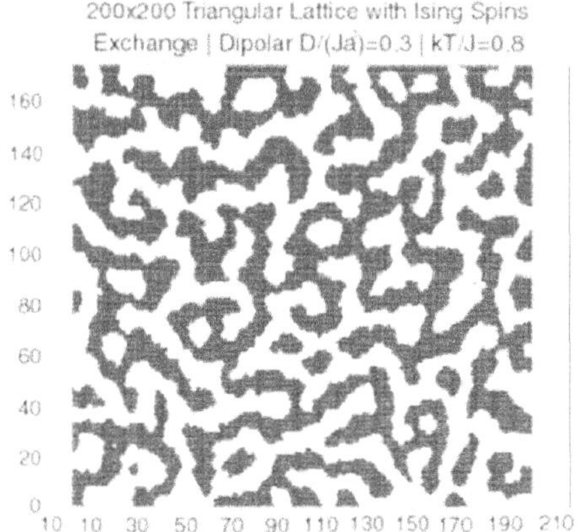

Fig. 11.6. Same as Fig. 11.3 with $k_BT/J = 0.8$

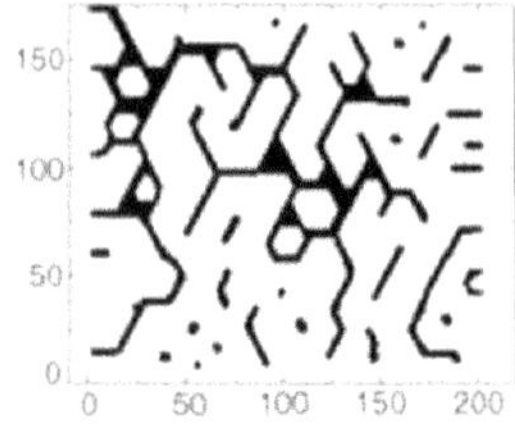

Fig. 11.7. Stability of pin shaped domains with $H/J = 0.4$, $k_B T/J = 0.2$

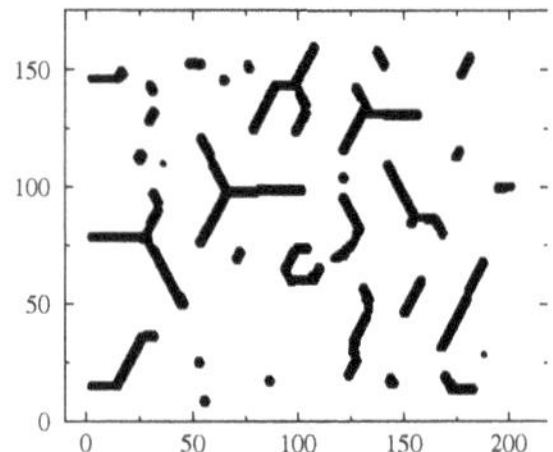

Fig. 11.8. Same as Fig. 11.7 with $H/J = 0.5$

increased is a proof of bent stripe stability. This is also in agreement with the previous considerations.

The stability of very narrow domains of Ising spins in presence of an external field is studied in Figs. 11.7-11.8 with the same dipolar effect as in Figs. 11.3-11.6 and moderate fields $H = 0.4\,J$ or $H = 0.5\,J$. These calculations were done with long Monte-Carlo relaxation between any variation of the external field. At this low temperature narrow linear domains are observed in well defined directions, and there are also some very restricted domains such as magnetic bubbles and dumb-bells which appear simultaneously, as well as branched linear domains. These features are evidence for nonlinear effects as briefly shown before.

The following examples are devoted to equilibrium spin configurations at low temperature of spherical hcp dots of two different sizes: 5947 spins and 47443 spins with different parameters. In order to give an idea of the geometric arrangement of spins, three sections of the sample, namely the equatorial plane normal to the c-axis and two parallel planes, six (resp. ten) monolayers above and below the equatorial plane are shown here. In the first example shown in Figs. 9-11, a small sphere of Heisenberg spins with strong anisotropy and strong dipolar effect is studied. These spins are close to Ising spins. Just a few domains appear and walls even if smooth are far from being rectilinear, neither in the cut plane nor in the transverse sections. This is well in agreement with the mazes observed in two-dimensional Ising samples. The average magnetization is practically zero. In Figs. 11.12-11.13 the magnetization process of a large sphere of Ising spins at low temperature is reported for strong dipolar interaction: $D/Ja^3 = 1$. Under a rather large field the sixfold branched core and the external part of the sphere are domains of opposite signs. in fact, the external part has a larger surface and is thus divided in six domains according to the underlying hexagonal lattice

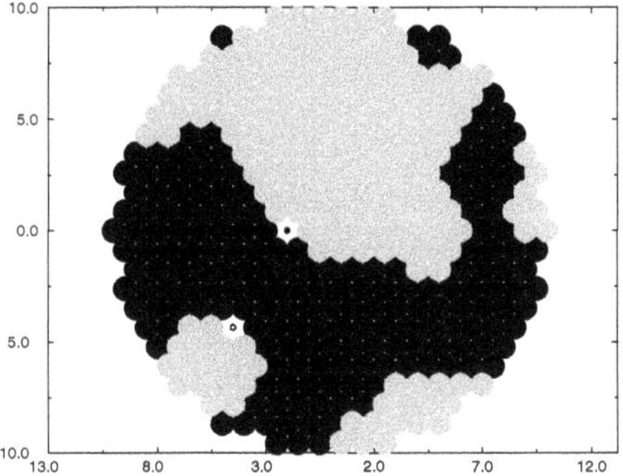

Fig. 11.9. Sphere of 5947 spins with $D/Ja^3 = 1.$, $k_BT/J = 0.01$, $H = 0$. Spins in the equatorial plane are partitioned in four classes; spins mainly up: black area, mainly down: gray area, weakly up (down): small black (open) circles

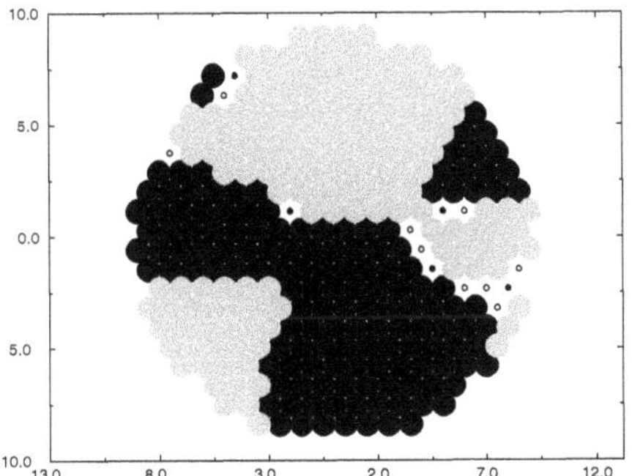

Fig. 11.10. Same sample and same symbols as in Fig. 11.9 for spins lying in the plane located 6 monolayers above the equatorial plane

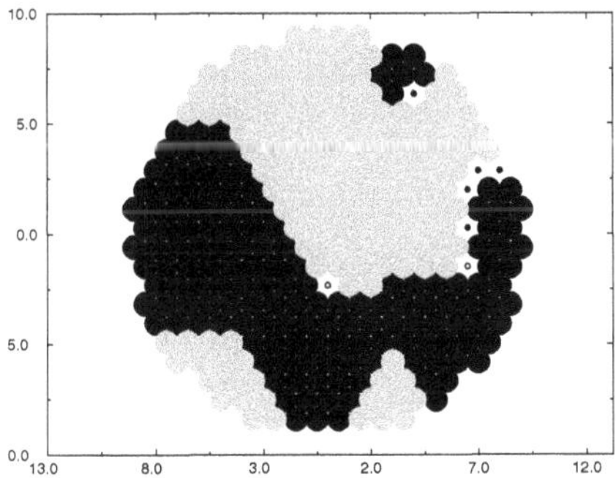

Fig. 11.11. Same sample and same symbols as in Fig. 11.9 for spins lying in the plane located 6 monolayers below the equatorial plane

symmetry. Under a higher field, the central part decreased and several external domains coalesce as it is seen from the equatorial cut. The occurrence of complex shapes of the domain walls is well in agreement with the previous considerations.

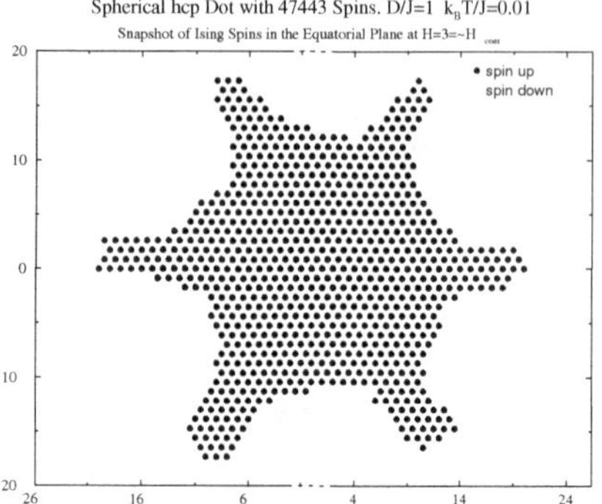

Fig. 11.12. Magnetization process of a large sphere of Ising-like spins: shrinking of the sixfold branched core domain: $D/Ja^3 = 1$, $k_B T/J = 0.01$, $H/J = -3. \simeq -H_{coer}$

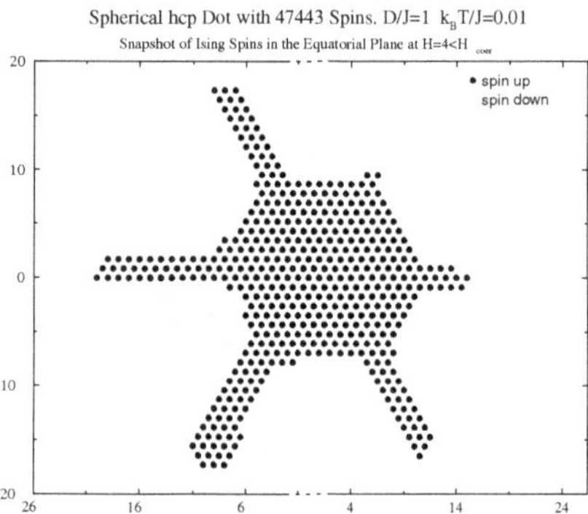

Fig. 11.13. Same as Fig. 11.12 with $H/J = -4$.

For a large sphere of Heisenberg spins with rather weak dipolar interactions characterized by $D/Ja^3 = 0.1$, there is only one vortex within a plane normal to the c-axis lying at some distance off the center as shown in the equatorial cut of Fig. 11.14. As observed for a disk in Fig. 11.1, this vortex is close to a spiral and is made of adjacent uniform domains. As mentioned above, we notice that the magnetization is parallel to the boundary. The magnetization process of a small sphere of Heisenberg spins during an hysteresis loop is shown in Figs. 11.15-11.18. First a very abrupt hysteresis loop is seen in Fig. 11.15. The three sections displayed in Figs. 11.16-11.18 at a field close to the coercive field show that a large uniform central domain appears and is surrounded by a vortex arrangement as well as by minor domains close to the surface where the magnetization is parallel to the surface. This demonstrates once more the stability of vortex states in magnetic nanostructures. The last example is given in Fig. 11.19 for the equatorial section of a small sphere with a strong dipolar effect: $D/Ja^3 = 4.$ and a strong anisotropy $A/J = 4.$ In that case in-plane vortices appear the core of which bears spins which are perpendicular to the plane. Thus vortices and mazes are quite complementary as discussed in the general part.

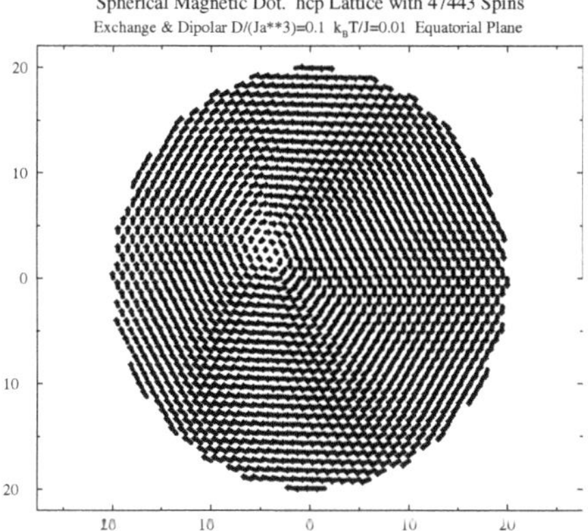

Fig. 11.14. An off-center vortex in the equatorial plane of a large sphere of vector spins at low temperature

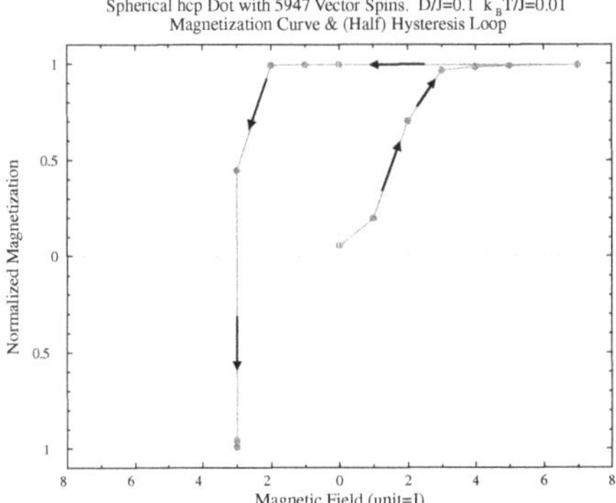

Fig. 11.15. First magnetization curve and (half) hysteresis loop for a small sphere of vector spins at low temperature

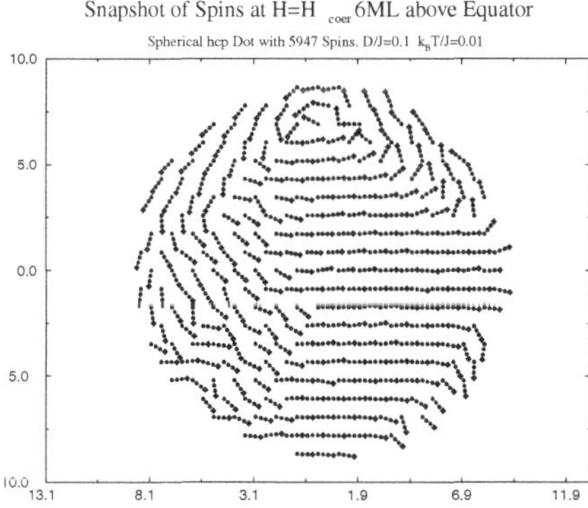

Fig. 11.16. Same sample as in Fig. 11.15. Snapshot of spins located 6 monolayers above the equatorial plane at $H = -H_{coer}$

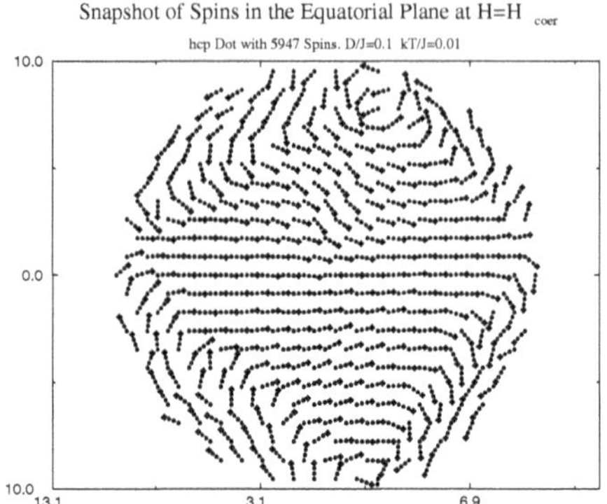

Fig. 11.17. Same as Fig. 11.16. Here the spins are lying in the equatorial plane

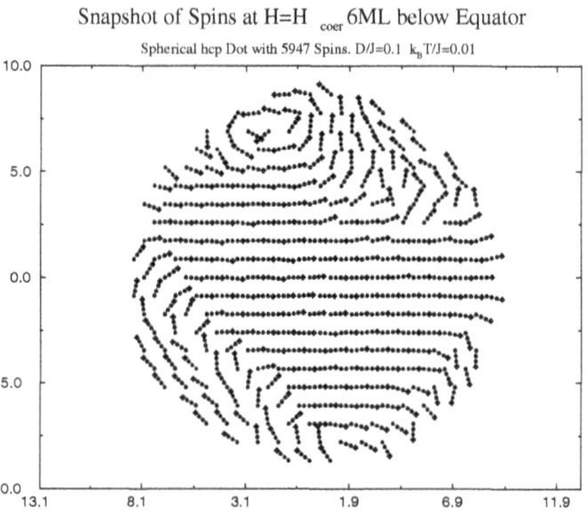

Fig. 11.18. Same as Fig. 11.16. Here the spins are lying 6 monolayers below the equatorial plane

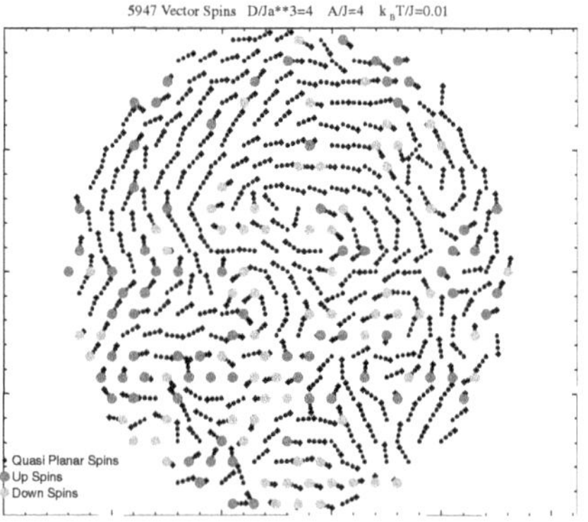

Fig. 11.19. Magnetization configuration in the equatorial plane of a small sphere of vector spins with exchange (J), dipolar (D) and anisotropy (A): $D/Ja^3 = 4.$, $A/J = 4.$, $k_B T/J = 0.01$. the arrows show the in-plane components while nearly vertical spins are shown in black (up) and gray (down)

As a conclusive remark about the stability of topological defects in experimental and simulated nanostructures, the effect of an extra magnetic field of convenient orientation must be considered as shown in the examples of Figs. 11.7-11.8, 11.12-11.13 and 11.16. Solitons were also observed when the external field is equal to the coercive field [11]. During this process, the part of singularities is essential. These results do confirm the importance of topological defects in nanostructure magnetization processes.

Acknowledgements

The authors are glad to acknowledge fruitful exchange and collaboration with Dr. Elena Vedmedenko, now at Hamburg University and with Christophe Jorssen from Paris-Sud Orsay University.

References

1. R. Allenspach, J. Magn. Magn. Mater. **129**, 160 (1994)
2. T. Duden, E. Bauer: Phys. Rev. Lett. **77**, 2308 (1996)
3. A. Vaterlaus, C. Stamm, U. Maier, M.G. Pini, P. Politi, D. Pescia, Phys. Rev. Lett. **84**, 2247 (2000)
4. M. Hehn, K. Ounadjela, J.-P. Bucher, F. Rousseaux, D. Decanini, B. Bartenlian, C. Chappert: Science **272**, 1782 (1996)
5. A. Lebib, S. Li, M. Natali et al.: J. Appl. Phys. 89, 3892 (2001) S. Li, D. Peayrade, M. Natali, A. Lebib, Y. Chen: Phys. Rev. Lett. **86**, 1102 (2001)
6. A. Hubert, R. Schäfer: Magnetic Domains (Springer, Berlin 1998)
7. W. Wulfhekel, J. Kirschner: Appl. Phys. Lett. **75**, 1944 (1999)
8. A.B. MacIsaac, J.P. Whitehead, K. De'Bell, P.H. Poole: Phys. Rev. Lett. **77**, 739 (1996)
9. A. Hucht, K.D. Usadel: J. Magn. Magn. Mater. **156**, 423 (1996)
10. J. Sasaki, F. Matsubara: J. Phys. Soc. Jpn. **66**, 2138 (1997)
11. E.Y. Vedmedenko, A. Ghazali, J.-C.S. Lévy: Surf. Sci. **402-404**, 391 (1998) E.Yu. Vedmedenko, A. Ghazali, J.-C.S. Lévy: Phys. Rev. B **59**, 3329 (1999) E.Yu. Vedmedenko, H.P. Oepen, A. Ghazali, J.-C.S. Lévy, J. Kirschner: Phys. Rev. Lett. **84**, 5884 (2000)
12. G.S. Kandaurova, A.E. Sviderskii: Sov. Phys. JETP **70**, 684 (1990)
13. M.V. Logunov, M.V Gerasimov: Phys. Met. Metallogr., to be published
14. T. Suzuki, L. Gal: Jpn. J. Appl. Phys. **18**, 1609 (1979)
15. J.M. Kosterlitz, D.J. Thouless: J. Phys. C **f6**, 1181 (1973)
16. J.N. Chapman: J. Phys. D **17**, 623 (1984)
17. A.C. Daykin, J.P. Jakubovics: J. Appl. Phys. **80**, 3408 (1996)
18. R.E. Dunin-Borkowski, M.R. McCartney, B. Kardynal, D.J. Smith: J. Appl. Phys. **84**, 374 (1998)
19. A.K. Petford-Long, B. Dieny: private communications
20. R.P. Cowburn et al.: Phys. Rev. Lett. **83**, 1042 (1999)
21. W. Schepper, H. Kubota, G. Reiss: 'Analysis of the disturbing influence of stray fields in very small MRAM cells by computer simulation', this conference
22. J.-C.S. Levy: Phys. Rev. B **63**, 104409 (2001)

23. M.M. Midzor, P.E. Wigen, D. Pelekhov, W. Chen, P.C. Hammel, M.L. Roukes: J. Appl. Phys. 87, 6493 (2000); P.E. Wigen et al.: 'Magnetic resonance force microscopy in microscopic magnetic systems', this conference
24. G.T. Rado: Phys. Rev. **83**, 821 (1951)
25. R.B. Dingle: *Asymptotic Expansion: Their Derivation, Interpretation* (Academic Press, London 1973)
26. Y. Yafet, E.M. Gyorgy: Phys. Rev. B 38, 9145 (1988)
27. P.Molho, J.L. Porteseil, Y Souche, J Gouzerh, J.-C.S. Lévy: J. Appl. Phys. **61**, 4188 (1987)
28. H.A.M. van den Berg: J. Appl. Phys. **60**, 1104 (1986)
29. M. Seul, D. Andelman: Science **267**, 476 (1995)

12 Novel Magnetic Materials Based on HTSC-Ferrite Heterostructures and Co/SiO_2 Nanocomposites

Galina A. Nikolaychuk[1], Alexey V. Lukashin[2], Vladimir V. Matveev[1], and Ivan V. Pleshakov[3]

[1] Institute of Physics of St. Petersburg State University, 1 Ulianovskaya, Petrodvorets, 198504, Saint-Petersburg, Russia
[2] Department of Chemistry, Moscow State University, Vorobievy Gory, 119899, Moscow, Russia
[3] A.F. Ioffe Physical–Technical Institute RAS, 21 Politechnicheskaya, 194021, Saint-Petersburg, Russia

Abstract. A brief survey of the present day investigations in the field of magnetically ordered nanocomposites is represented. The special attention is paid to magnetometry and NMR of cobalt multilayers and dispersed structures. In the original part of work the fabrication methods and physical properties of two kinds of materials are considered: (i) thin ferrimagnet/superconductor structures and (ii) dispersed cobalt nanocomposites impregnated into porous matrices. For the latter case novel radiospectroscopic data (high-frequency peaks, time transformation of NMR spectrum) are presented.

12.1 Introduction

The search of new magnetic substances carried out during the last two decades has resulted today in the creation of a whole direction in material science dealing with compound systems. One or more constituents of such a system can be a magnetically ordered material, and in such a case it is called magnetic composite. Essentially this way was chosen owing to the fact that the traditional trends – preparing of different ferrites, growing of single crystals, etc. – in some sense have exhausted their ressources. But, what is most important, one hopes to obtain materials with novel physical properties. The systems where at least one of the characteristic dimensions is of the order of a few nanometers attract a special attention. These may be, for example, thin films and film structures with the thickness of several monolayers, ultra-dispersed powders, or more complicated dispersed substances with very small magnetic particles. Being extremely diversed in manufacturing and morphology, they are similar in one feature – it is possible to observe, at least in principle, the quantum size effects in them. Following the generally accepted terminology further we shall call such structures nanocomposites.

There are two main types of nanostructures: regular structures and chaotic dispersed systems. The first class includes the different kinds of films, multilayetrs and heterostructures.[1] The second class consists of nanocrystalline powders, clusters, nanocomposites, etc.

In order to understand their unique properties one needs to have some background on the bulk materials used for nanostructure preparation as well as the essentials of the main physical methods applied for the investigation of nanostructures.

There is a variety of physical phenomena in magnetic films and their sandwiches, but the necessary initial data at the first stages of investigation is the knowledge on the peculiarities of layer formation, the sharpness of interface, crystal lattice parameters, and the lattice strains. Together with x-ray analysis, nuclear magnetic resonance (NMR) belongs to the main experimental techniques giving such an information. NMR is a local method providing microscopic data in addition to the results of conventional magnetometry. The magnetometry and NMR are also of the same efficiency in the case of chaotic magnetic nanostructures. Further we shall give a background on magnetic materials as well as a brief summary of magnetometry and NMR methods. After that some literature data on the application of these two methods to magnetic nanostructures will be presented. Results related to the film structures of a superconductor/magnetic type, and dispersed nanocomposites based on porous media will be given.

12.1.1 Magnetic Ordering and Magnetic Materials

In the present paper we shall deal mainly with magnetically ordered substances. The latter term generally belongs to the vast class of materials with different kinds of magnetic ordering - ferromagnetic, antiferromagnetic, and ferrimagnetic. In the first case the electronic magnetic moments of ions are parallel, in the second there are two magnetic sublattices with the magnetizations equal in value and oppositely directed in the ways, which means the complete compensation of total magnetic moment, and, at last, there are two magnetic sublattices in the third case, different in value and directed in opposite ways, which signifies the existence of an uncompensated total magnetic moment. Of course, this is only a rough classification and it does not conclude a great number of possible magnetic structures. But here we shall not be engaged in magnetic structure details. Moreover, as we shall see further, in one of the special types of samples we shall interested in – nanostructured composites – the simplest magnetic materials are predominantly used as a magnetic constituent. What is important, is the existence of a certain ordering in the system of magnetic moments of a crystal. The source of this phenomenon is the exchange interaction between magnetic ions which arises either as a result of immediate overlap of their electron wave functions or indirect interplay through intermediate nonmagnetic ions.[2] In the

[1] Here we should mention that the term 'heterostructure' used in this paper differs from the common one: here we regard it like a simple contact between two different materials without epitaxial growing requirement.

[2] It should be noted here that the exchange interaction is of electrostatic nature, and magnetic dipole–dipolar interaction between magnetic moments does not play any considerable role in their alignment.

case of metals the conduction electrons take part in exchange process. For the description of the exchange interaction the Heisenberg Hamiltonian is usually used:

$$\hat{H}_{ob} = -2J\hat{S}_i\hat{S}_k \tag{12.1}$$

where J is the exchange integral, connected with the electron wave functions overlap, $\hat{S}_i$, $\hat{S}_k$ are the spin operators of interacting ions. When $J > 0$, the minimal energy will be for the parallel magnetic moments, and the ferromagnetism takes place. In the opposite case we deal with the antiferromagnetic ordering.

At the careful examination of the expression (12.1) under different realistic conditions various kinds of magnetic ordering can be obtained. The simplest but representative class of magnetic materials is ferromagnetic. Their main macroscopic property is the spontaneous magnetic moment M_s (i.e. a moment which exists without of an external magnetic field) at temperatures lower than the so-called Curie point, T_c. The latter is the temperature of ferromagnetic phase transition when the energy of thermal agitation destroys the spin alignment. Naturally, ferromagnetic materials should include the ions with partially filled electron shells. First of all these are elements of the iron group with partially filled 3d shell, and the rare-earth ions with partially filled 4f shell. Ferromagnetism observed in three elements of the iron group: iron, cobalt (below we will pay special attention to this element), nickel, and in a few heavy rare-earth ions like gadolinium, terbium, dysprosium, etc. Only four elements have relatively high Curie temperatures - iron ($T_c \simeq 1100\,\mathrm{K}$), nickel ($T_c \simeq 650\,\mathrm{K}$), cobalt ($T_c \simeq 1400\,\mathrm{K}$), and gadolinium ($T_c \simeq 290\,\mathrm{K}$). At the same time a number of different iron alloys exists. The class of nonmetallic ferromagnets is not numerous, these are EuS, analogous compounds with sulfur, selenium, and chalcogenides ($CdCr_2S_4$ for example).

The magnetization tends to align along preferred directions of an anisotropic system. Phenomenologically this behavior can be described by the introduction of the effective field of anisotropy, H_a, acting along the axis. The direction where the energy of magnetic anisotropy becomes minimal is called the easy direction, and the direction with maximal magnetic anisotropy energy is called the hard one. The crystals with only one easy axis are called "crystals of an easy axis type". For metallic iron the easy directions are the edges of unit cell, for nickel they are body diagonals. When the magnetization process is equally or almost equally easy in all directions of any plane one calls such a system as "easy plane".

For uniaxial crystals the simplest expression for the anisotropy energy per a unit volume is

$$E_a = K_a \cdot \sin^2\theta,$$

where K_a is the anisotropy constant and θ is the angle between the easy direction and magnetization. The situation with $K_a > 0$ corresponds to the easy axis system (because the energy is minimal when $\theta = 0$), and in the absence of an external magnetic field the equilibrium direction of the spontaneous magnetization coincides with the easy axis. For $K_a < 0$ the energy minimum reaches at $\theta = \pi/2$, which corresponds to the easy plane type. It should be noted that this case is more complicated because small in-plane anisotropy is possible. The anisotropy fields in real ferromagnetic crystals vary from hundreds to thousands of oersteds.

It is well known that magnetically ordered materials form the inhomogeneous structure, dividing into macroscopic domains. The mechanism of this effect is associated with the minimization of magnetic energy. In the domains the spontaneous magnetization is oriented in different directions. The domains are separated by the relatively narrow walls, where magnetization rapidly changes its orientation. Such a structure lowers the magnetostatic energy of the field surrounding a sample, but, at the same time, it increases the energy of magnetic anisotropy in domain walls. The competition of these two energies produces an equilibrium domain state. Domain walls can move in an external magnetic field, they affect the details of the magnetization process of a sample .

Some Properties of Cobalt

Some Co/SiO_2 composites will be discussed below. There are two crystalline modifications of metallic cobalt: hexagonal closed parked (hcp) and face centered cubic (fcc). Thermodynamically the hcp phase is more stable at lower temperatures (below 690 K) and the fcc phase above 690 K. However, the fcc structure normally remains partly in bulk material below 690 K as a metastable phase. As to nanodispersed cobalt the fcc phase becomes more stable than hcp phase for all temperatures.

Both modifications are ferromagnetic. The hcp phase belongs to the easy axis crystal type, while the fcc has cubic anisotropy. The energy of anisotropy of fcc cobalt structure is significantly lower that in hcp case.

Superparamagnetic State of Fine Particles

Chaotic magnetic nanostructures belong to dispersed systems of magnetically ordered fine particles. Physics of such systems exhibits a number of specific effects, e.g. so-called superparamagnetism. When the size of particles decreases it is possible to reach a single-domain state, and the magnetic properties of the system change significantly. Such systems at a temperature higher than the so-called blocking temperature, T_B, in the approximation of noninteracting magnetic particles, behave like a gas of magnetic moments, and their properties may be described as superparamagnetic, i.e. paramagnetic with the atomic magnetic moment substituted by the spontaneous moment of a particle. Below T_B the characteristic time of thermal fluctuations of magnetic moment is lower than the time constant of measuring device, and the system looks “frozen”, i.e. the sample in the whole can be regarded as a ferromagnet. Of course the blocking temperature should not be considered as a physical parameter of the system because it depends on the device used. Therefore, T_B describes the physical system together with measuring technique rather than only the physical system.

A Few Quantitative Expressions

According to the simple activation model the superparamagnetic relaxation time obeys the Arrhenius low

$$\tau = \tau_0 \cdot exp(K_a V/k_B T). \qquad (12.2)$$

Here V is a particle volume, and $K_a V$ is a height of the potential barrier which the magnetic moment of an easy axis type particle overcomes under the action of thermal fluctuations; k_B is the Boltzmann constant, and τ_0 is the relaxation one, associated commonly with the electron Larmor frequency (the rough estimation of the τ_0 upper limit is 10^{-9} sec).[3]

Let V and T be given values, and τ_i is the measuring time, i.e. the characteristic time of a device (of magnetometer, for example). In that case if $\tau > \tau_l$ the system from the experimenter point of view is ferromagnetic. When $\tau < \tau_i$, the spontaneous magnetic moment of every particle fluctuates thermally, overcoming an energy barrier many times during the period of measurement. In this case the particles are in superparamagnetic state.

As it is clear from above and the expression (12.2) T_B, separating superparamagnetic and ferromagnetic states, depends on the instrumental time. *Thus we would like to underline once more that T_B it is not a physical constant of the material and characterizes the system like "magnetic particles together with measuring device"*. A series of experiments was carried out by methods with very different measuring times. They may be, for instance, the standard magnetometry (τ_i is of the order of 10–100 sec) and Mössbauer spectroscopy ($\tau_i \sim 10^{-8}$ sec). Of course, the resulting values of T_B were quite different for these two kinds of measurements. This property of T_B is often used for determination of real parameters of a system under investigation. For instance, it is easy to obtain from (12.2) an expression for the "critical volume" of a particle for given T_B

$$V_c = (k_B T_B/K_a) \cdot ln(\tau_i/\tau_0). \qquad (12.3)$$

The estimation by (12.3) for cobalt yields a linear particle size of a several nanometers for T_B at room temperature. It should be noted that significant variations in the τ_i and τ_0 values are not of great importance because of the properties of logarithmic function.

The magnetic response of a superparamagnetic particle (the magnetization M in external magnetic field) is described by the well known Langevin formula, and under the condition $K_a V \ll k_B T$ it may be written as

$$M - [NV^2 \cdot {M_s}^2(T) \cdot H]/3k_B T, \qquad (12.4)$$

where N is a number of the particles in a unit volume. It is clear from (4) that the magnetic response of a substance in the superparamagnetic state is much bigger than in the paramagnetic one due to the substitution of an atomic magnetic moment by a spontaneous moment of the particle.

[3] It is useful to emphasize that K_a may be significantly changed in fine particles in comparison with the bulk material because of various reasons; usually for dispersed systems it is bigger by one or two orders of magnitude.

12.1.2 High-T_c Superconductors

The other important class of materials that will be considered later is the superconducting copper oxides with the high T_c.[4] Further we shell discuss the composite systems of magnetic/superconductor type, and here we have to give a brief description of high-T_c superconductors (HTSC). We shall use the well known substance $YBa_2Cu_3O_{7-x}$ (1-2-3 system), which has a critical temperature near 90 K. Its superconducting properties depend on the oxygen deficiency x.

The orthorhombic unit cell has a perovskite-like characteristic structure with CuO_2 planes. Such structural peculiarities determine the anisotropic properties of the material. For example, the difference between the conductivity measured parallel and perpendicular to c-axis reaches 30 (in the normal state near T_c).

The other transport and magnetic properties also differ considerably in these directions. The system 1-2-3 is a type II superconductor, i.e. the external magnetic field can penetrate into the sample in the form of magnetic vortices. The samples that will be discussed in this paper are polycrystalline thin films. In such systems the Josephson junctions between the grains play an important role providing the transport current percolation through the sample.

12.1.3 NMR in Magnetically Ordered Materials

The NMR in magnetic materials for the first time was observed by Portis and Gossard in 1959 on the ^{59}Co nuclei in bulk metallic cobalt. Already at that time the essential difference of the NMR in magnetically ordered substances from the conventional NMR spectroscopy were pointed out. Here we mention two distinctive features of this NMR:

i) it is possible to observe the NMR signal without a static external magnetic field, because there are big and fairly homogeneous magnetic fields acting on the nuclei of magnetic ions due to hyperfine interaction (sometimes the NMR in magnetic materials called "zero field NMR");

ii) the nuclear spins interact with the external radio-frequency (rf) field indirectly through the magnetic moments of electrons, which results in a considerable enhancement of the resonance intensity.

The local field is given by

$$H_{loc} = A_0 \langle M \rangle + \langle H_{dip} \rangle, \tag{12.5}$$

where A_0 is the hyperfine interaction constant, $\langle\rangle$ is the designation of averaging, and H_{dip} is the dipolar part of the local field (which is small and can be neglected in the first approximation). In (5) M may be the spontaneous magnetization for a ferromagnetic material, or the sublattice magnetization in the case of ferri- and antiferromagnets.

Hence the local field at the nuclei mainly is of a hyperfine nature originating from the electron shell. In comparing with the standard laboratory fields it

[4] In this subsection T_c means the critical temperature of the superconducting transition.

is extremely strong. For the elements of iron group it is of the order of tens, and for the rare-earth ions at about of hundreds teslas. The NMR frequency is determined by the expression

$$\omega = \gamma \cdot H_{loc},$$

where γ is the gyromagnetic ratio. So, the frequencies vary in the wide range from tens to thousands MHz. The detail of the hyperfine field origin is complicated. It arises from the so-called Fermi contact interaction by the immediate interplay of the nuclear magnetic moment with the electron. But the electron density at a nucleus site is nonzero only for s-electrons, which results in zero field for an electron pair with the antiparallel spins. But it turns that under the action of partially filled external shells the wave functions of s electrons become slightly asymmetric (the effect of shell polarization) with nonzero differences of the electron densities on a nucleus. Thus, the local field mainly originates from the Fermi contact interaction. We have to note here that for the ions of iron group the effect of orbital moment quenching exists, and the total magnetic moment of an ion is determined predominantly by the spins.

As is well known the resonance transitions in NMR are driven by the rf field, but in the case of NMR in ordered magnetics the direct effect of this field is negligibly small. Instead of this the resonance transitions are excited by the alternating part of a local field. Let us consider the simplest case of the single domain ferromagnet with easy axis, placed in the external magnetic field H_a applied along this axis. The rf field applied perpendicular to the easy axis deflects the magnetization from its equilibrium direction, and produces the transverse alternating magnetization. If h is the magnitude of the rf field, and it is small in compared to H_a, then the declining angle α is equal to $h/(H + H_a)$. The magnetization is related to the local field by (5). Neglecting dipolar fields H_{dip} it is easy to obtain for the transverse alternating part of the local field:

$$h^t_{loc} = H_{loc} \cdot [h/(H + H_a)].$$

The enhancement factor η is defined as a ratio of the magnitude of the alternating part of the local field to an external rf field. Thus for the enhancement factor we obtain:

$$\eta = H_{loc}/(H + H_a) \qquad (12.6)$$

In the case when frequency of rf field coincides with the NMR frequency the nuclear magnetic moments in the magnetic material are excited by the $\eta \cdot h$ rf field. Taking into account the orders of H_{loc} and H_a given above we can estimate the enhancement factor to be in the range of $10^1 \div 10^4$. It should be mentioned that the reverse effect also exists, which means that the oscillating nuclear magnetization induces the electron magnetic moment movement through the hyperfine interaction. Finally, the intensity of NMR signal in ferromagnets is proportional to η^2.

The excitation of NMR signal in domain walls requires a lower rf amplitude than in domains because it is driven by the magnetic moment rotation in the wall during its linear displacement under the action of the rf field. Normally the enhancement factors in the wall and the domain differ by one order of magnitude.

The observation of NMR in magnetic materials can be carried out by conventional radiospectrometers, moreover, it is easier than the usual NMR experiment because it is not necessary to use homogeneous permanent fields and powerful rf oscillators. The difficulty arises, however, that the NMR frequency now is unknown and depends on the material properties. One of the purposes of the experiment is to determine this value.[5].

To summarize this subsection one can conclude that NMR in magnetic materials in a certain sense is the measuring of magnetic susceptibility of electron system in the vicinity of the frequencies of nuclear Zeeman splittings. The position of a resonance peak is fixed by the intrinsic fields. The signal intensity depends on magnetic anisotropy and external field, while the latter practically does not affect the resonance frequency.

12.1.4 Other Methods Used for the Investigation of Magnetically Ordered Systems

There are many different experimental techniques used for the investigation of magnetic structure. Almost all the facilities of solid state physics can be applied to the investigation of magnets: the x-ray diffraction, electron microscopy, etc. The microscopic magnetic structure and its correlation with the crystal structure are established by neutron scattering, because it is sensitive to the mutual orientation of the magnetic moments of neutrons and ions. The macroscopic (i.e. domain) structure is investigated by the methods of powder patterns, magnetooptics and many others. We shall not discuss every method and will concentrate on NMR (see subsection 1.3) and magnetometry as the techniques directly related with the general properties of the magnetic state on the one hand, and giving the unique information on the microscopic properties on the other hand.

The Magnetometry deals with measuring a magnetic parameter (very often it is a sample magnetization) at different conditions. In the magnetometry it is possible to observe the hysteresis magnetization process, making conclusions on the magnetization details, on such parameters as residual magnetization, coercitivity, and the spontaneous magnetization. The data on the static magnetic susceptibility χ (we remember that $M = \chi \cdot H$) can be extracted from the magnetization curves. The measurements of temperature dependencies of M_s (or χ) play here the key role. The characteristic features of $M_s(T)$ or $\chi(T)$ can help to investigate the properties of such a structure as the system of nanosize magnetically ordered particles. The experiments in this field are carried out by utilizing one or another standard magnetometer especially fitted to the concrete measurement task.

[5] NMR in magnetic materials possesses many other interesting and important peculiarities which were not considered here. More detail on the problem can be found in the monograph [1]

12.2 Properties of Nanostructured Objects

As it has been mentioned above there is a variety of new physical phenomena in the magnetic thin films and multilayers. Recently many important effects have been observed in nanodispersed systems as well. In the sections following below we represent a brief review of some unusual physical properties of both type nanoobjects as obtained by magnetometry and NMR.

12.2.1 Ultrathin Films and Dispersed Nanocomposites

The samples of a thickness of a few monolayers conventionally can be regarded as ultrathin films. At the early stage of investigation such objects attracted the attention of the researchers as a model of two-dimensional (2D) magnetic formation, and they have been investigated from this point of view. It is well known that the magnetic ordering in ideal isotropic 2D spin system with Heisenberg interaction (1) does not occur (Mermin–Wagner theorem, [2]). In reality it turns out, however, that the most of the theorem conditions are broken, because the films of finite thickness possess the anisotropy, etc. Thus, the magnetic ordering in ultrathin films exists [3], and the goal of experimental study of such a system consists of the investigation of a transformation of magnetic properties under the transition from 3D systems to objects with minimal thickness. During the last years fairly many works were accomplished in this field. For example, the magnetometric measurements of ultrathin Co films were carried out [4] in order to specify 2D-magnetic behavior. For the 2D-system the values of critical exponent near the phase transition have been established, and distinctive features of the magnetic moment behavior at low temperatures were revealed. In [3] the possibility of the formation of a peculiar 2D-superparamagnetic in the vicinity of Curie point because of the film fragmentation was mentioned. Simultaneously with physical studies, the ultrathin film and layered superlattices technology is being developed. In the latter study the novel phenomena, that have been mentioned before, were revealed.

In multilayer structures with alternating ultrathin films of magnetic and non-magnetic metals the dimensional effects [5], magnetic anisotropy peculiarities [6], and giant magnetoresistance [7] have been observed. Among the thin film systems the structures of a superconductor/magnet type have a special significance. They represent a model of the coexistence of magnetically ordered and superconducting states, and also are promising materials for tunable elements of spin-wave high frequency microelectronics.

As to chaotic dispersed nanocomposites one can notice the ordering in the system of interacting particles [8,9], quantum tunneling between them [10,11], leading to a nonthermal relaxation process or peculiarities of the magnetic susceptibility, etc. It is interesting to mention that a nanosized structure of the natural (biological) origin was investigated [11], namely a protein molecule containing the antiferromagnetic core with the diameter of 7.5 nm.

12.2.2 Cobalt Nanostructures

There are many different ways to fabricate the chaotic magnetic nanostructures like, for instance, using dispersion alloys, cosputtering metal and chemically neutral substances, or impregnating magnetic nanoclusters into the medium with the cavities of an appropriate size. In every case the system of fine metallic particles embedded into a matrix forms a composite material with unusual magnetic properties. The particles are supposed to be single–domain due to their nanoscale dimensions.

The systems containing metallic cobalt as a magnetic part belong to a specific type of nanostructures. This class is very vast because Co, on the one hand, is an extensively investigated magnetically ordered 3d-conductor, and on the other hand the ^{59}Co nucleus can be very well observed by NMR method. The content of a magnetic material in nanostructures may be vanishingly small and in this case the problem of low signal-to-noise ratio in NMR becomes very serious. Just because of this almost exclusively Co-containing materials are utilized in works, where the NMR method is used.

NMR Results

The main subject in the analysis of NMR data is the information on the fixed frequencies and relative intensities of resonance peaks, giving a knowledge about the atomic position. Potentially the NMR method is able to give another information, for instance, as it follows from (6) that it is possible to estimate the anisotropy field H_a.

Let us consider the results of [12] as an example of NMR application to the investigation of multilayers. The widespread structure Co/Cu was studied in this work. It was possible to observe not only the NMR from the internal part of the Co film, but also to separate for the first time a signal from the transition region between Co and Cu. In the spectrum the broad line was revealed, which was positioned in the wide frequency range below the main peak (215 MHz at 2 K) . Its integrated relative intensity decreased with increasing of the number of layers, and its position corresponds to the additional peaks of a dilute alloy Co–Cu. According to these findings the low frequency spectrum was attributed to the Co atoms near the interface, and the rate of Cu diffusion into Co was estimated. Also the content of fcc and hcp phases of Co was determined by NMR, and the degree of Co crystal lattice expansion was calculated. A great number of other investigations have been carried out on Co/Cu multilayers (see, for example, [13–15]). In [13,14] the low frequency tail of spectrum was decomposed into a series of resolved peaks, and it was shown that it can be used for the determination of the interface quality. In [15] the NMR data were correlated with the magnitude of the giant magnetoresistance effect. Similar approaches were applied to the cobalt sandwiches with different metallic layers (Au, for instance [16]). In all the works on NMR in superlattices essential information was obtained, supplementing the data of x-ray analysis and magnetometry. We have to emphasize that there were no works where NMR peaks were detected at frequencies distinctly higher than the hcp frequency (near 230 MHz at liquid helium temperatures).

In the case of magnetic nanocomposites the NMR gives the same information as for film structures, supplemented by the data on the effective anisotropy constant and blocking temperature. The NMR spectrum has been measured for metallic Co isolated in nanosize cages of zeolite, [17]. It consisted of a broad line (150÷260 MHz at low temperatures) and was interpreted in terms of fcc- and hcp-structures. An unusual dependence of the NMR signal in the external magnetic field was revealed for the system. The spectrum obtained for the cobalt nanoparticles embedded in the SiO_2 matrix [18] demonstrates such a distinctive feature like a well pronounced high-frequency line centered at about 243 MHz. In the authors [18] opinion this is because of the magnetic moment increasing in ultrafine particles.

In conclusion of this paragraph we have to mention the work [19] where the unique observation of the NMR spectrum of Co nanowires at room temperature has been achieved.

12.3 Ferrite/Superconductor Thin Film System

One of the possibilities to create a new compound structure is the fabrication of superconductor/magnetic systems. It is well known that thin magnetic films are used as working cells of tunable filters and delay lines, operating on the principle of spin wave excitation in the magnetic material. Usually such an element of functional electronics is a sandwich of dielectric/magnetic material/metal layers (the dielectric constitutes a substrate for the main structure grown on top of it). It forms a waveguide for the coupled modes of spin and electromagnetic waves propagating along the structure. The substitution of a normal metal by a superconductor (in our case HTSC) imparts additional possibilities to the device [20]. In the latter work the dielectric/ferrimagnetic/HTSC system was analyzed by the examination of Maxwell–Landau–Lifshits equations (for the magnetic part of structure) and Maxwell–London equations (for the superconducting part). It was shown that the system has potential for efficient control of the dispersive characteristics of spin waves by the change of boundary conditions at the ferrimagnetic/HTSC interface. An essential point here is the degree of proximity between the layers. There are particular difficulties in the creation of such layers related to the chemical influence of the magnetic substrate on HTSC. Therefore, for carrying out the experimental check of the theory a special technique was developed for the monolithic structure fabrication [21,22].

The superconductors in sandwich compositions were prepared from the nitrate solutions with adding of a polymeric binder and subsequent pyrolysis on the surface of the magnetic material. The additional method to obtain the monolithic film structures was verified, which consists of doping the initial HTSC powder by silver, lowering the sintering temperature, and introducing the double-stage cooling regime. Such a technique provides the HTSC films of the yttrium 1-2-3 system with the transition temperature to superconducting state T_c not lower than 85 K. As a magnetic substrate the nickel-zinc, manganese-zinc spinels, yttrium, and yttrium-gadolinium garnets were used. Also single crystal films of yttrium garnet, epitaxially grown on the gadolinium-gallium garnet substrate were employed.

An experimental check of the theory was carried out on the structures with the ceramic and single crystal films of $YB_2Cu_3O_{7-x}$ [20]. In the range of 2–5 HHz the amplitude and phase characteristics of the dielectric/ferrimagnetic/HTSC structure were investigated. Significant peculiarities of the systems with the HTSC material instead of metal were revealed. Two frequency ranges were distinguished with substantially different behavior of the system below T_c. In the lower range the damping from HTSC material practically did not depend on temperature, while the group velocity of waves increased by a factor of two when increasing the temperature to 77 K. Vice versa, for the high frequencies a strong increase of the damping was observed. Such a behavior qualitatively corresponds to the theoretical predictions.

So, it is possible to use these structures as filters with controlled characteristics. The control can be realized not only by temperature, but also by transport current [20].

12.4 Magnetically Ordered Nanocomposites Based on Porous Media

12.4.1 Synthesis of Cobalt Nanocomposites

The use of purely nanocrystalline systems is limited because of their low stability and tendency to form aggregates. A well-known solution of the problem is an encapsulation of nanoparticles in a chemically inert matrix [23]. Moreover such an approach enables one to protect nanoparticles from the effect of external factors (for example, from the oxidation in air).

In the present research we report a simple and inexpensive synthesis of cobalt nanoparticles embedded in silica framework using Co^{2+}–poly(vinylalcohol), further CoPVA, complex as a precursor [24]. It should be noted that the matrix, used for preparation of Co nanoparticles, determines basic properties of the nanocomposite.

The CoPVA/SiO_2 composite, N2 was obtained by mixing 100 ml of SiO_2 sol and 100 ml of CoPVA complex. CoPVA–containing sol was gelled in the thermostat at 80°C for a several hours with following drying in air at 120°C. After that the sample was annealed in O_2 at 500°C for 2 h (heating rate =1 min^{-1}) to remove organic components and reduced in hydrogen at 600°C for 2 hours to ensure Co nanoparticles formation.[6]

Another sample, N1 (SiO_2-(TEOS)-Co) was prepared using the TEOS gelation (here TEOS means tetraethyloxisilan). CoPVA complex has been prepared by the same technique. Alcogel was obtained by hydrolysis of TEOS in water–ethanol medium. Gelation was performed in a hermetic plastic container in the

[6] The similar procedure was a few months earlier used for the N3 sample preparation, however the gelation was reached by chemical way.

thermostat at 80°C. After gelation, the gel was aged 2–3 times longer than necessary for gelation, in order to ensure complete condensation.

The cobalt content in the fresh samples was preliminarily estimated to about 2.5 per cent by weight for N2 and N3 samples and as not more than 6 per cent for N1 one.

Another possible way to fabricate the magnetic nanocomposites was tried in present work. As matrices we used the porous glasses with well defined size of internal cavities (the diameter at about 40 nm). Metallic cobalt was impregnated into the glass by the electrolysis process from the water solution of cobalt chloride. According to our preliminary data a nanonocomposite material has been obtained with the strong magnetic response to the static magnetic field.

12.4.2 Characterization of the Nanocomposites

Studies of the surface properties of Co/SiO_2 nanocomposites as well as SiO_2 matrices by the method of capillary adsorption of nitrogen at 77 K showed that all samples are characterized by high specific surface area and narrow pore size distribution. Pure SiO_2 matrix has surface area about 410 m^2/g and pore radius 20–60 Å. By other hand, Co/SiO_2 nanocomposite has the surface area about 320 m^2/g and pore radius 28–30 Å. The change of the pore size distribution and the surface area is probably due to the formation Co-nanoparticles directly in the pores. TEM image of the N2 sample after annealing is shown in Fig. 12.1.

For the N1 sample the picture is similar. Nanoparticles are well seen on both micrographs. The electron diffraction studies confirm the formation of cobalt oxide together with metallic cobalt. Preliminary analysis of the TEM micrographs showed that the average size of Co nanoparticles is between 5 and 10 nm for the sample N2 and between 10 and 15 nm for the sample N1. The size of the particles is probably determined by the amount of Co atoms in polymeric clusters incorporated into silica network.

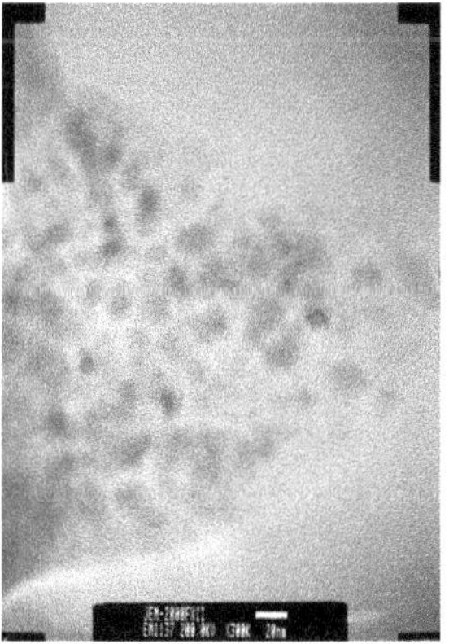

Fig. 12.1. TEM picture for the N2 sample.

12.4.3 NMR in Co-SiO_2 Nanocomposites

Frequency-sweep zero-field spectra of the samples N1 and N2 have been presented earlier [25] and are similar. The three groups of lines are clearly resolved (Fig. 12.2). The central and most intensive one at about 221 MHz is typical for cobalt thin films and nanostructures and may be attributed to fcc and/or hcp cobalt structures. More detailed investigation using field sweeping allow us to conclude that for both samples the field-sweep spectrum consists of fcc and hcp overlapping signals [26].

The second group in the lower frequency range is common for cobalt crystal structure with defects or admixtures. It is essential that the spin-echo spectra of the nanocomposites are sensitive to the details of preparation procedure (Fig. 12.3). There is a difference in the low-frequency part of the spectra, in particular

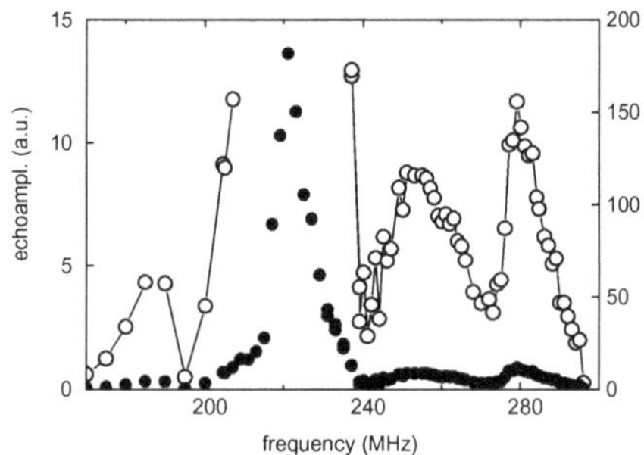

Fig. 12.2. Echo amplitude vs frequency for N2 sample. Bullets – a whole spectrum (right scale), open circles – an enhanced part of the spectrum outside the main line (left scale)

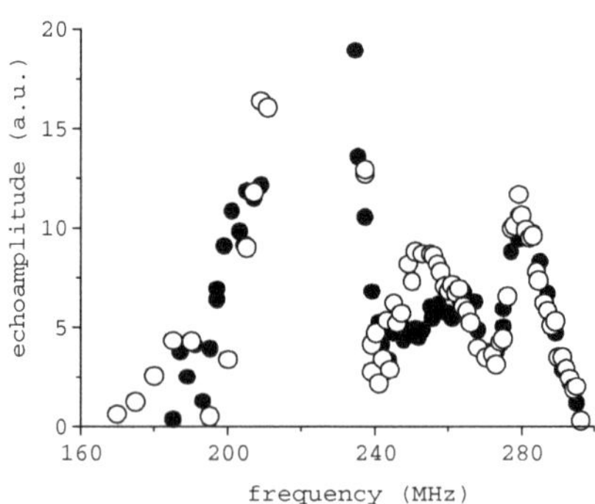

Fig. 12.3. Echo amplitude vs frequency for two samples prepared by different variants of the sol-gel method: a comparison of the minor line groups. Bullets – N1 sample, open circles – N2 sample

the absence of a line near 200 MHz for the N2 sample. A similar effect appears at the left side of the high-frequency part of the spectra as well.

At the same time both samples have practically identical high-frequency lines near 280 MHz (Fig. 12.2, 12.3). All high-frequency peaks from 245 up to 290 MHz never have been observed in cobalt-containing structures. The main line in the group appears at 280 MHz and this frequency is the highest one as compared to ^{59}Co NMR signals in bulk metal as well as in other Co films and nanostructures. The high frequency lines correspond to a hyperfine field enhancement. It means, in turn, the enhancement of the hyperfine constant and/or magnetic moment of individual atom (μ_{Co}). The latter was predicted theoretically for free cobalt clusters [27]. An experimental confirmation of the μ_{Co} enhancement was made for a few systems by magnetometry [28–30]. As to NMR in cobalt nanostructures there were only one observation of a weak signal at 243 MHz [18] and one short remark about a signal at 250 MHz [31]. According to [18] the 243 MHz line may be attributed to cobalt atoms with enhanced magnetic moments. Following such a point of view we have estimated the μ_{Co} value as $2.17\pm\ 0.02\,\mu_B$ in our case.

The Physical Nature of the μ_{Co} enhancement is not clear enough. One of the possible explanations is given below. Every particle may be considered as a core with a structure similar to bulk cobalt, transition part, and thin oxide layer (see [32] and refs therein). The particles under investigation are large enough to consider the transition layer and oxide one as a Co–CoO thin bilayer. The transition layer undergoes the strong influence from the oxide. It is well known that AF–FM exchange interaction can change the magnetic properties of the FM part of a system [33,34]. As to chaotic Co-nanostructures the AF–FM exchange appeared in both cases when an enhanced magnetic moment was observed by NMR [18,31]. *Therefore, the interaction with the AF CoO cover shell can lead to the enhancement of cobalt magnetic moment in the transition layer*

This conclusion is consistent to the data on Co-Mn granular films, where the cobalt spectrum expands up to 270 MHz [35]. The system has the similar AF–FM transition layer.

Aging of the Sample. The effect of the sample aging has been observed in our experiments. As it is seen from the Fig. 12.4 a spectrum of the N3 sample dramatically changed in a few months [36].

The spectrum of the fresh sample was similar to the N2 sample spectrum. However the main line after six months disappears and the previous minor peaks become well pronounced. A strong aging effect is detected for the relaxation curves as well (Fig. 12.5).

A mechanism of the phenomenon may be preliminary attributed to a permanent oxidation of the metallic cobalt. According to the authors knowledge, a similar effect in NMR spectra was not detected before. Moreover, the aging effect was not detected earlier for the our samples during their characterization by TEM, adsorption curves, etc.

It is necessary to note that the effect observed may lead to a controversy in the explanation of the high-frequency lines (at least in principle). Namely

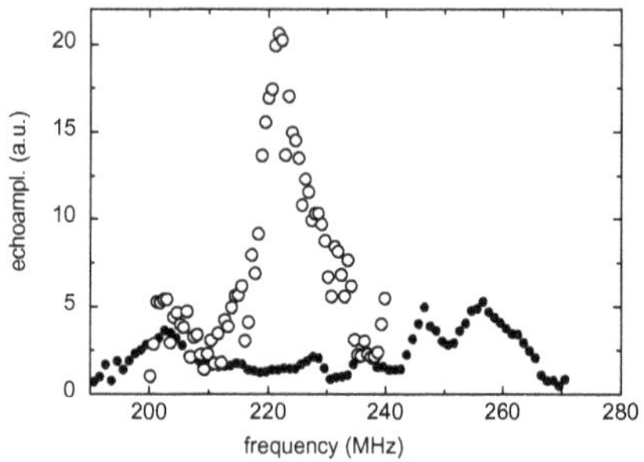

Fig. 12.4. Echo amplitude vs frequency for the N3 sample at different time: open circles – soon after preparation; bullets – about six months later

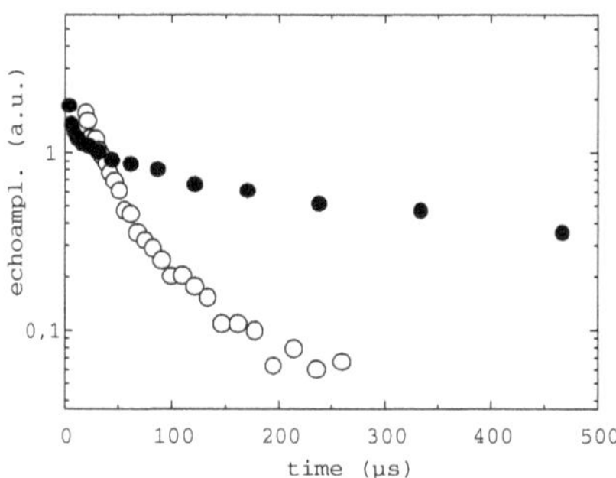

Fig. 12.5. Relaxation curves (T_2) for the N3 sample versus time of storage. Open circles – soon after preparation; bullets – about six months later

one can attribute the lines to any cobalt oxide. This explanation looks like very natural but it is very doubt in reality. CoO oxide is pure antiferromagnet and its zero field NMR spectrum consists of one low-intensity line at about 500 MHz [37], that is far away from range under investigation. The Co_2O_3 oxide does not exist in an unhydrous state [38]. As to antiferromagnetic Co_3O_4 [39,40] nobody observed NMR signal for it even in a bulk state (at least according to the authors knowledge).

Acknowledgements

The work has been partly supported by Russian Federation Program "Integration", project # A0142/01/8 and GCNS grant #E00-3.5-483. One of the authors (V.V. Matveev) acknowledges the support of the Saint-Petersburg – Hamburg Universities exchange program. We thank Frank Wiekhorst very much for NMR spectra recording and very grateful to Dr. G.P. Roskova and Dr. T.S. Tsekhomskaya for supplying porous glass matrices.

References

1. E.A. Turov, M.P. Petrov: *Nuclear Magnetic Resonance in Ferro- and Antiferromagnets* (Halstead Press/Wiley, New York 1972)
2. N.D. Mermin, H. Wagner: Phys. Rev. Lett. **17**, 1133 (1966)
3. R Allenspach: J. Magn. Magn. Materials **129**, 160 (1994)
4. J. Kohlhepp, H.J. Elmers et al.: Phys. Rev. B **45**, 12287 (1992)
5. W. Durr, M. Taborelli et al.: Phys. Rev. Lett. **62**, 206 (1989)
6. P.J. Garcia, A.D. Meinhaldt, A. Suna: Appl. Phys. Lett. **47**, 178 (1985); F.J.A. den Broeder, H.C. Doncersloot et al.: J. Appl. Phys. **61**, 4317 (1986)
7. M.N. Baibich, J.M. Broto et al.: Phys. Rev. Lett. **62**, 206 (1989)
8. M.J. Carey, F.T. Parker et al.: J. Appl. Phys. **69**, 5282 (1991)
9. S.Morup, G.Christiansen: J. Appl. Phys. **73**, 6955 (1993)
10. J. Tejada, L.I. Balcelles et al.: J. Appl. Phys. **73**, 6952 (1993)
11. D.P. Awschalom, J.F. Smyth et al. J. Appl. Phys. **73**, 6715 (1993)
12. K. Le Dang, P. Veillet et al.: Phys. Rev. B **41**, 12902 (1990)
13. H.A.M. de Gronckel, K. Kopinga et al.: Phys. Rev B **44**, 9100 (1991)
14. C. Meny, P. Panissod, R. Loloee: Phys. Rev. B **45**, 12269 (1992)
15. T. Thomson, P.C. Riedi, D. Greig: Phys. Rev B **50**, 10319 (1991)
16. C. Christides, S. Stavroyannis et al.: Phys. Rev. B **59**, 8812 (1999)
17. Gang Xiao, S.H. Liou et al.: Phys. Rev. B **34**, 7573 (1986)
18. T. Thomson, P.C. Riedi et al.: J. Appl. Phys. **81**, 5549 (1997)
19. V. Scarani, B. Doudin, J.-Ph. Ansermet: J. Magn. Magn. Mater.**205**, 241 (1999)
20. A.B. Altman, B.M. Lebed et al.: Sverkhprovodimost – Fizika, Himia, Tekhika **3**, 2205 (1990), in Russian
21. G.A. Nikolaychuk, S.M. Dunaevsky et al.: Russian Federation patent (1991), # 2030817
22. G.A. Nikolaychuk, S.V. Yakovlev et al.: Inorganic materials **34**, 718 (1998), pp. in Russian edition
23. X.H. Jin, L. Gao: J. Inorg. Mater. **16**, 200 (2001)
24. A.A. Eliseev, A.V. Lukashin et al.: Mat. Res. Innov. **3**, 308 (2000)
25. F. Wiekhorst, S. Skwirblies et al.: "^{59}Co NMR in Nanoparticles of Metallic Cobalt". In: *Abstracts of NATO ARW "Magnetic Resonance in Colloid and Interface Science" at St. Petersburg, Russia, June 26–30, 2001*, p. 65
26. F. Wiekhorst, J. Koetzler et al.: *to be published*
27. Z. Li, B. Gu: Phys. Rev. B **47**, 13611 (1993)
28. J.P. Bucher, D.C. Douglass et.al.: Phys. Rev. Let. **66**, 3052 (1991)
29. J.A.C. Bland, A.D. Johnson et. al.: JMMM **104–107**, 1798 (1992)
30. J.P. Chen, C.M. Sorensen et al.: J. Appl. Phys. **76**, 6316 (1994)
31. K. Yoshida, T. Takayama et. al.: J. Appl. Phys. **69**, 5184 (1991)
32. X.G. Li, A. Chiba et.al.: J. Appl. Phys. **83**, 3871 (1998)
33. W.H. Meiklejohn, C.P. Bean: Phys. Rev. **102**, 1413 (1956)
34. M. Gruyters, D. Riegel: Phys. Rev. B **63**, 052401 (2000)
35. K. Le Dang, P. Veillet et al.: J. Phys. F **16**, 93 (1986)
36. F. Wiekhorst: *private communication*
37. K. Okada, H. Yasuoka: J. Phys. Soc. Japan **43**, 34 (1977)
38. *Brief Chemical Encyclopedia*. ed. by I.L. Knunyants (Sovetskaya Entsiklopediya, Moscow, 1963) v.2, p. 619
39. *Tables of Physical Quantities*. ed. by I.K. Kikoin (Atomizdat, Moscow 1976) p. 620
40. W.Z. Roth: Bull. Amer. Phys. Soc. **8**, 213 (1963)

13 Influence of Nanocrystalline Ferrite Particles on Properties of Magnetic Systems

Robert Müller, Tobias Habisreuther, Robert Hiergeist, Hanna Steinmetz, Matthias Zeisberger, and Wolfgang Gawalek

Institut für Physikalische Hochtechnologie e.V., POB 100239, D-07702 Jena, Germany

Abstract. Nanocrystalline mainly superparamagnetic ferrite particles $\approx 10\,\text{nm}$ are used for the preparation of magnetic fluids. Barium hexaferrite $BaFe_{12-2x}Ti_xCo_xO_{19}$ powders with mean particle sizes $< 30\,\text{nm}$ show the transition to single domain Stoner-Wohlfarth behaviour. Hysteresis parameters, losses and the initial susceptibility versus temperature were obtained by VSM. Ba-ferrite ferrofluids have been prepared using Isopar M or dodecane as carrier liquid. Small Angle Neutron Scattering curves lead to a bimodal size distribution consisting of single magnetic particles and aggregated magnetic particles. Particle size investigations were done by TEM.

13.1 Introduction

In a stable dispersion of magnetic particles (ferrofluid) the rheological properties of the fluid can be influenced by magnetic fields or a heating due to magnetic losses is achieved. The particles have to be superparamagnetic to avoid agglomeration and sedimentation.

In the first section the preparation of nanocrystalline particles is described. We focus on the glass crystallization method (GCM) for Ba- hexaferrite preparation. The second section reveals investigations on Ba-ferrite ferrofluid (Ba-ferrite-FF).

13.2 Powder Preparation by Glass Crystallization Method (GCM)

Usual powder preparation methods supply in the case of ferrites particles with mean sizes above 50 nm, which results in a fast sedimentation of the particles in a liquid. Using a GCM, which was originally investigated for single domain Ba-ferrite particles $< 0.5\,\mu\text{m}$ for recording applications [1], we were able to prepare non-sintered single-crystalline Ba-ferrite particles with mean sizes $< 15\,\text{nm}$ [2]. From a melt, supersaturated with components of the desired powder material, an amorphous solid is prepared by quenching the melt with $\approx 10^4\,\text{K/s}$ to amorphous flakes. During a temperature treatment a phase separation and crystallization of the particles in a matrix of glass network former compounds occurs. The powder can be prepared by dissolving the soluble matrix (borate) by means of acetic acid. The powder particles are usually single crystalline because of the separated growth in the solid matrix. There is no sintering of particles.

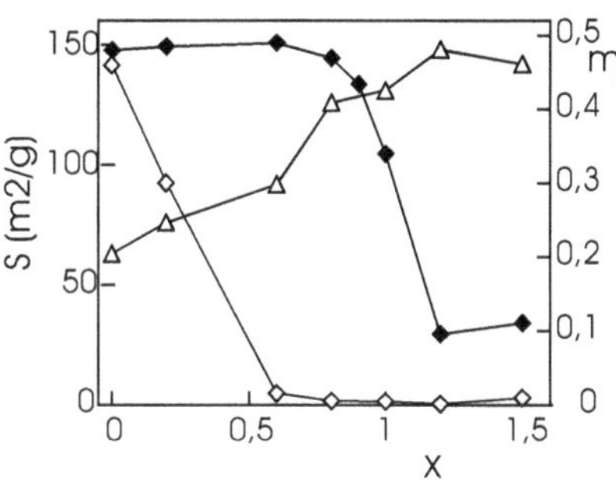

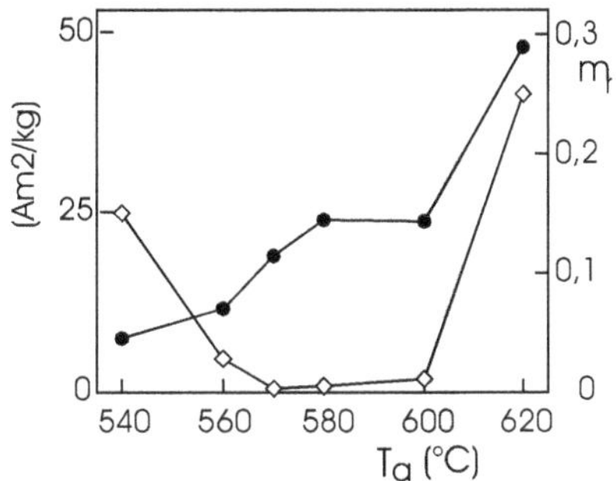

Fig. 13.1. Influence of the degree x of substitution (left) and the annealing temperature T_a (right) on specific saturation magnetization σ_∞ (circle) and relative remanence m_r (square) of substituted ferrite (x=0.8) and specific surface S (triangle). Black squares indicate remanence values of non-superparamagnetic powders (T_a=780°C)

The phase composition (formation of hexagonal ferrite or spinel ferrite) and the particle size can be influenced by the melt composition (e.g. use of nucleation agents) as well as temperature treatment (temperature, time). For the preparation of nanocrystalline $BaFe_{12-2x}Ti_xCo_xO_{19}$ particles (x=0.8) glass flakes from the system $40BaO\text{-}33B_2O_3\text{-}27(Fe_2O_3\text{+}CoO\text{+}TiO_2)$ (mol %) were heat treated between 540 and 620°C for long times (h or d). The influence of the annealing temperature T_a on the magnetic properties is shown in Fig. 13.1 for a substituted ferrite with x=0.8 (annealing time: 10 d). The curves reveal an increasing specific magnetization σ_∞ due to hexaferrite formation and an increasing relative remanence m_r at T_a¿570°C due to particle growth.

The mean particle size is ≈ 10 nm at T_a= 580°C (estimated from specific surface by BET-method). The samples are nearly superparamagnetical ($H_c \approx 0$, m_r ≈0) at T_a=570-600°C.

Substituting Fe by Co-Ti leads to smaller particle sizes. Ti has an effect as nucleation agent. The Co-Ti incorporation in microcrystalline, non-superparamagnetic single domain ferrite particles leads to a decrease of the magnetocrystalline anisotropy [3] resulting in the strong decrease of m_r at x≈1.0 (Fig. 13.1) caused by the transition from uniaxial (m_r=0.5) to planar anisotropy (m_r <0.5 [4]). In the case of nanoparticles this effect is superimposed by superparamagnetism in the range of substitution x = 0.6 – 1.0 due to the small particle size.

13.3 Magnetic Ba-ferrite Particles in Ferrofluids

13.3.1 Motivation for Ba-ferrite Containing Ferrofluids

With decreasing size magnetic particles show the transition from stable single-domain behaviour to superparamagnetism (thermally activated magnetization reversal). The transition strongly depends on temperature and magnetocrystalline anisotro-py. Barium hexaferrite $BaFe_{12-2x}Ti_xCo_xO_{19}$ has an intrinsic anisotropy one order of magnitude higher than that of magnetite, the usual material for commercial ferrofluids (FF). For magnetite based fluids the anisotropy barrier – which determines the Néel relaxation – is not defined by the particle volume only. Small deviations from spherical particle shape will induce an additional shape anisotropy. In contrast, for Ba-ferrite FFs the influence of shape anisotropy is less severe. Néel relaxation is mainly defined by the size distribution of the core particles. It is expected, that Néel relaxation is significantly suppressed by the high intrinsic anisotropy. Magneto-rheological effects will be much more significant in the case of Ba-ferrite FFs. Numerical calculations [5] [6] show for individual spherical particles a much higher specific viscosity gain for Ba-ferrite FFs compared to magnetite FFs for the same particle volume, but the difference in the specific viscosity gain can be reduced due to shape effects or chain formation.

13.3.2 Ferrofluid Preparation

Ba-ferrite ferrofluids have been prepared using oleic acid as surfactant and Isopar M or dodecane as carrier liquid [6]. The ferrite powder (x=0.8) was prepared at 580°C. K-oleate was added to the ferrite to adhere at the particle surface. The oleate was converted to oleic acid by adding mineral acid. All the mixing steps were done in a ball mill. The substance was washed (removing of water, salts and surplus of oleic acid), the coated particles were separated and carrier liquid was added. Aggregates in the FF were reduced by centrifugation at 5000 g. The supernatant liquid is the initial ferrofluid, which usually will be diluted.

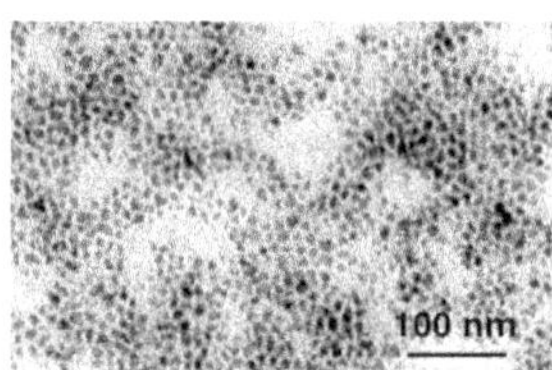

Fig. 13.2. TEM picture of a Ba-ferrite ferrofluid

A TEM picture of a concentrated suspension (Fig. 13.2) shows isolated particles and only small agglomerates. From this picture a particle size distribution with a mean particle size in the FF of about 8 nm could be determined.

13.3.3 Magnetic and Structure Investigations

The initial susceptibility (magnetic moment at H=4 kA/m measured by vibrating sample magnetometry (VSM)) vs. temperature after zero field cooling shows a maximum typical for small particle systems in the transition region between superpara- and stable ferrimagnetic behaviour (Fig. 13.3) [7]; the irregularity (see arrow) is caused by melting of the carrier liquid. From the initial susceptibility in the superparamagnetic state we calculated a mean magnetic diameter

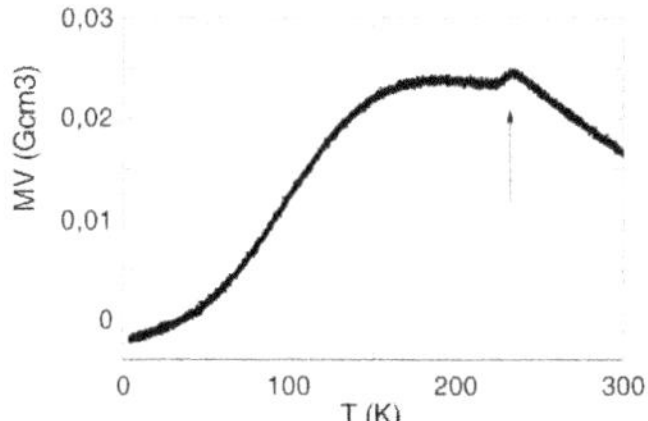

Fig. 13.3. Initial susceptibility (magnetic moment) vs. temperature after ZFC measured at H=4 kA/m

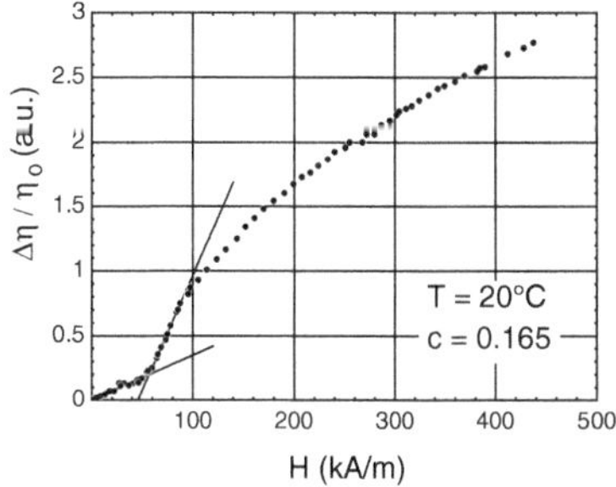

Fig. 13.4. $\Delta\eta/\eta_0$ vs. field stength of a concentrated fluid. Note the transition from linear increase at low fields (a) towards an linear behaviour at medium field strength (b)

of the particles of 4.7±0.5 nm (FF in Fig. 13.2). Measurements of the rheological properties were performed with a rheometer of concentric cylinder double gap geometry [6]. Results of field dependent measurements of a concentrated fluid with magnetic field perpendicular to the hydrodynamic vortex-axis are shown in Fig. 13.4. Measurements of dilute fluids (c = 0.013) reveal even higher values up to $\Delta\eta/\eta_0 = 7$ at the field, which cannot be explained by the model mentioned above. Collective phenomena as e.g. the formation of particle chains with increasing field have to be considered in oder to explain these results.

An analysis of the size distribution of the magnetic cores of the ferrite particles was done by magnetic granulometry as described in [8] by calculating the superparamagnetic volume fraction at different temperatures according to

$$V_{core} = \ln(\tau_o/\tau)\frac{k_B T}{K(T)}. \tag{13.1}$$

Stoner-Wohlfarth behaviour was assumed for non-superparamagnetic particles. For estimation of the anisotropy $K(T)$ and τ_o see [9]. Deviations of an effective K is due to differences in the particle shape (coarse particles are platelets) and surface anisotropy of the nanocrystalline powder. Magnetic granulometry data are shown in Fig. 13.5(a). The first value (for $d_{core} < 1.63$ nm) is uncertain because (i) the temperature range below 4.2 K is experimentally not accessible and therefore assumptions for the 0 K-values have to be made. (ii) The relative remanence M_r/M_s is lower than 1/2 (= 0.41) even at 5 K, in contrast to the Stoner-Wohlfarth theory.

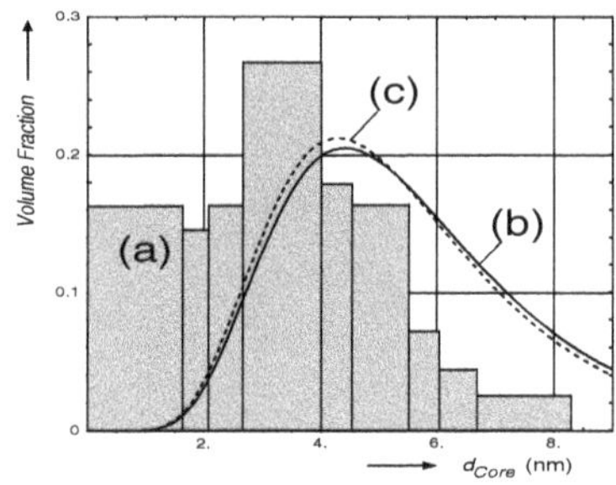

Fig. 13.5. Core size disribution of Ba-ferrite nanopowder obtained by magnetic granulometry (a) compared with the lognormal distribution calculated from magnetization data for a FF from the same batch of powder (b) and powder suspended in gel (c).

Reconstruction of the particle core size distribution $f(d_{core})$ from magnetization data $M_{fluid}(H)$ at room temperature was as well done by solving the Langevin integral equation

$$M_{fluid}(H) = M_{s\ fluid} \cdot \int_0^{\infty} f(d_{core}) \cdot L\left(\mu_0 \frac{M_{s\ bulk} \cdot \pi \cdot d_{core}^3}{6 \cdot k_B T} \cdot H\right) d\, d_{core} \tag{13.2}$$

with $L(\alpha) = \coth\alpha - 1/\alpha$ under the assumption of a lognormal core size distribution

$$f(d_{core}) = \frac{1}{\sigma\sqrt{2\pi}d_{core}} \cdot \exp\left[-\frac{(\ln d_{core} - \ln d_1)^2}{2 \cdot \sigma^2}\right] \tag{13.3}$$

(cf. reference [10]). Note that the quantities σ and d_1 in Eq.(13.3) are just parameters of the lognormal core size distribution and are not identical with the mean particle core diameter and the standard deviation of the particle core size distribution respectively. Results of this reconstruction are also given in Fig. 13.5 for a fluid prepared from the same batch of Ba-ferrite nanopowder (b) and for the nanopowder suspended in gel (c). Since there is no significant difference between these two distributions, we conclude that a significant loss of larger particles due to sedimentation during the FF preparation does not occur.

Real (χ') and imaginary part (χ'') of the ac-susceptibility were measured in dependence on frequency of the oscillating magnetic field in a frequency range between 20 Hz and 100 kHz. Fig. 13.6 shows specific ac-susceptibility data

$$\tilde{\chi} = \chi \cdot M_{S\ bulk}/M_{S\ fluid} \tag{13.4}$$

i.e. ac-susceptibility with respect to the volume concentration of magnetic material in the fluid – of a water based FF. (The fluid was prepared from Ba-ferrite nanopowder by N. Buske, Berlin Heart AG, Wiesenweg 10, D-12247 Berlin, Germany.)

A linear decrease of (χ') in dependence on log f could be observed experimentally for the most ferrofluid samples. According to Kramers-Kronig relations, it can be shown that from this linear dependence a frequency independent value of (χ') can be derived [10] as can be seen in Fig. 13.6. A nearly constant value

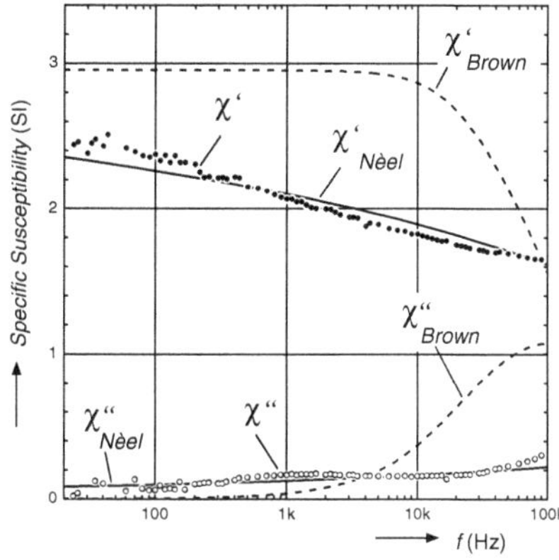

Fig. 13.6. Real and imaginary part of the specific ac-suceptibility of a water based Ba-ferrite FF in dependence on frequency measured at room temperature with an amplitude of the oscillating magnetic field of 98.6 A/m.

of (χ'') over a broad frequency range can be explained by a very broad distribution of relaxation times τ which can be due to either Brownian or Néelian relaxation mechanism [11]. In the case of the water based ferrofluid discussed here, we obtained from the reconstruction of the lognormal core size distribution from magnetization data a mean particle core diameter of $\langle d_{core} \rangle = 7.1\,\text{nm}$ and a broad distribution with $FWHM = 6.6\,\text{nm}$.

By use of the lognormal size distribution data the specific ac-susceptibility spectra according to

$$\tilde{\chi} = \int_0^\infty f(d_{core}) \cdot \mu_0 \cdot \frac{M_{s\ bulk}^2}{3 \cdot k_B T} \cdot \frac{\pi \cdot d_{core}^3}{6} \cdot \frac{1}{1 + i\omega\tau} d\, d_{core} \tag{13.5}$$

for the case of Néel relaxation with

$$\tau_{Neel} = \tau_0 \cdot \exp\left(\frac{K \cdot \pi d_{core}^3/6}{k_B T}\right) \tag{13.6}$$

and for the case of Brownian relaxation

$$\tau_{Brown} = \frac{3\pi(2.5 \cdot d_{core} + 1nm)^3 \cdot \eta}{6 \cdot k_B T} \tag{13.7}$$

(here a dead layer of 1.5 times the magnetic particle core diameter and in addition a surfactant layer of 0.5 nm thickness was taken into account) were calculated by numerical integration over the particle core diameter range from 0 up to 15 nm. Within this range 97.5 % of all particles of the lognormal distribution are taken into account. Because the concentration of Ba ferrite in the fluid was only 0.72 vol. % – i.e. the fluid was very dilute – we used in our calculations for the viscosity of the fluid $\eta = 1.01 \cdot 10^{-3}\,\text{kg/ms}$ (i.e. the viscosity of water). Temperature was set to $T = 295\,\text{K}$ and τ_o in Eq.(13.6) to $10^{-9}\,\text{s}$. For the anisotropy constant we used the value $K = 40 \cdot 10^3\,\text{J/m}^3$ which provides the best fit to the experimental data in Fig. 13.6. This value is lower then the value for bulk material of the same composition ($K = 70 \cdot 10^3\,\mathrm{J/m^3}$ [12]). This is reasonable since the crystal lattice of those small particles is expected to be distorted leading to a reduced anisotropy.

The results of both calculations are shown together with the experimental data in Fig. 13.6, for the case of Néel mechanism as full lines and as dashed lines for the case of Brownian relaxation. In contrast to the calculated data for the Brownian mechanism, we observe a very good agreement of the experimental data with calculated data for the Néelian mechanism. This is due to the small particle core size: The transition towards blocking is expected to occur for particles with core diameter larger than 10.8 nm (for $K = 40 \cdot 10^3\,\mathrm{J/m^3}$ [i.e. $\tau_{Neel} = \tau_{Brown}$ according to Eqs.(13.6) and (13.7)]). Because of the small mean particle core diameter of 7.1 nm most of the particles are not blocked in spite of the high anisotropy of the Ba-ferrite.

Ferrofluids with a partly deuterated carrier liquid ($C_{12}H_{26}$ and $C_{12}D_{26}$) and a particle content between 1 and 4 vol. % were investigated in the fluid state by SANS. SANS allows to distinguish between single magnetic particles and agglomerates in the range up to about 50 nm. The deuteration causes the scattering contrast between the organic surfactant and the carrier liquid. SANS curves lead to a bimodal size distribution (Fig. 13.7) consisting of single magnetic particles (mean diameter 8 nm corresponding with the TEM Fig. 13.2) with an organic shell and agglomerates (mean diameter about 22 nm) consisting of ferrite particles with an incomplete organic layer. Further details on SANS investigations are given in [13].

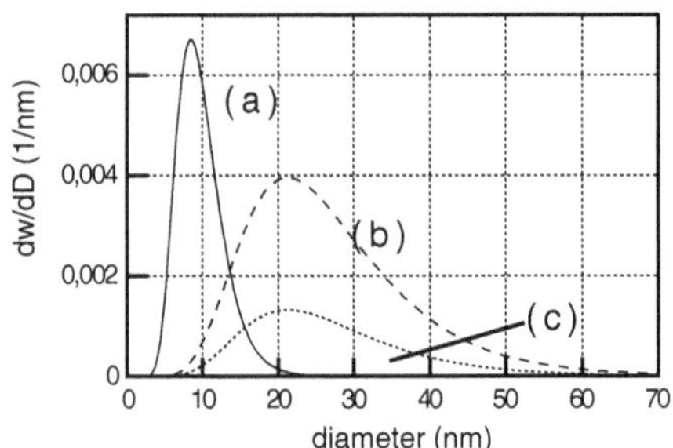

Fig. 13.7. Relative volume size distribution of a FF by SANS of single particles without organic layer (a) and agglomerates including organic material (b). (c) shows the magnetic particle fraction in the agglomerates estimated from the scattering length density.

Acknowledgement

The authors thank Dr. Th. Höche (FSU Jena) for the TEM investigations. This work was supported by DFG Ga 622/2-1.

References

1. Toshiba Review No. 154, Winter, 1985, 5-31
2. R. Müller et al. Proceed. ELECTROCERAMICS IV 1994 Aachen, vol. II, 1183
3. E.C. Stoner and E.P. Wohlfarth, Phil. Trans. Roy. Soc. 240A (1948) 599
4. Th. Klupsch, Phys. Stat. Sol. 113 (1989) 561-572
5. A. Cebers, Magnetohydrodyn. 11 (4) (1975) 439
6. R. Müller, et al. J. Magn. Magn. Mater. 201 (1999) 34
7. H. Pfeiffer, R. W. Chantrell, J. Magn. Magn. Mater. 120 (1993) 203
8. E. Kneller, "Ferromagnetismus", Springer, Berlin, 1962, 427
9. R. Müller et al., J. Magn. Magn. Mater. accepted
10. E. Pytte, Y. Imry, Phys. Rev. B 35 (1987) 1465
11. Yu. L. Raikher, A.F. Pshenichnikov JETP Lett., 41 (1985) 132
12. M. Kishimoto, Shin-ichi Kitahata, IEEE Trans. Magn. 25 (1989) 4063
13. A. Hoell et al., J. Magn. Magn. Mater. accepted

14 Phase Stability, Structural and Magnetic Properties of Some CMR Manganites

R. Ganguly[1], I.K. Gopalakrishnan[1], V. Siruguri[2], and J.V. Yakhmi[3]

[1] Novel Materials and Structural Chemistry Division,
Bhabha Atomic Research Centre, Mumbai – 400 085, India
[2] Inter University Consortium for DAE Facilities – Mumbai Centre,
Bhabha Atomic Research Centre, Mumbai – 400 085, India
[3] Technical Physics and Prototype Engineering Division,
Bhabha Atomic Research Centre, Mumbai – 400 085, India

Abstract. Discoveries of colossal magneto-resistance (CMR) in the perovskite manganites, $Ln_{1-x}A_xMnO_3$ (Ln = La, rare earth; A = divalent ions), as well as in the layered manganites, $(La,Sr)_3Mn_2O_7$, has created a renewed interest in them. Their physical characteristics point to the existence of a fine interplay between spin, charge and lattice interactions. Substitutional studies, particularly at the rare earth and Mn sites in these compounds, can throw light on important crystal chemical aspects. Through systematic site-selective substitutions we have tried to evaluate the phase stability of some of these layered and perovskite manganites. In particular, our study on the layered manganites, $La_{1.2}(Sr_{1-x}Ca_x)_{1.8}Mn_2O_7$, and $Ca_{3-x}La_xMn_2O_7$, establishes that the compositions $Ca_{3-x}La_xMn_2O_7$ ($1.2 \leq x \leq 2.3$) do not form layered compounds with $Sr_3Ti_2O_7$ - structure, in contrast to literature reports. Instead, they form multiphase mixtures with $La_{1-x}Ca_xMnO_3$ as the majority phases which, in fact, determine their magnetic and transport behavior. By careful structural and magnetic studies on $La_{0.7}Ce_{0.3}MnO_3$ and $La_{0.7}MnO_3$ we have also proved that the compositions $La_{1-x}Ce_xMnO_3$ do not form electron-doped manganites, but actually yield multiphase mixtures comprising $La_{1-x}MnO_3$ and CeO_2. Lastly, we demonstrate that both Co and Cr as dopant ions destabilize the double-exchange mechanism driven ferromagnetism and metallic characters in $La_{0.67}Ca_{0.33}Mn_{1-x}Co_xO_3$ and $La_{0.67}Ca_{0.33}Mn_{1-y}Cr_yO_3$, and highlight the differences in the nature of the magnetic exchange interaction of Mn ions with Co, and with Cr ions in these hole-doped perovskite manganites.

14.1 Introduction

Discoveries of high temperature superconductivity in the layered cuprates [1] and colossal magneto-resistance in the perovskite manganites [2,3] have created a tremendous interest in them. The hole-doped manganites draw special attention as they show a rich variety of magnetic and electrical properties due to the presence of a fine interplay between spin, charge and lattice interactions in them. Site-selective substitutions can modify such interactions and have, therefore, been carried out in detail in the case of the manganites $R_{1-x}A_xMnO_3$ (R = Rare Earth, A = divalent ion) [4–23]. These compounds have perovskite structure and can have varying amount of lattice distortions, determined by the value of the tolerance factor, t. The tolerance factor is defined as $R_{(R-O)}/\sqrt{2}R_{(Mn-O)}$, where $R_{(R-O)}$ and $R_{(Mn-O)}$ stand for the rare earth-oxygen and manganese-oxygen bond distances, respectively. A decrease of t, from its ideal value of $t = 1$, leads to a decrease in the Mn-O-Mn bond angle from 180°. As magnetic and electrical properties of these compounds depend critically on the Mn-O-Mn bond angle, substitution of ions at the rare earth site, which can modify the value of the tolerance factor, has significant influence on those physical properties [4–6]. The compounds with $t \approx 1$ exhibit simultaneous onset of ferromagnetism and metallic character mediated by the double exchange (DE) process [24]. The e_g electrons of the Mn^{3+} ions which are coupled strongly to the spin of the t_{2g} electrons by Hund's coupling, undergo hopping between Mn^{3+} ($t_{2g}^3 e_g^1$) and Mn^{4+} ($t_{2g}^3 e_g^0$) ions, with their spins remaining parallel to the core spin, and mediate ferromagnetism. A decrease in the one-electron bandwidth and the electron transfer probability of the e_g electrons, as a result of a decrease in the value of t and consequently the Mn-O-Mn bond angle, leads to suppression of DE interaction. This in turn suppresses ferromagnetism and insulator to metal (I-M) transition, and leads to the stabilization of other competing interactions like charge-ordering and antiferromagnetism [4–6].

After the discovery of CMR in the hole-doped layered manganites, $(La,Sr)_3$-Mn_2O_7 [25], extensive studies have been carried out in order to understand the influence of the exchange interaction between Mn ions on the magnetic and transport properties of these compounds. Due to a reduced dimensionality of the DE interaction in the layered structure, these compounds show lower values of ferromagnetic transition temperature (T_C) than the corresponding perovskite manganites, $(La,Sr)MnO_3$. The maximum value of T_C exhibited, for instance, is 124 K by the compound, $La_{1.2}Sr_{1.8}Mn_2O_7$ [25], which is much lower than the maximum value reported for the $(La,Sr)MnO_3$ compounds, namely 380 K [5]. In both cases, of course, a decrease in the average size of the substituent ions at the La-site results in a decrease in the ferromagnetic transition temperature as a result of a decrease in the one-electron bandwidth of the e_g electrons. It has been reported that in the single phase region of $La_{1.2}Sr_{1.8-x}Ca_xMn_2O_7$, i.e. in the composition range $0.0 \leq x \leq 0.8$, the values of T_C and the insulator-to-metal transition temperature (T_P) decrease monotonically with an increase in the concentration of Ca [26]. Similar is the case for the $La_{1.4}(Sr_{1-x}Ca_x)_{1.6}Mn_2O_7$ [27] system, for which the T_C for the parent compound (x=0) is 100 K. However, some reports in the literature, claiming that the samples of nominal compositions, $Ca_{3-y}La_yMn_2O_7$ ($1.0 \leq y \leq 2.3$), form with layered $Sr_3Ti_2O_7$ structure, and show ferromagnetic transitions in the temperature range 160 K to 240 K [28–32], are in contravention with this trend and are, therefore, intriguing. Through systematic studies on $La_{1.2}Sr_{1.8-x}Ca_xMn_2O_7$ and $Ca_{3-y}La_yMn_2O_7$ ($0.0 \leq y \leq 2.3$) we have made efforts to check whether these reports are correct.

Quite similarly to the influence of doping holes in $LaMnO_3$, double exchange driven I-M transitions and CMR behavior can also be observed through electron doping in $LaMnO_3$ and in $CaMnO_3$ [33–36]. The $Ca_{1-x}Ln_xMnO_3$ compounds show I-M transitions with smaller range of 'x' values and with smaller average size of dopant ions at the rare earth site (r_A), than the hole-doped perovskite manganites $La_{1-x}A_xMnO_3$. The nature of carriers in these compounds are electrons, which confirms the electron-doped state in them. It may be mentioned, however, that the nature of carriers in the other electron-doped compounds, $La_{1-x}Ce_xMnO_3$ [37–42], have been found to be holes. Our studies have shown that the samples $La_{1-x}Ce_xMnO_3$ actually form multiphase mixtures, comprising La-deficient $LaMnO_3$ and CeO_2, the magnetic and transport properties being determined by the La-deficient $LaMnO_3$ phases.

Substitution of Cr, Co, Ni and Ru ions in the charge-ordered insulating compounds $R_{0.5}Ca_{0.5}MnO_3$ (R = Rare Earth) destroy charge ordering, and induce ferromagnetic as well as and insulator to metal (I-M) transitions [7–14]. It has been suggested that substitution of these ions results in coexistence of ferromagnetism with short-range charge ordering, the ferromagnetic phase being the majority phase [13,14]. A completely different behavior i.e. suppression of ferromagnetic and metallic character is observed when Ti, Fe, In, Co and Ru [15–22] are substituted for Mn in ferromagnetic $La_{1-x}Ca_xMnO_3$, indicating that these ions do not take part in the DE process. Gundakaram et al [43] have observed that the compounds $LaMn_{1-x}Cr_xO_3$ (x=0.2 and 0.3) show ferromagnetic behavior but do not show any I-M transition near the T_C. It has been suggested that Cr^{3+} (d^3) ion, which is iso-electronic with Mn^{4+} ion, does not take part in the DE process and ferromagnetism in these compounds appears due to strong super-exchange interaction between the Cr^{3+} and the Mn^{3+} ions. We have carried out systematic studies on the magnetic and transport properties of the two series of compounds $La_{0.67}Ca_{0.33}Mn_{1-x}M_xO_3$ (M = Co, Cr). For both Co and Cr substitutions, the double exchange driven ferromagnetic and I-M transitions are destabilized leading to a decrease in T_c and the disappearance of the I-M transition. However, due to the stronger ferromagnetic exchange interaction between the Cr^{3+} and Mn^{3+} ions, the ferromagnetic character could be observed in Cr-substituted compounds for higher doping level (50 %) than in the case of the Co-substituted compounds (30 %). The difference in the nature of the ferromagnetic state of the Co-substituted and Cr-substituted compounds has been explained on the basis of anisotropic nature of ferromagnetic exchange interaction induced by the substitution of Co.

14.2 Experimental

The samples with nominal compositions $La_{1.2}(Sr_{1-x}Ca_x)_{1.8}Mn_2O_7$, $Ca_{3-y}La_yMn_2$ $La_{0.7}Ce_{0.3}MnO_3$, $La_{0.7}MnO_3$, $La_{0.67}Ca_{0.33}Mn_{1-x}Co_xO_3$ and $La_{0.67}Ca_{0.33}Mn_{1-y}Cr$ were prepared by the solid-state reaction between stoichiometric mixtures of respective oxids, carbonates or nitrates. Room temperature powder x-ray diffraction (XRD) patterns of the samples were recorded using either a 12 kW rotating anode based Rigaku powder diffractometer or a Phillips (model PW 1820) diffractometer with Cu-K_α radiation and a graphite monochromator. The lattice parameters of the compounds were calculated and refined by least square fitting of the XRD data. Room temperature powder neutron diffraction (ND) patterns were collected at a neutron wavelength of 1.278 Å using the position-sensitive detector based Hi-Q diffractometer at Dhruva reactor, Trombay. A Cu (111) monochromator was used which gave a second-order contribution with an intensity ratio of 0.004. Rietveld refinement was carried out using the program FULLPROF. The parameters to be refined were first varied individually, and then in groups and care was taken to avoid refinement of highly correlated parameters at the same time. DC electrical resistance measurement was done by the standard four-probe method using a Datron (model 1071) multimeter. AC susceptibility data were obtained using an APD CCR with Meissner coil assembly in conjunction with an EG & G model 5208 lock-in amplifier. Measurements were done at 80 Hz and 0.5 Oe. DC magnetization vs. temperature data both under zero field cooled (ZFC) and field cooled (FC) conditions were taken using PAR EG & G (model 4500) vibrating sample magnetometer. The ZFC data were taken by first cooling the sample below T_C to required temperature and then applying the field while the FC data were taken by cooling the sample from above T_C at the same field. For the magnetization measurement as a function of field the samples were first cooled to the temperature of measurement in zero field, the magnetic field was then applied and hysteresis loop generated.

14.3 Results and Discussions

14.3.1 $La_{1.2}(Sr_{1-x}Ca_x)_{1.8}Mn_2O_7$ and $Ca_{3-y}La_yMn_2O_7$

The x-ray diffraction patterns of the samples $La_{1.2}(Sr_{1-x}Ca_x)_{1.8}Mn_2O_7$ (x=0.00, 0.25, 0.50, 0.65 and 1.00) are shown in the Fig. 14.1. The neutron diffraction patterns recorded on some of these compositions are also shown in Fig. 14.2. The XRD patterns confirm that the layered structure of the parent compound is maintained only up to x=0.50, because a noticeable change occurs in the pattern at $x \geq 0.65$ indicating a change of phase as one goes from x=0.50 to 0.65. A drastic change occurs in the ac-susceptibility behavior, too, when one goes from x=0.50 to x=0.65 as shown in the Fig. 14.3, which supports the conjecture that a phase change occurs in this composition window. The ferromagnetic transition temperature, T_C, as well as the size of the ac-susceptibility signal decreases with an increase in 'x' up to x=0.50. But for x=0.65, and for x=0.80, a strong ferromagnetic signal, with a probable ferromagnetic transition above room temperature, appears. For x=1.0, a sharp signal corresponding to a ferromagnetic transition is observed at about 250 K. Since the value of T_C for x=1.0 is very close to those of the hole-doped perovskite manganites, $La_{1-x}Ca_xMnO_3$ ($0.3 \leq x \leq 0.4$), we recognized a possibility that this sample may be a multiphase mixture with hole-doped perovskite manganite as a major constituent. The Rietveld refinement of the XRD pattern for this sample (x=1.0), conducted assuming the sample as a mixture of $La_{0.6}Ca_{0.4}MnO_3$ and CaO, gives an excellent fit to the observed data. The neutron diffraction patterns, too, exhibit a change in the nature of the samples with an increase in 'x', from a predominantly layered phase to a predominantly perovskite phase (Fig. 14.2). Thus, we can conclude that the sample with the composition x=1.0 does not form with the layered $Sr_3Ti_2O_7$-structure. We tend to believe that the magnetic and transport properties reported in literature [30,31] for this composition actually arise from the presence of hole-doped perovskite phases.

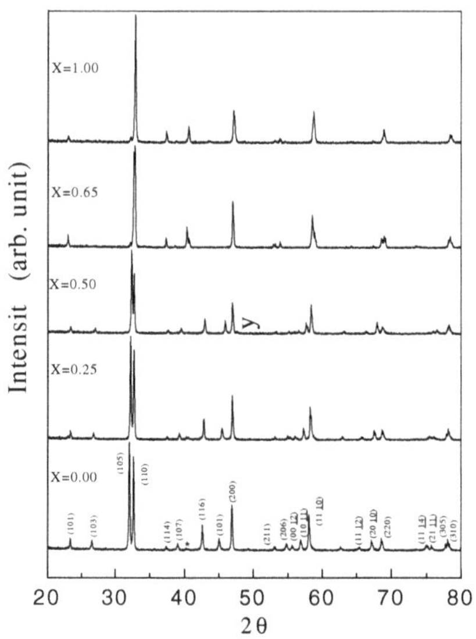

Fig. 14.1. X-ray diffraction patterns for the series $La_{1.2}(Sr_{1-x}Ca_x)_{1.8}Mn_2O_7$.

A number of reports based on neutron diffraction studies [44–47] have indicated that the sample with composition x=0.00, i.e., $La_{1.2}Sr_{1.8}Mn_2O_7$ forms a single-phase compound with layered $Sr_3Ti_2O_7$-type structure (***I4/mmm*** space group). But, none of these reports have presented the observed neutron diffraction pattern for this composition, i.e. x=0.0. Repeated attempts made by us to synthesize a single-phase compound with this structure were not successful and there always were impurity phases present which could be identified to be perovskite (La,Sr)MnO_3 phases (***Pbnm*** space group). The neutron diffraction pattern recorded by us on this composition (x=0.0) was similar to neutron patterns reported by Moritomo *et al* [48] and Chi *et al* [49] for the compounds $(La_{0.9}Nd_{0.1})_{1.4}Sr_{1.6}Mn_2O_7$ and $La_{1.4}Sr_{1.6}Mn_2O_7$, respectively, both of which had a co-existence of the layered and the perovskite phases. The phase-purity of the layered compounds is extremely sensitive to the synthesis conditions. We believe that the existence of the perovskite phases in $La_{1.2}Sr_{1.8}Mn_2O_7$ is not noticeable in XRD pattern for two reasons. Firstly, these phases are present with low volume fraction, and secondly, their most intense peaks, (020, 200), overlap very closely with the characteristic (105, 110) peaks of the layered phases.

X-ray diffraction patterns of the samples with nominal compositions $Ca_{3-y}La_yMn_2O_7$ are shown for y=0.0, 0.2, 0.4 and 0.8, in Fig. 14.4 and for y=0.8, 1.2, 1.5 and 2.3 in Fig. 14.5. No extra peaks indicative of the presence

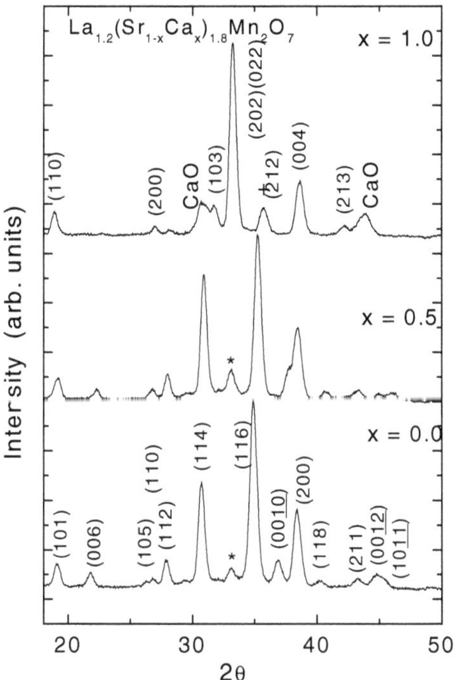

Fig. 14.2. Neutron diffraction patterns for the series $La_{1.2}(Sr_{1-x}Ca_x)_{1.8}Mn_2O_7$.

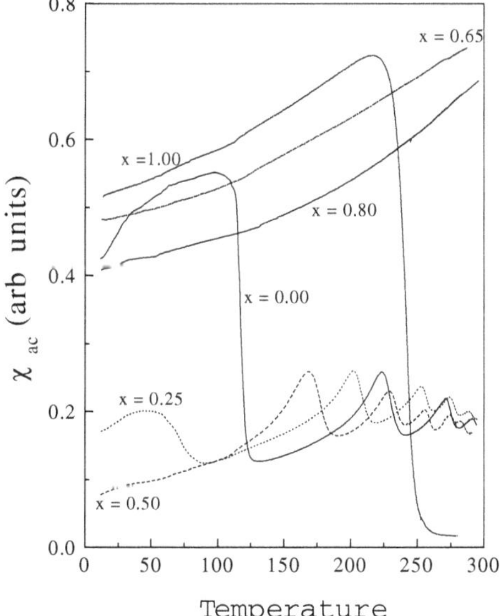

Fig. 14.3. AC-susceptibility versus temperature for $La_{1.2}(Sr_{1-x}Ca_x)_{1.8}Mn_2O_7$.

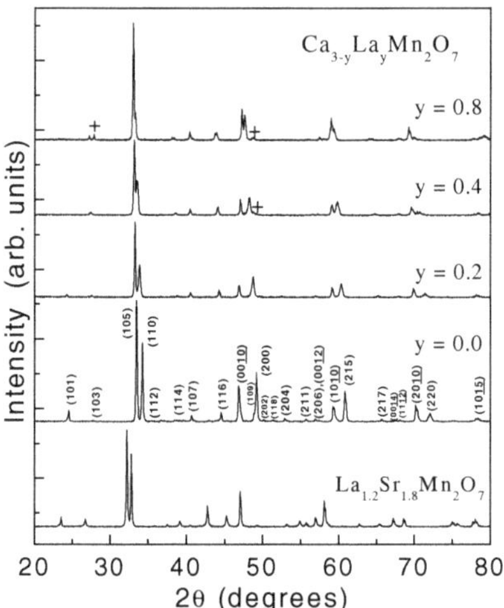

Fig. 14.4. X-ray diffraction patterns for $Ca_{3-y}La_yMn_2O_7$ (y=0.0, 0.2, 0.4 and 0.8).

of any phase other than the layered phase are observed for y=0.0 and y=0.2 (Fig. 14.4). Appearance of extra peaks in the case of samples with y=0.4 and 0.8 (marked as "+") indicates the presence of a second phase, the layered phase still being the majority phase (Fig. 14.4). A significant change is observed in the nature of the XRD pattern at y=1.2 (as highlighted in Fig.14.5 with dotted lines), indicating that this sample forms a multiphase mixture with perovskite manganite as the majority phases. In other words, as one proceeds from y=0.8 to y=1.2, the nature of the product changes from a composition with layered phases as the majority phases to a composition with perovskite phases as the majority phases. The sample with y=1.5 shows similar behavior to that of the sample with y=1.2. The minor phase in these two samples, which could be detected from their XRD patterns, is CaO, as indicated in the Fig. 14.5. In the case of the y=2.3, although the perovskite phases remain as the majority phases, the minor phase which could be detected is not CaO but La_2O_3, as shown by asterisks in Fig. 14.5.

The neutron diffraction patterns of all samples belonging to the series $Ca_{3-y}La_yMn_2O_7$ are shown in Fig. 14.6 and Fig. 14.7. The samples with y=0.0 and 0.2 are single-phase compounds with layered structure without the presence of any impurity phases (Fig. 14.6). The perovskite phase, as the second phase, starts appearing as minority phase at y=0.4 and its fraction increases with increase in 'y' up to y=0.8 (marked with asterisk in Fig. 14.6). At y=1.2, the perovskite becomes majority phase with CaO and layered phases as the minority phases (Fig. 14.7). These neutron patterns thus essentially support the conclusions made by us from the x-ray diffraction patterns (Fig. 14.4 and 14.5), with even more clarity. For instance, the impurity phase indicated in the XRD

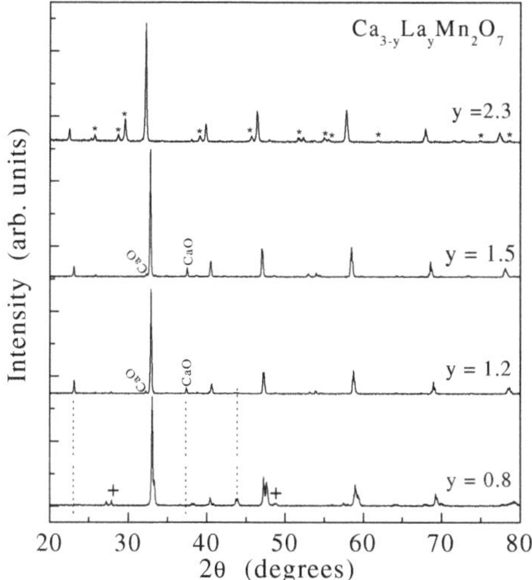

Fig. 14.5. X-ray diffraction patterns for $Ca_{3-y}La_yMn_2O_7$ (y=0.8, 1.2, 1.5 and 2.3).

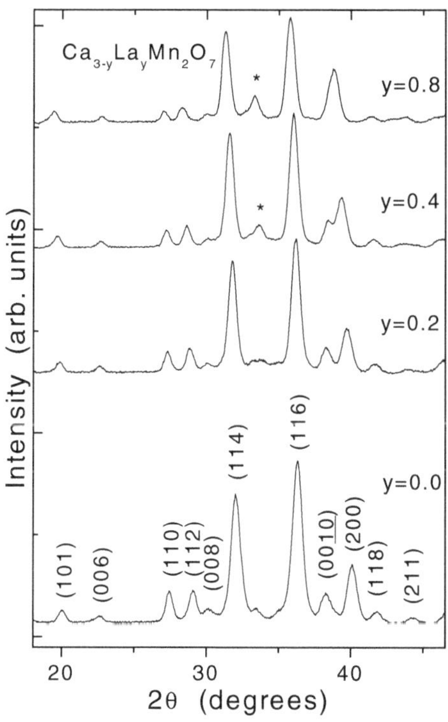

Fig. 14.6. Neutron diffraction patterns for $Ca_{3-y}La_yMn_2O_7$ (y=0.0, 0.2, 0.4 and 0.8).

patterns of the samples with y=0.4 and 0.8 (Fig. 14.4) can be unambigously recognized from their ND patterns as the perovskite phase. Similarly, the change in the nature of the product at y=1.2, from a composition with predominantly layered phase to a composition with predominantly perovskite phase can be appreciated clearly from the neutron diffraction patterns. This is because the most intense reflections of the layered phase are well separated from the most intense reflections of the perovskite phase in the neutron diffraction patterns. For both y=1.5 and 2.3, the layered phase is not observable within the resolution limits of our instrument, and the minor phases in these samples are CaO and La_2O_3, respectively.

Our neutron diffraction studies on the series of $Ca_{3-y}La_yMn_2O_7$ samples do not support the literature reports [28–32] that the samples with nominal compositions, $La_{2-2x}Ca_{1+2x}Mn_2O_7$ ($0.0 \leq x \leq 0.5$) form single-phase compounds with layered $Sr_3Ti_2O_7$-structure. This composition range when considered in our representation, becomes $1.0 \leq y \leq 2.0$. With the synthesis conditions followed by us, the samples y=1.2 and 1.5, falling in this composition range, form multiphase mixtures with perovskite manganates as the majority phases. We concur with Asano *et al* [31] that $Ca_3Mn_2O_7$ (y=0 in the present case) forms with layered $Sr_3Ti_2O_7$-structure. We believe that a detailed ND study is essential for an unambiguous understanding of the structural characteristics of $La_{2-2x}Ca_{1+2x}Mn_2O_7$ ($0.0 \leq x \leq 0.5$).

14.3.2 $La_{0.7}Ce_{0.3}MnO_3$ and $La_{0.7}MnO_3$

Recently, a number of reports have been published on the magnetic, electrical and structural studies of the samples with nominal compositions $La_{1-x}Ce_xMnO_3$[37–42]. Das and Mandal [37–39] have reported that $La_{0.7}Ce_{0.3}MnO_3$ forms single-phase compound with $GdFeO_3$-type structure. These authors have explained the origin of ferromagnetic and insulator-to-metal (I-M) transitions in $La_{0.7}Ce_{0.3}MnO_3$ to arise from a double exchange (DE) type interaction between $Mn^{3+}(t_{2g}^3e_g^1)$ and $Mn^{2+}(t_{2g}^3e_g^2)$ [37,38]. Specifically, the Mn^{2+} ions are generated in this compound through the substitution of Ce^{4+}-ions for La^{3+}-ions, which at constant level of oxygen content leads to doping of electrons. Thermo-power studies on this compound by the same authors, however, showed that it is not in electron-doped state but in hole-doped state. Philip and Kutty [42] have reported detailed structural, magnetic and electrical studies on the series $La_{1-x}Ce_xMnO_3$($0.05 \leq x \leq 0.8$). Hole-like nature of the carriers in the $La_{1-x}Ce_xMnO_3$($0.05 \leq x \leq 0.8$) was also concluded by Philip and Kutty [42] from thermo-power studies. They observed that the ferromagnetic and metallic properties of these compounds arise from the DE interaction between the Mn^{3+} and Mn^{4+} ions. The presence of Mn^{4+} ions in the Ce-substituted samples was explained by them on the basis of increased oxygen content in the presence the Ce^{4+} ions at the La^{3+}-sites [42].

Our study, however, suggests that the system $La_{1-x}Ce_xMnO_3$ does not yield single-phase compounds with perovskite structure, contrary to the published reports [37–42]. Instead, they form multi-phase mixtures comprising La-deficient lanthanum manganates ($La_{1-x}MnO_3$) with perovskite structures and CeO_2. The ferromagnetic and I-M transitions exhibited by these samples results from

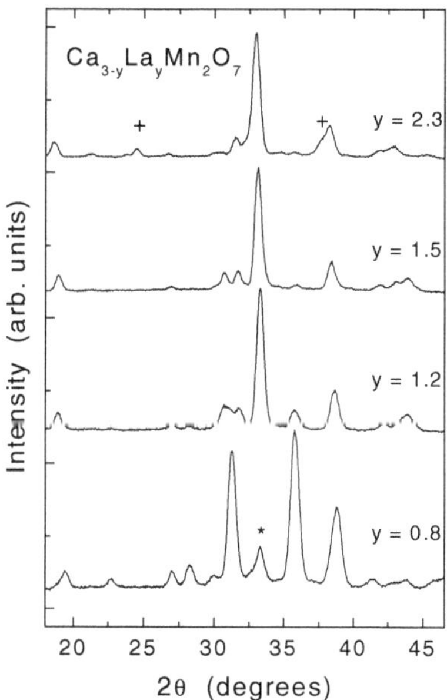

Fig. 14.7. Neutron diffraction patterns for $Ca_{3-y}La_yMn_2O_7$(y= 0.8, 1.2, 1.5 and 2.3).

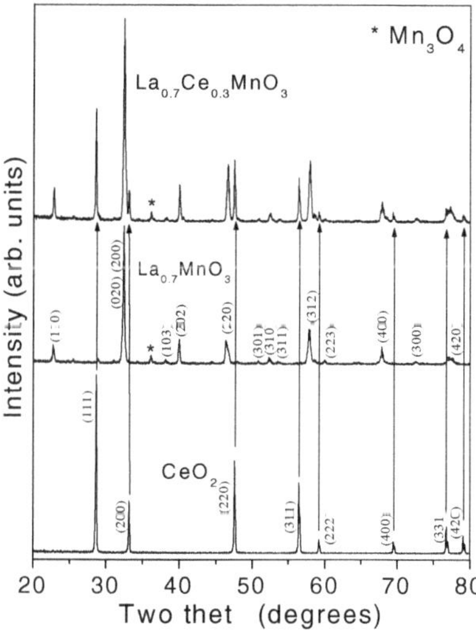

Fig. 14.8. X-ray diffraction patterns for $La_{0.7}Ce_{0.3}MnO_3$, $La_{0.7}MnO_3$ and CeO_2.

the presence of the hole-doped $La_{1-x}MnO_3$ phases [50]. Fig. 14.8 shows the XRD patterns for the samples with nominal compositions $La_{0.7}Ce_{0.3}MnO_3$, $La_{0.7}MnO_3$ and for CeO_2. The pattern for the $La_{0.7}MnO_3$ sample indicates that it is a mixture of orthorhombically distorted perovskite ($GdFeO_3$-type) lanthanum manganate and a small amount of Mn_3O_4 [50]. An inter-comparison of these three patterns clearly shows that the sample $La_{0.7}Ce_{0.3}MnO_3$ can be considered as a mixture of orthorhombically distorted lanthanum deficient perovskite manganates and CeO_2. The XRD pattern for this sample appears quite different from those reported by Mandal et al [37,38] and Philip and Kutty [42]. Both groups reported patterns looking significantly different from each other, but nonetheless containing the three most intense peaks due to the presence of the CeO_2-phase (occurring at 28.55^o, 47.62^o and 56.35^o). These peaks cannot be indexed on the basis of the reported lattice parameters values [37,38,42] considering the structure belonging to $\boldsymbol{Pbnm}$ space group. To confirm the multi-phase nature of this sample, we have carried out Rietveld refinement of the XRD pattern of the sample $La_{0.7}Ce_{0.3}MnO_3$, presuming it to be a mixture of $La_{1-x}MnO_3$ ($\boldsymbol{Pbnm}$ space group) and CeO_2 phases, leaving the Mn_3O_4 phase as the impurity phase. A reasonably good fit (Fig. 14.9) accounting for all the observed diffraction peaks (excluding the peak due to the Mn_3O_4 phase) could be obtained considering the sample as a mixture of CeO_2 and two perovskite manganate phases with compositions close to $La_{0.7}MnO_3$, and with slightly different lattice parameter values. It is clear from Fig. 14.9 that the three most intense peaks of the CeO_2 phase cannot be indexed on the basis of the orthorhombic structure with $\boldsymbol{Pbnm}$ space group, using lattice parameter values close to those reported in literature [37–42]. Hence, we conclude that $La_{0.7}Ce_{0.3}MnO_3$ does not form a single-phase compound with orthorhombically distorted perovskite structure. The XRD patterns for the samples $La_{1-x}Ce_xMnO_3$ ($0.05 \le x \le 0.8$), provided by Philip and Kutty (Fig. 1 of Ref. 42) show that the relative intensity of the peaks due to the CeO_2 phase increases monotonically with an increase in the value of 'x'. A closer look at these patterns shows that the sample with $x = 0.05$ contains perovskite phase as the majority phase and CeO_2 phase as the minority phase, whereas the opposite is true for the sample with $x = 0.8$.

Further support to our conjectures comes from our magnetization studies. Very similar ac-susceptibility signatures were observed by us for $La_{0.7}Ce_{0.3}MnO_3$ and $La_{0.7}MnO_3$, as shown in Fig. 14.10, which suggests that the magnetic behavior of the $La_{0.7}Ce_{0.3}MnO_3$ is determined by the presence of the hole-doped $La_{1-x}MnO_3$ phases with compositions close to $La_{0.7}MnO_3$ [50]. Broader transition exhibited by $La_{0.7}Ce_{0.3}MnO_3$ vis-à-vis the sample $La_{0.7}MnO_3$ could result from the formation of two different $La_{1-x}MnO_3$ phases in it in the presence of CeO_2 as the impurity phase (Fig. 14.9). Our results, therefore, suggest that the ferromagnetic behavior exhibited by the samples with nominal compositions, $La_{1-x}Ce_xMnO_3$ ($0.05 \le x \le 0.8$) [37–42], arises from the presence of hole-doped $La_{1-x}MnO_{3-\delta}$ phases in them.

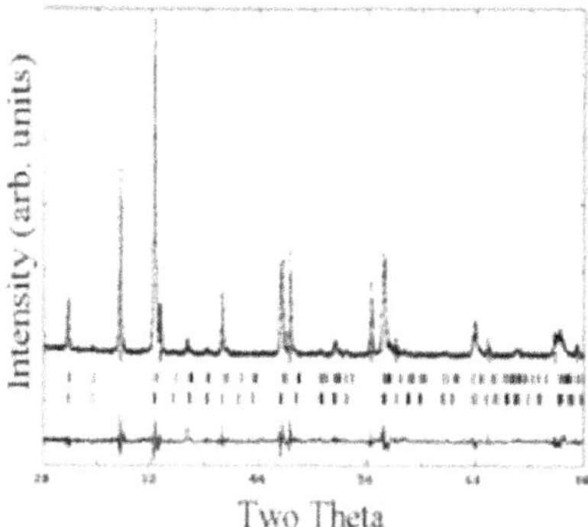

Fig. 14.9. Rietveld refinement of the XRD pattern of $La_{0.7}Ce_{0.3}MnO_3$presuming it to be a mixture of $La_{1-x}MnO_3$ (*Pbnm* space group) and CeO_2 phases. Mn_3O_4 is treated as impurity.

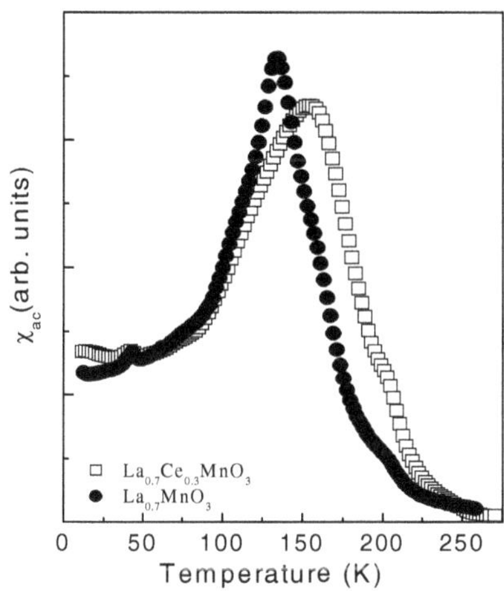

Fig. 14.10. AC-susceptibility versus temperature data for $La_{0.7}Ce_{0.3}MnO_3$ and $La_{0.7}MnO_3$.

14.3.3 $La_{0.67}Ca_{0.33}Mn_{1-x}Co_xO_3$ and $La_{0.67}Ca_{0.33}Mn_{1-y}Cr_yO_3$

X-ray diffraction patterns shows that all these compositions form single-phase compounds with orthorhombically distorted perovskite structures. All the diffraction peaks could be indexed using ***Pbnm*** space group. The lattice parameters did not show any systematic variation in view of the small difference in sizes between the substituent Co and Cr ions and the substituted Mn ions [51]. Figs. 14.11 and 14.12 show the ac-susceptibility vs. temperature plots for the Co and Cr substituted compounds, respectively. For both Co and Cr substituents, the ferromagnetic transition temperature and magnitude of the magnetic signal decreases steadily with an increase in the concentration of the dopant ion which indicates that both Co and Cr ions do not take part in the DE interaction. The difference in the nature of the Co and Cr ions is to the extent that whereas in the case of Cr substitution the ferromagnetic signal is observed up to y=0.50, it is observed only up to x=0.30 for Co. This difference probably arises from the stronger ferromagnetic super-exchange interaction between the Mn^{3+} and Cr^{3+} ions [43] as compared to the ferromagnetic exchange interaction between the Mn and Co ions. Besides, the decrease in the value of T_C, a change in the shape in magnetization plot is observed in the case of Co substitution. The compositions x=0.00 and 0.05 show typical ferromagnetic behavior, whereas the compounds $0.10 \leq x \leq 0.30$ shows ferromagnetic onset followed by a cusp after which magnetization reduces sharply. The broadness of the cusps increases as the Co concentration increases. The x=0.075 compound shows a bit anomalous behavior. We have measured ac-susceptibility of the compounds with x=0.00 and 0.30 annealed under high pressure (50 atm.) oxygen and argon environments and observed the same T_c values as for the as-prepared compounds. This indicates that oxygen content in these compounds is stable and remains close to three in $0.00 \leq x \leq 0.30$. Hwang et al. [4] have also observed that for pure manganites oxygen content remains stable irrespective of the heat treatments. The susceptibility plots for the Cr-substituted compounds with $y \geq 0.10$ show a change in slope below 50 K indicating a change in the magnetic nature. This can result from: i) canting of ferromagnetically aligned spins; or, ii) onset of mictomagnetic (spin glass or cluster glass) behavior; or, iii) a change in the pinning strength of magnetic domains as a result of change in the domain wall configuration. The observation of cusps in the AC susceptibility is indicative of either mictomagnetic behavior or ferromagnetism. Such cusps are symmetric in shape for mictomagnetic behavior and asymmetric in shape for ferromagnetic compounds. The asymmetric nature of cusps observed by us for $0.10 \leq x \leq 0.30$, therefore, points to ferromagnetic ordering. Similar cusps in the AC susceptibility measurement were observed for $La_{1-x}Sr_xCoO_3$ [52] and for UCu_2Ge_2 [53,54] and were explained on the basis of ferromagnetism, with anisotropic exchange interaction. The initial susceptibility $\chi(T)$, is proportional to $M_s(T)^2/K(T)$, where $M_s(T)$ is saturation magnetization at the temperature T and K(T) is the anisotropy constant. The increase in the value of K(T) with decrease in temperature results in a reduction in magnetization with decreasing temperature [52] below T_C. This gives rise to cusp in the susceptibility versus temperature plot.

Since the ac-susceptibility study cannot distinguish between these three types of behaviors, DC magnetization study was carried out. ZFC and FC magnetization vs. temperature plots for the compounds are shown in Figs. 14.13, 14.14, 14.15 and 14.16. The plots for the x=0.0, i.e. $La_{0.67}Ca_{0.33}MnO_3$ in Fig.14.13 show small thermomagnetic irreversibility that is typical of a ferromagnetic compound. Much higher magnetic irreversibility occurs at much lower temperature at 3000 Oe than at 50 Oe of field for $La_{0.67}Ca_{0.33}Mn_{0.80}Co_{0.20}O_3$ (Fig.14.14). Rubinstein et al. [55] and Gayathri et al [56] have also reported an increase

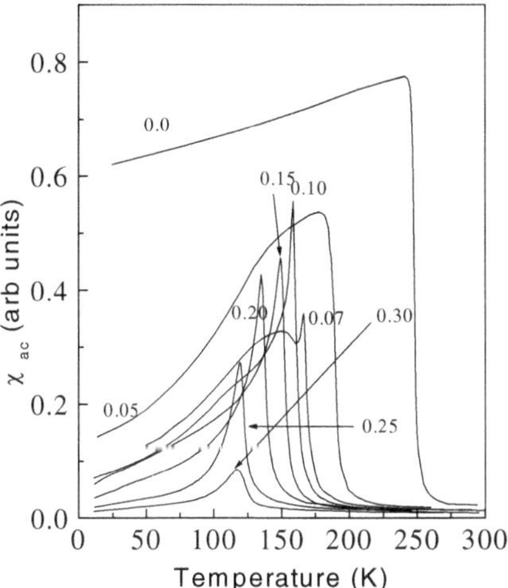

Fig. 14.11. AC-susceptibility versus temperature data for $La_{0.67}Ca_{0.33}Mn_{1-x}Co_xO_3$ for different values of x as shown.

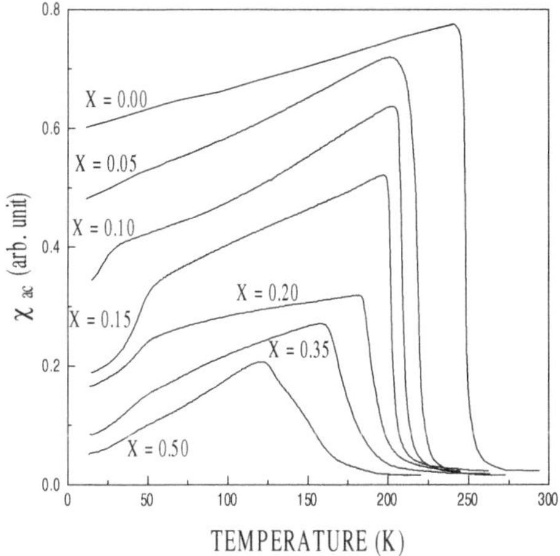

Fig. 14.12. AC-susceptibility versus temperature data for $La_{0.67}Ca_{0.33}Mn_{1-y}Cr_yO_3$ for different values of y as shown.

in the thermomagnetic irreversibility with increase in the Co concentration in $La_{0.67}Ca_{0.33}MnO_3$ and $La_{0.7}Ca_{0.3}MnO_3$, respectively. They have attributed it to mictomagnetic behavior [57]. However, such behavior can also be a signature of ferromagnetism with hysteresis and anisotropy of exchange interaction [58–61], giving rise to high domain wall pinning.

The observation of an increase in the value of T_C-onset with an increase in the applied field (H_{app}) for $La_{0.67}Ca_{0.33}Mn_{0.80}Co_{0.20}O_3$ is in good agreement

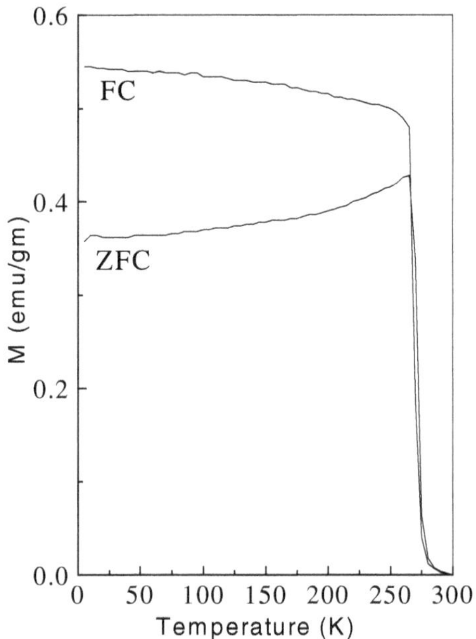

Fig. 14.13. DC magnetization (ZFC and FC) plots for x=0.0, i.e. $La_{0.67}Ca_{0.33}MnO_3$ taken at 50 Oe.

with the results of De Teresa et al [62], who have presented evidence for the existence of magnetic polarons above the ferromagnetic ordering temperature, T_C, by using a combination of volume thermal expansion (with and without field), magnetic susceptibility and small angle neutron scattering measurements on $La_{0.67}Ca_{0.33}MnO_3$.

For the Cr-substituted compounds $La_{0.67}Ca_{0.33}Mn_{1-y}Cr_yO_3$ with y=0.15 and 0.50, too, similar change in slope was observed in the ZFC magnetization plots at low field (50 Oe) (Figs. 14.15 and 14.16, respectively). Absence of any such loss in the FC magnetization plots for the compounds with y=0.15 and 0.50 rules out the possibility of onset of canting of ferromagnetically aligned spins and antiferromagnetic behavior. The ZFC plot for y=0.50 taken at 3000 Oe also do not show any change in slope below 50 K (Fig.14.16), which rules out the possibility of any mictomagnetic transition. In general, low field DC magnetization plots under ZFC conditions always show a decrease in magnetization with decrease in temperature below the ferromagnetic transition temperature. AC magnetization versus temperature plots for ferromagnetic compounds also show similar behavior as measurements are taken at very low fields. This typical behavior of ferromagnetic compounds has been explained by Hietzfield et al [63] to be due to presence of hysteresis and anisotropy. For ZFC magnetization measurements, the field is applied after cooling the samples. At low temperature, the movements of the magnetic domains along the magnetic field is restricted due to pinning of domain walls thus resulting in an incomplete magnetization. With increase in the temperature, an increase in the thermal energy allows more and more domains

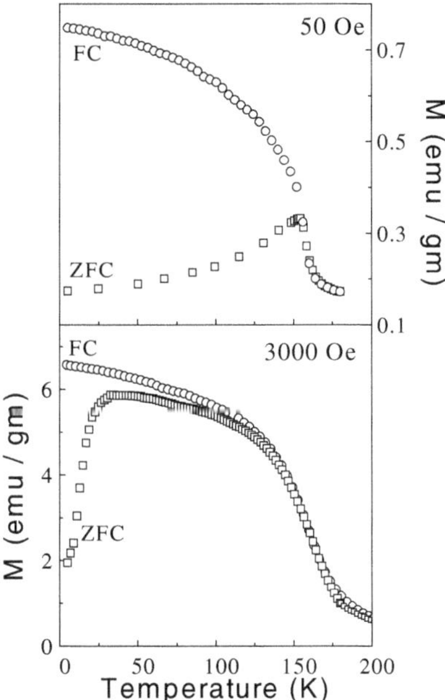

Fig. 14.14. DC magnetization (ZFC and FC) plots for $La_{0.67}Ca_{0.33}Mn_{0.80}Co_{0.20}O_3$, i.e., x=0.2 for applied fields: (a) 50 Oe, and (b) 3000 Oe.

to align along the field and the magnetization increases. Compounds with larger hysteresis will have higher domain wall pinning and thus will show larger fall in magnetization after the ferromagnetic onset. Higher applied field helps the domains to overcome the pinning, and a decrease in magnetization in the ZFC plot region starts appearing at lower temperature. Our observation of disappearance of the change in slope in the ZFC magnetization vs. temperature plots at high applied field, therefore, indicates that the observed low magnetic transition temperature is most probably related to change in domain wall configuration. Our experimental data does not throw light on the origin of this probable change in domain wall configuration.

Fig. 14.17 shows the hysteresis loops for compounds $La_{0.67}Ca_{0.33}Mn_{1-x}Co_xO_3$ with composition x=0.025, 0.15 and 0.25, whereas Fig. 14.18 shows those for the compounds $La_{0.67}Ca_{0.33}Mn_{1-y}Cr_yO_3$ with y=0.00, 0.15 and 0.50 taken at 5 K and field ranging between -10 kOe and 10 kOe. It can be seen that as we go from x=0.025 to 0.15 there is an increase in the area under the hysteresis loop, as well as the coercive field (H_c). The compound with x=0.25, however, does not show any sign of saturation up to the maximum applied field. These results indicate that the domain wall pinning increases with increase in Co concentration. Fig. 14.19 shows the plot of coercive field (H_c) versus temperature for $La_{0.67}Ca_{0.33}Mn_{0.80}Co_{0.20}O_3$ x=0.2. In the low temperature region the value of

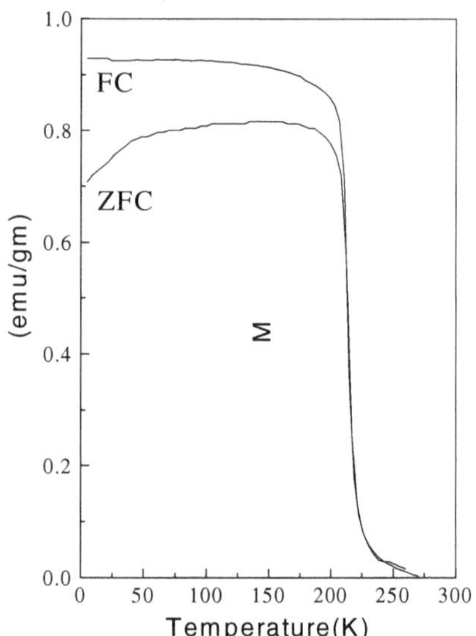

Fig. 14.15. DC magnetization (ZFC and FC) plots for $La_{0.67}Ca_{0.33}Mn_{0.85}Cr_{0.15}O_3$, i.e., y=0.15 taken at an applied field of 50 Oe.

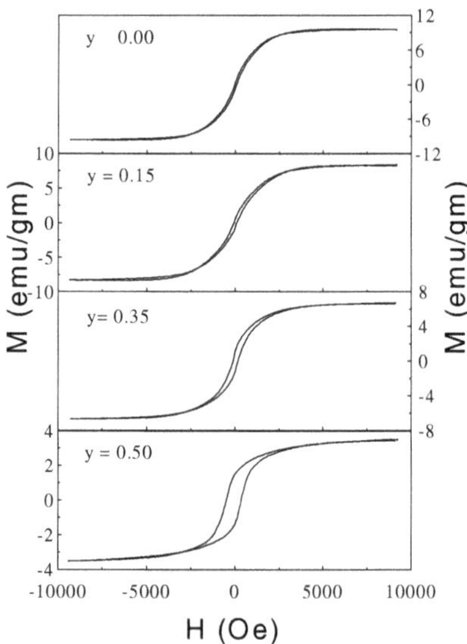

Fig. 14.16. DC magnetization (ZFC and FC) plots for $La_{0.67}Ca_{0.33}Mn_{0.50}Cr_{0.50}O_3$, i.e., y=0.5 taken at applied field values as shown.

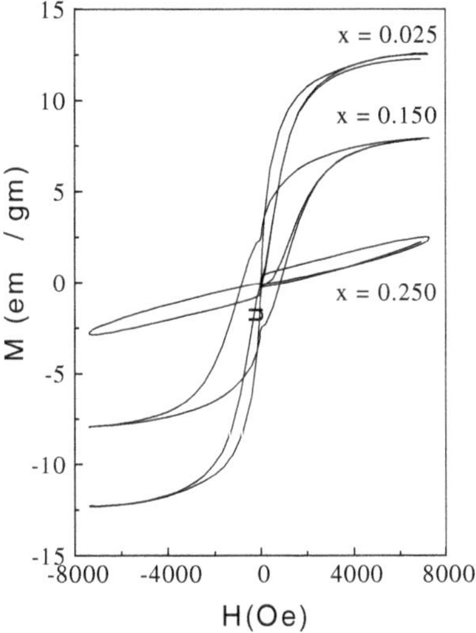

Fig. 14.17. Hysterisis loops for $La_{0.67}Ca_{0.33}Mn_{1-x}Co_xO_3$ for x=0.025, 0.15 and 0.25 recorded at 5 K.

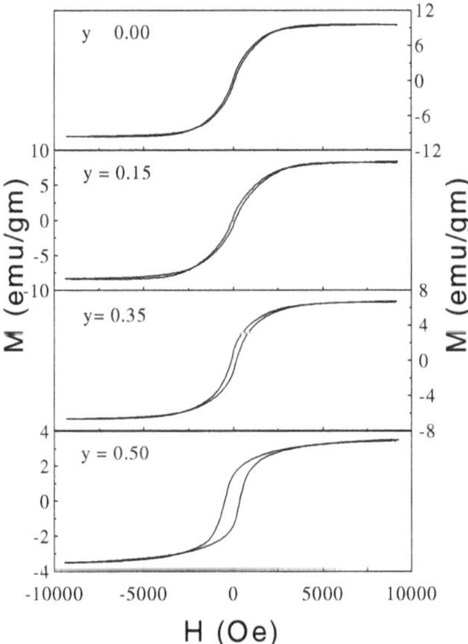

Fig. 14.18. Hysterisis loops for $La_{0.67}Ca_{0.33}Mn_{1-y}Cr_yO_3$ for y=0.00, 0.15 and 0.50 recorded at 5 K.

H_c decreases sharply to a very low value with increase in the temperature. In general, for a ferromagnetic compound the coercive field is a sum of intrinsic and extrinsic contributions:

$$H_c = H_c^{int} + H_c^{ext} \tag{14.1}$$

It has been shown [58] that highly anisotropic ferromagnetic systems have narrow domain walls since the domain wall width, $\delta \quad \alpha \ (J/K)^{1/2}$ where J and K are exchange and anisotropy constants, respectively. Hilzinger and Kronmuller [59] have shown that H_c^{int} increases exponentially with decrease in δ and dominates only at low temperature. This shows that H_c and subsequently the domain wall pinning increases with decrease in δ and increase in K. Our observation of a large drop in the value of H_c for the x=0.2 compound in the low temperature region shows that the compound is an anisotropic ferromagnet. Thus, the increase in the thermomagnetic irreversibility with increase in Co concentration occurs because of an increase in the domain wall pinning arising out of an increase in the anisotropy of exchange interaction. The suggestion of Rubinstein et al. [55] and Gayathri et al. [56] that substitution of Co leads to mictomagnetic behavior is debatable. Hitzfeld et al. [63] have also observed that ferromagnetic compounds with hysteresis and anisotropy show thermomagnetic irreversibility due to the presence of large domain wall pinning. At low temperature, when the value of H_c is larger than the applied field, ZFC magnetization is much less as domains movements are restricted due to large domain wall pinning. With increasing temperature $H_c$decreases drastically, the domain starts moving and the ZFC magnetization increases, finally becoming equal to the FC magnetization at the irreversibility temperature. A decrease in the irreversibility temperature upon increasing the applied field points to the lesser restriction to the movement of magnetic domains as difference between the values of H_c and the applied field decreases.

The explanation for the probable source of anisotropy is as follows. In general, there are two sources of the anisotropy of exchange interaction: i) magnetocrystalline anisotropy; and, ii) spin-orbit coupling which can arise from an asymmetry in charge distribution [64,65]. In our case, the XRD analyses show that the crystal symmetry does not change with substitution of Co. Therefore, the substituent Co ions must have asymmetric charge distribution and are involved in the exchange interaction. Recently, it has been reported that the hole-doped cobaltates show a large decrease in the value of H_c at low temperature [66] and also show huge anisotropic magnetostriction [67] both of which have been reported to be due to the presence of spin orbit coupling of the Co ions.

The nature of the hysteresis loops taken at 5 K for $La_{0.67}Ca_{0.33}Mn_{1-y}Cr_yO_3$ with y=0.00, 0.15, 0.35 and 0.50 (Fig. 14.18) indicates that these compounds are ferromagnetic at 5 K. Though the area under the loop increases with increase in y, the effect is much less than what is observed in the case of Co-substituted compounds. This clearly shows the difference in the nature of the exchange interactions, of Co and Cr ions with the Mn ions. Figs. 14.20 and 14.21 show normalized DC resistance vs. temperature plots for the Co- and Cr-substituted

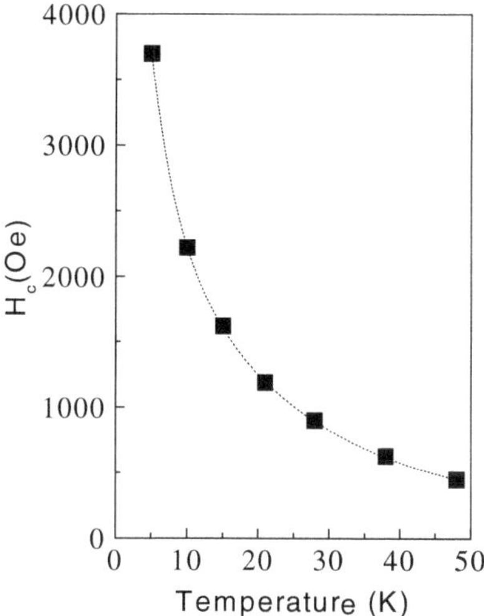

Fig. 14.19. Coercive field H_c versus temperature plot for $La_{0.67}Ca_{0.33}Mn_{0.80}Co_{0.20}O_3$, i.e., $x = 0.2$.

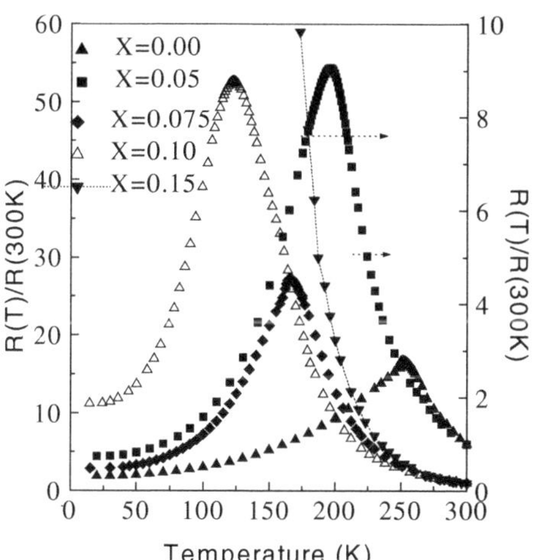

Fig. 14.20. Normalized resistance versus temperature plots for $La_{0.67}Ca_{0.33}Mn_{1-x}Co_xO_3$ for different values of x as shown.

compounds, respectively. In both the cases the metal-insulator transition temperature, T_p, decreases with dopant content (x or y) up to 0.10, but for dopant

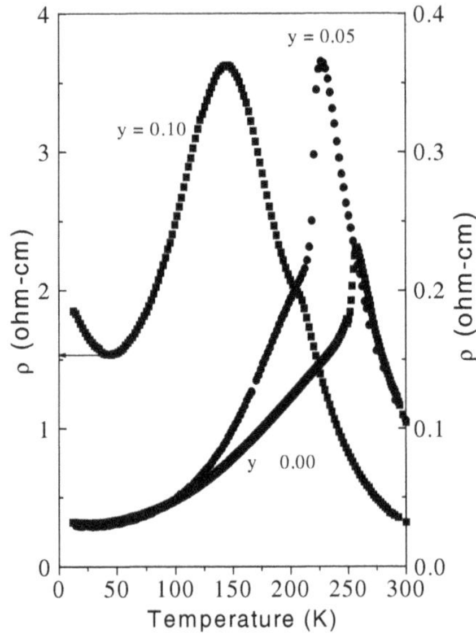

Fig. 14.21. Resistivity versus temperature plots for $La_{0.67}Ca_{0.33}Mn_{1-y}Cr_yO_3$ for different values of 'y' as shown.

content higher than 0.10, no metal-insulator transition is observed and instead one continues to see an increase in resistance upon lowering the temperature down to 10 K. The two transition temperatures, T_p and T_C are quite close for all the compounds except for x=0.10 and for y=0.10, respectively, for which T_p is observed at much lower temperature than T_C. The resistivity behavior of the compounds studied by us are similar to that of Cr-substituted $LaMnO_3$ [43] which are ferromagnetic due to the super-exchange interaction Cr^{3+}-O-Mn^{3+} but are insulating due to the absence of any double-exchange interaction. In the present series of compounds, the super-exchange interaction between Co and Cr ions with the Mn ions may be the reason for the observed behavior. Room temperature resistivity values are of nearly same magnitude up to substitution levels of 10 % in the case of both Co- and Cr-substitutions.

14.4 Conclusions

In view of the potential of hole-doped CMR manganites for use in sensors and devices, it is essential that one establishes the stability parameters of the different phases in these materials. It is in this context that the present study was undertaken. We have been able to throw light on some controversial issues in this respect. In particular, we have been able to establish that samples of nominal compositions $Ca_{3-y}La_yMn_2O_7$ ($1.0 \leq y \leq 2.3$) do not lead to compounds with layered $Sr_3Ti_2O_7$-structure. Our studies on layered manganites $La_{1.2}(Sr_{1-x}Ca_x)_{1.8}Mn_2O_7$ and $Ca_{3-y}La_yMn_2O_7$ give new insights into this aspect. In view of our observation that the sample of nominal composition $La_{0.7}Ce_{0.3}MnO_3$ forms a multi-phase mixture comprising hole-doped lanthanum deficient manganite phases and CeO_2. This provided us an opportunity to look into the phase stability of samples prepared from nominal compositions, $La_{1-x}Ce_xMnO_3$ ($0.05 \leq x \leq 0.8$), and establish that they do not form single phase compounds with $GdFeO_3$-type structure. This implies that electron-doping of $LaMnO_3$ by the substitution of Ce at Mn sites is not easy.

Lastly, our studies the perovskites $La_{0.67}Ca_{0.33}Mn_{1-x}Co_xO_3$ and $La_{0.67}$ $Ca_{0.33}Mn_{1-y}Cr_yO_3$ showed that the substitution of Co and Cr for Mn in $La_{0.67}Ca_{0.33}MnO_3$ destabilizes the double-exchange (DE) driven ferromagnetic as well as insulator-to-metal transition temperatures.

References

1. J. G. Bednorz and K. A. Muller, Z. Phys. B **64**, 189 (1986)
2. S. Jin, T.H. Tiefel, M. McCormack, R. A. Fastnacht, R. Ramesh and L.H. Chen, Science **264**, 413 (1994)
3. R. Von Helmholt, J. Wecker, B. Holzapfel, L. Schultz and K. Samwer, Phys. Rev. Lett. **71**, 2331 (1994)
4. H.Y. Hwang, S-W. Cheong, P.G. Radaelli, M. Marezio and B. Batlogg, Phys. Rev. Lett. **75**, 914 (1995)
5. A. Urushibara, Y. Moritomo, T. Arima, A. Asamitsu, G. Kido and Y. Tokura, Phys. Rev. B **51**, 14103 (1995)
6. R. Mahendiran, R. Mahesh, A.K. Raychaudhuri and C. N. R Rao, Solid State Commun. **94**, 515 (1995)
7. A. Barman, M. Ghosh, S. Biswas, S.K. De and S. Chatterji, Solid State Commun. **106**, 691 (1998)
8. A. Barnabe, A. Maignan, M. Hervieu, F. Damay, C. Martin and B Raveau, Appl. Phys. Lett. **71**, 3907 (1997)
9. B. Raveau, A. Maignan and C. Martin, J. Solid State Chem. **130**, 162 (1997)
10. P.V. Vanitha, R.S. Singh, Sreenivasan Natarajan and C.N.R. Rao, J. Solid State Chem. **137**, 365 (1998)
11. C. Martin, A. Maignan and B. Raveau, J. Mater. Chem. **6**, 1245 (1996)
12. A. Maignan, F. Damay, C. Martin and B. Raveau, Mat. Res. Bull. **32**, 965 (1997)
13. B. Raveau, A. Maignan and C. Martin, R. Mahendiran, and M. Hervieu, J. Solid State Chem. **151**, 330 (2000)
14. R. Mahendiran, M. Hervieu, A. Maignan C. Martin and B. Raveau, Solid State Comm. **114**, 429 (2000)
15. Y. Moritomo, A. Machida, S. Mori, N. Yamamoto and A. Nakamura Phys. Rev. B **60**, 9220 (1999)
16. P.V. Vanitha, A. Arulraj, A.R. Raju and C.N.R. Rao, C. R. Acad. Sci. **2**, 595 (1999)
17. N. Gayathri, A.K. Raychaudhuri, S.K. Tiwari, R. Gundakaram, A. Arulraj and C.N.R. Rao, Phys. Rev. B **56**, 1345 (1997)

18. M. Rubinstein, D.J. Gillespie, J.E. Snyder and T.M. Tritt, Phys. Rev. B **56**, 5412 (1997)
19. R. Ganguly, I.K. Gopalakrishnan and J.V. Yakhmi, Physica B **266**, 332 (1999)
20. K.H. Ahn, X.W. Wu, K. Liu and C.L. Chien, Phys. Rev. B **54**, 1 (1996)
21. S.B. Ogale, R. Shreekala, S.I. Patil, B. Hannoyer, F. Pettit and G. Marest, Phys. Rev. B **57**, 7841 (1998)
22. M. Sahana, K.M. Satyalakshmi, M.S. Hegde, V. Prasad and S.V. Subramanyam, Mat. Res. Bull. **32**, 831 (1997)
23. M.C. Sanchez, J. Blasco, J. Garcia, J. Stankiewicz, J. M. De Teresa and M. R. Ibarra, J. Solid State Chem. **138**, 226 (1998)
24. C. Zener, Phys. Rev. **82**, 403 (1951)
25. Y. Moritomo, A. Asamitsu, H. Kuwahara, and Y. Tokura, Nature **380**, 141 (1996)
26. R.S. Liu, C.H. Shen, J.G. Lin, J.M. Chen, R.G. Liu and C.Y. Huang, Int. J. Inorg. Mater. **1**, 61 (1999)
27. C.F. Chang, P.H. Chuo, H.L. Tsay, S.S. Weng, S. Chatterji, H.D. Yang, R.S. Liu, C.H. Shen and W-H Li, Phys. Rev. B **58**, 12224 (1998)
28. H. Asano, J. Hayakawa and M. Matsui, Appl. Phys. Lett. **68**, 3638 (1996)
29. R. Suryanarayanan, I. Zelenay and B. Berthon, Mat. Res. Bull. **32** 595 (1997)
30. K.V. Kamenev, M.R. Lees, G. Balakrishnan and D. Mck. Paul, Phys. Rev. B **56**, R12688 (1997)
31. H. Asano, J. Hayakawa and M. Matsui, Phys. Rev. B **56**, 5395 (1997)
32. H. Asano, J. Hayakawa and M. Matsui, Phys. Rev. B **57**, 1052 (1997)
33. H. Chiba, M. Kikuchi, K. Kusaba, Y. Muraoka and Y. Sayono, Solid State Commun. **131**, 144 (1997)
34. I.O. Troyanchuk, N.V. Samsonenko, H. Szymczak and A. Nabialek, J. Solid State Chem. **131**, 144 (1997)
35. C. Martin, A. Maignan, F. Damay, M. Hervieu and B. Raveau, J. Solid State Chem. **134**, 198 (1997)
36. A. Maignan, C. Martin, F. Damay, and B. Raveau, Chem. Mater. **10**, 950 (1998)
37. P. Mandal and S Das, Indian J. Phys. **71**, 231 (1997)
38. P. Mandal and S Das, Z. Phys. B Cond. Matter **7**, 104 (1997)
39. P. Mandal and S Das, Z. Phys. Phys. Rev. B **56**, 15073 (1997)
40. J.R. Gebhardt, S. Roy and N.S. Ali, J. Appl. Phys. **85**, 5390 (1999)
41. P. Raychaudhuri, S. Mukherji, A.K. Nigam, J. John, U.D. Vaisna and R. Pinto J. Appl. Phys. **86**, 5718 (1999)
42. J. Philip and T.R.N. Kutty, J. Phys. Cond. Matter **11**, 8537 (1999)
43. R. Gundakaram, Anthony Arulraj, P.V. Vanitha and C.N.R. Rao, J. Solid State Chem. **127**, 354 (1996)
44. J.F. Mitchell, D.N. Argyriou, J.D. Jorgensen, G.D. Hinks, C.D. Potter and S.D. Bader, Phys. Rev. B **55**, 63 (1997)
45. D.N. Argyriou, J.F. Mitchell, J.B. Goodenough, O. Chmaissem, S. Short and J.D. Jorgensen, Phys. Rev. Lett. **78**, 1568 (1997)
46. S. Rosenkranz, R. Osborn, J.F. Mitchell, L. Vasiliu-Doloc, J.W. Lynn, S.K. Sinha and D.N. Argyriou, J. Appl. Phys. **83**, 7348 (1998)
47. D.N. Argyriou, H.N. Bordallo, J.F. Mitchell, J.D. Jorgensen, and G.F. Strouse, Phys. Rev. B **60**, 6200 (1999)
48. Y. Moritomo, K Ohoyama and M Ohashi, Phys. Rev. B, **59**, 157 (1999)
49. E. O. Chi, K.P. Hong, Y.U. Kwon, N.P. Raju, J.E. Greedan, J.S. Lee, and N.H. Hur, Phys. Rev. B **60**, 12867 (1999)
50. A. Gupta, T.R. McGuire, P.R. Duncombe, M. Rupp, J.Z. Sun, W.J. Gallagher and Xiao Gang, Appl. Phys. Lett. **67**, 3494 (1995)

51. R.D. Shanon, Acta Crystallogr. A **32**, 751 (1976)
52. R. Mahendiran and A.K. Raychaudhuri, Phys. Rev. B **54**, 16044 (1996)
53. S.B. Roy and B.R. Coles, Phil. Mag. B **64**, 741 (1991)
54. S.B. Roy, A.K. Pradhan and P. Chaddah, Phil. Mag. B **71**, 97 (1995)
55. M. Rubinstein, D.J. Gillespie, J.E. Snyder and T.M. Tritt, Phys. Rev. B **56**, 5412 (1997)
56. N. Gayathri, A.K. Raychaudhuri, S.K. Tiwari, R. Gundakaram, A. Arulraj and C.N.R. Rao, Phys. Rev. B **56**, 1345 (1997)
57. Ronald B. Goldfarb and Carl E. Patton, Phys. Rev. B **24**, 1360 (1981)
58. K.M.B. Alves, L.C. Sampaio, A.P. Guimaraes and S.F. Da Cunha, J. Alloys and Compounds, **210**, 325 (1994)
59. H.R. Hilzinger and H. Kronmuller Phys. Stat. Sol. (b) **58**, 593 593 (1972); ibid **59**, 71 (1973)
60. T.V. Chandrasekhar Rao, P Raj, Sk. Mohammad Yusuf, L. Madhav Rao, A. Sathyamoorthy and V.C. Sahni, Phil. Mag. B **74**, 275 (1996)
61. W. Zhong, J. Lan and Z. Liu, J. Magn. magn. Mater, **68**, 197 (1987)
62. M. De Teresa, M.R. Ibarra, P.A. Algarabel, C. Ritter, C. Marquina, J. Blasco, J. Gracia, A. del Moral and Z. Arnold, Nature **386**, 256 (1977)
63. M. Hitzfeld, P. Ziemann, W. Buckel, and H. Clus, Phys. Rev. B **29**, 5023 (1984)
64. C.M. Hurd, Contemp. Phys. **23**, 469 (1982)
65. S. Chikazumi: *Physics Of Magnetism.* (Wiley, New York, 1964) p. 153
66. R. Ganguly, I.K. Gopalakrishnan and J.V. Yakhmi, Physica B **271**, 116 (1999)
67. M.R. Ibarra, R. Mahendiran, C. Marquina, B. Gaarcia-Landa and J. Blasco, Phys. Rev. B **57**, R3217 (1998)

15 Colossal Magnetoresistance in Screen Printed Manganite Films

A.K.M. Akther Hossain

Department of Physics, Bangladesh University of Engineering & Technology, Dhaka 1000, Bangladesh

Abstract. Thick films of $La_{0.67}Ca_{0.33}MnO_3$ were fabricated on single crystal (100) $LaAlO_3$ (LAO), single crystal (100) yttria stabilised zirconia (YSZ) and polycrystalline Al_2O_3 by a screen printing technique. The films were sintered at 1200, 1300 and 1400°C, in air, oxygen and nitrogen atmospheres to explore the optimum annealing conditions of manganite thick films. The thick films on LAO exhibit a M-I transition at T_{p1}. T_{p1} varies from 180 to 258 K depending on sintering temperature and atmospheres for these films. The thick film sintered at 1300°C in air shows highest (80%) MR at 8 T near T_{p1} (180 K). The thick films on YSZ also show a M-I transition in the measured temperature range except for the film sintered at 1400°C in nitrogen. The Tp1s for the thick films on YSZ varied from 187 to 265 K. The film sintered at 1400°C in oxygen exhibits 94 % MR around T_{p1} (187 K) in an 8 T applied field. The MR sensitivity as a function of field at $T \ll T_{p1}$ is highest for the oxygen sintered samples for the films on both LAO and YSZ substrates. The thick films fabricated on Al_2O_3 substrates reacted very badly for the films sintered at 1300 and 1400°C and are not suitable for further investigations. However, the films sintered at 1200°C in air, oxygen and nitrogen atmospheres show a M-I transition. The T_{p1} varies for these films from 253 to 26 K. The film sintered in air shows 65% MR around T_{p1} (252 K) in an 8 T applied field.

Thick films of $La_{0.63}Y_{0.07}Ca_{0.30}MnO_3$ were fabricated on (100) LAO and polycrystalline Al_2O_3 substrates. The films were sintered at 1200°C in air and oxygen. All these thick films show a M-I transition at T_{p1} ~220 K. For both substrates, the films sintered in oxygen flow have slightly higher T_{p1}. The film on Al_2O_3 sintered in oxygen flow shows 71 % MR at 220 K in presence of an 8 T applied field.

The salient feature of the CMR observed in these thick films are that MR is not limited to a small temperature window near T_{p1} (e.g. epitaxial thin films, single crystal). The MR peak is very broad and for some films a temperature independent CMR is observed at $T \ll T_{p1}$.

15.1 Introduction

Screen printing is one of the oldest forms of graphic art reproduction and the technique used for generating images on T-shirts, mugs, pencils etc is very similar to that used in the production of modern day, precision thick film circuits. The capital and running costs are low for screen printing. This technique offers a wide range of circuit complexities, that is, from simple conductor patterns to highly complex hybrid integrated circuits, using the same basic materials and equipment. We have fabricated some patterned thick films, which have six contact pads for electrical connections. In most cases, we investigate the effect of post-annealing and oxidation on our samples. For this, we needed to use a definite set of electrical contacts. For applications of manganite materials, we need to make the technology simple and cheap. The thick film offers a good opportunity for future sensor and read head applications. For this reason we are motivated to make manganite thick films by a screen printing technique on different substrates. The substrates play an important role in the properties of materials. The thick films were fabricated from two different manganite compositions (e.g. $La_{0.67}Ca_{0.33}MnO_3$ and $La_{0.63}Y_{0.07}Ca_{0.30}MnO_3$) on three different substrates (e.g. polycrystalline Al_2O_3, single crystal (100) LAO and single crystal (100) YSZ).

15.2 Processing of $La_{0.67}Ca_{0.33}MnO_3$ Thick Films

15.2.1 Calcination of the Chemical Composition

The $La_{0.67}Ca_{0.33}MnO_3$ compositions were synthesised using a conventional solid state reaction technique. Powders of La_2O_3 (Aldrich 99.99 %), $CaCO_3$ (BDH 99.9 %) and $MnCO_3$ (Aldrich 99.9 %) were used to make the required compositions. We followed a slightly deviated calcination route. We pre-heat treated the La_2O_3 powder at 1000°C for 6 h before mixing with the other ingredients, as this is a hygroscopic chemical. The ingredients were weighed with appropriate proportions and then dry ball milled for 24 h. The calcination was done in 4 steps. The mixed powders were calcined at 950°C for 6 h, the temperature ramps being 300°C/h in both warming and cooling. This step was repeated three more times with intermediate grindings to obtain a homogeneous and phase pure composition.

15.2.2 Fabrication of Thick Films

Fine Particle and Ink Preparation

A fine powder is necessary to fabricate thick films by screen printing [1,2], otherwise the ink will not pass through the screen. As we are unable to apply any mechanical pressure on thick films the packing of particles will remain poor if we do not start from powders of small particle size. To obtain fine powders two different routes were followed. After a total period of 24 h of calcination, the final calcined powders were divided into two halves. One portion was ball milled with Zirconia beads in water media for 7 days and the other portion was ultrasonically dispersed for 30 minutes. The particle sizes were measured using Mastersizer. It was found that ultrasonically dispersed powders are finer than ball milled powders. Therefore, it was decided to make thick films from ultrasonically dispersed powders. The terpineol based Blythe vehicle was used for ink preparation. The ratio of powder to Blythe vehicle was 70 % to 30 % by weight respectively. The ink was mixed thoroughly until adequate printability was obtained. Sometimes tiny air bubbles were trapped in the ink. We needed to make sure that bubbles are completely removed from the ink by repeated mixing.

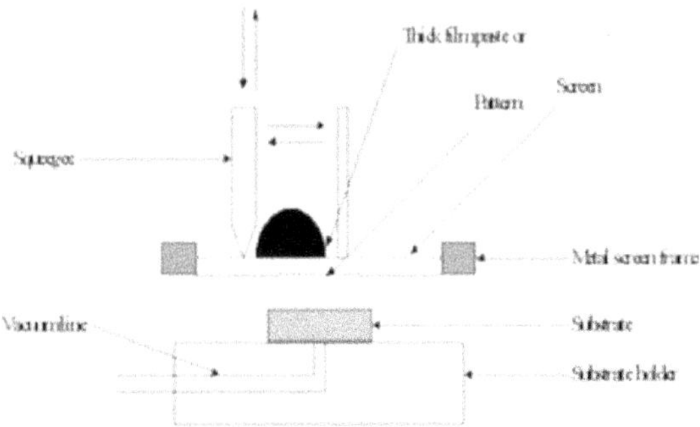

Fig. 15.1. Schematic representation of screen printing system.

Screen Printing Process

Fabrication of Thick Films. The basic components of a screen printing machine (shown in Fig. 15.1) are: (i) The screen, together with its mounting framework, (ii) The substrate carrier and feed system, (iii) The squeegee mechanism and pressuring system, (iv) The adjustment mechanism for the precise positioning of the screen relative to the substrate.

The Screen. The screen defines the pattern of the printed film and also meters the amount of paste which is used. It is a very important part of the screen printing equipment and is essentially a stencil through which the ink is forced during the printing process. The screen was made by DEK Precision Limited., UK. The pattern of the film is shown in Fig. 15.2. The frame size is 1200/1202. The mesh size of the screen is 325/inch, the material is stainless steel and the emulsion thickness is 30 mm. The stainless steel screen was mounted on a metal frame. The screen mounting is designed to clamp the screen frame rigidly and securely so that it does not move during the printing cycle. It is provided with means for adjusting the X and Y position of the screen. This allows adjustment of the pattern on the screen relative to the substrate on the substrate carrier.

Substrate Holder. The function of the substrate holder is to register the substrate accurately below the screen and also to keep it firmly fixed in position during printing. The substrate carrier contains a number of holes in it, which are connected to the vacuum line. The substrate was put on top of the holes and hence fixed rigidly by air suction.

Squeegee. The squeegee is essentially a flexible blade whose function is to transfer the paste through the screen and onto the substrate. During printing the squeegee forces ink through the open areas of the mesh and by virtue of the surface tension between the film and substrate, the required pattern is transferred to the substrate as the screen and substrate separate. The squeegee

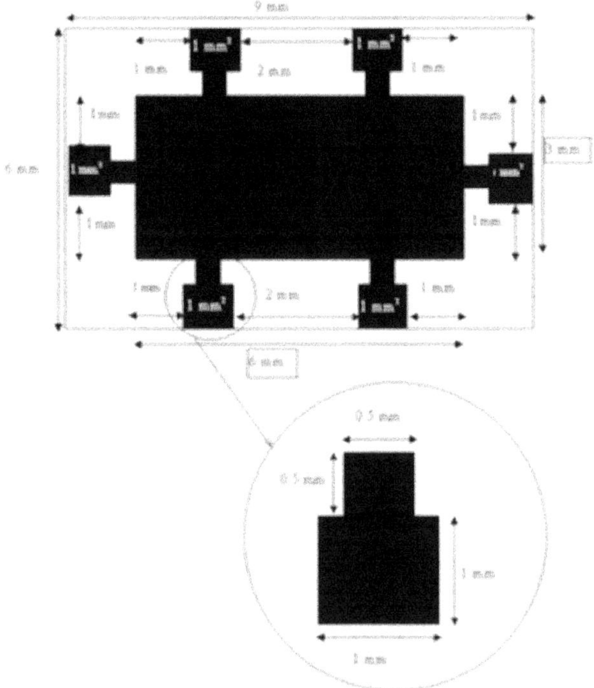

Fig. 15.2. The pattern of the film. Black represents the fabricated material on the substrate.

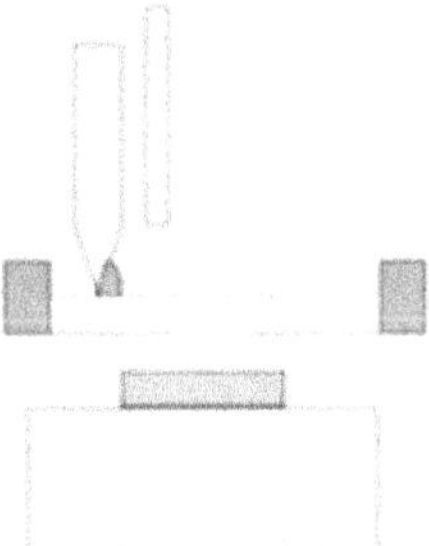

Fig. 15.3. Screen printing process. At the start of print stroke; the squeegee is in contact with screen and flood blade is raised.

mechanism and pressuring system is automated. The squeegee remains in contact with the screen only on the forward stroke (Fig. 15.3) and is raised out of contact on the reverse stroke (Fig. 15.4). In this case, a flood blade is provided to return the excess ink to the starting point during the reverse stroke prior to the next printing cycle.

Finally, the wet print will be laid down upon the substrates (Fig. 15.5).

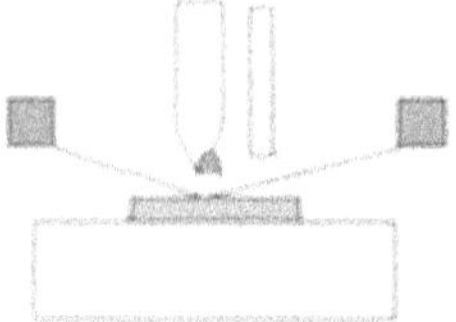

Fig. 15.4. Print stroke: Screen depressed.

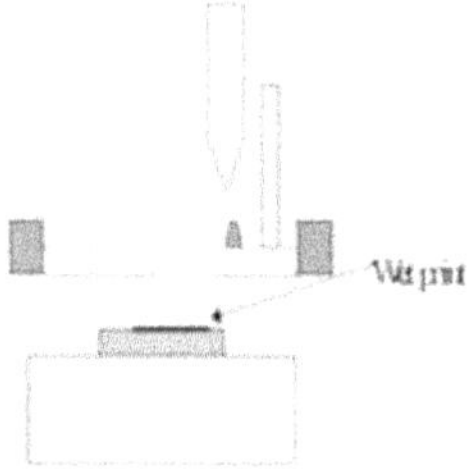

Fig. 15.5. End of print stroke; Squeegee and flood blade reverse relative positions.

Burning Out of Organic Vehicle

After fabrication of thick films we leave them in a desiccator to settle overnight. The next step is to burn out the organic vehicle from the fabricated thick films. The burning out of the organic vehicle was done at 300°C for 3 h and the temperature ramp was 30°C/h in both warming and cooling. The burn-out thick films are immune to smudging and subjected to high temperature sintering in different atmospheres.

15.2.3 Sintering of $La_{0.67}Ca_{0.33}MnO_3$ Thick Films

Sintering is defined as the process of obtaining a dense, tough body by heating a compacted powder for a certain time at a temperature high enough to significantly promote diffusion. There are many variables in a sintering process e.g. composition, average particle size, heating rate, temperature, dwell time, cooling rate, atmosphere of sintering, etc. Once a thick film has been fabricated onto the substrate and is immune to smudging (after burning out), the particles of manganite powder must be encouraged to coalesce to form a uniform and dense thick film. The driving force for sintering is the reduction of surface energy by the reduction of surface area and this can be optimised with the above variables of sintering. The optimum sintering conditions for growing manganite thick films are yet to be known. To explore the optimum sintering conditions, the films prepared from a specific composition were sintered in different temperatures and atmospheres. The sintering conditions of $La_{0.67}Ca_{0.33}MnO_3$ thick films on different substrates are given in Table 15.1.

Table 15.1. First sintering conditions of thick films of $La_{0.67}Ca_{0.33}MnO_3$ on single crystal (100) LAO, single crystal (100) YSZ and polycrystalline Al_2O_3 substrates. Heating rate is 100°C/h.

Sample No	Substrates	T_{max} (°C)	Time at T_{max}(h)	Atmos-phere	Cooling rate (1) °C/h	Cooling rate (2) °C/h
LCL5, LCY5	LAO, YSZ, Al_2O_3	1400	1	Air	100	-
LCL6, LCY6	LAO, YSZ	1400	5	Air	100	-
LCL8, LCY8	LAO, YSZ, Al_2O_3	1400	1	N_2	100 down to 600 and then halt 12 h	100
LCL9, LCY9, AHCa4	LAO, YSZ, Al_2O_3	1300	1	N_2	100 down to 600 and then halt 12 h	100
LCL10, LCY10, AHCa5	LAO, YSZ, Al_2O_3	1200	1	N_2	100 down to 600 and then halt 12 h	100
LCL11, LCY11, AHCa6	LAO, YSZ, Al_2O_3	1200	1	O_2	100 down to 600 and then halt 12 h	100
LCL12, LCY12, AHCa7	LAO, YSZ, Al_2O_3	1300	1	O_2	100 down to 600 and then halt 12 h	100
LCL13, LCY13, AHCa8	LAO, YSZ, Al_2O_3	1400	1	O_2	100 down to 600 and then halt 12 h	100
LCL14, LCY14, AHCa9	LAO, YSZ, Al_2O_3	1300	1	Air	100 down to 600 and then halt 12 h	100
LCL15, LCY15 AHCa10	LAO, YSZ, Al_2O_3	1200	1	Air	100 down to 600 and then halt 12 h	100
AHCa2	Al_2O_3	1200	24	Air	100	-

15.2.4 Microstructures of $La_{0.67}Ca_{0.33}MnO_3$ on Different Substrates

SEM pictures of some representative thick films of $La_{0.67}Ca_{0.33}MnO_3$ on different substrates are given below.

Thick Films on LAO Substrates

SEM pictures of a few thick films of nominal composition $La_{0.67}Ca_{0.33}MnO_3$ on single crystal (100) LAO substrates, sintered at different temperatures and atmospheres are shown in Figs. 15.6, 15.7, 15.8, 15.9. It was observed that the film sintered at 1400°C in nitrogen flow is densely packed with a larger grain size compared to the film sintered in an oxygen atmosphere at the same temperature. This result agrees well with the literature [3]. The lower temperature (e.g.1200°C) sintering results in a smaller grain size sample and a relatively more porous structure.

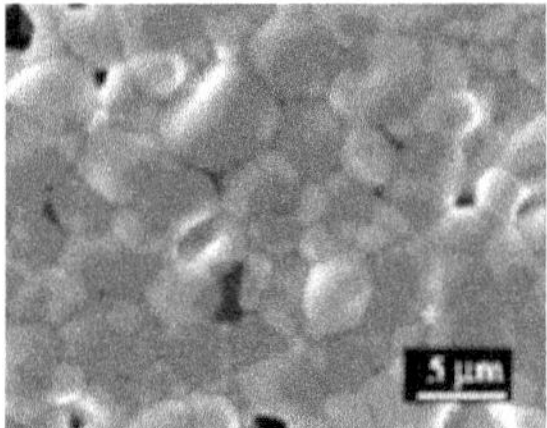

Fig. 15.6. SEM picture of LCL8, film on LAO, sintered at 1400°C in N_2 flow for 1 h.

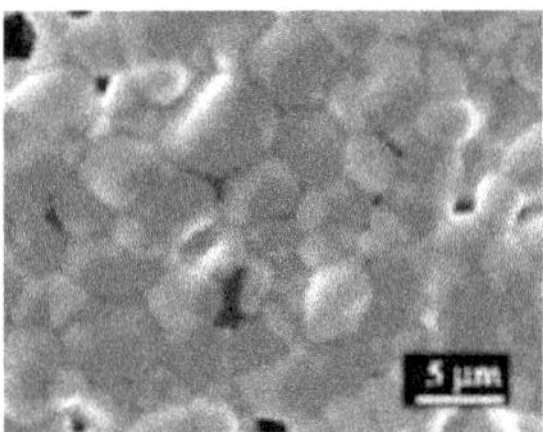

Fig. 15.7. SEM picture of LCL13, film on LAO, sintered at 1400°C in O_2 flow for 1 h.

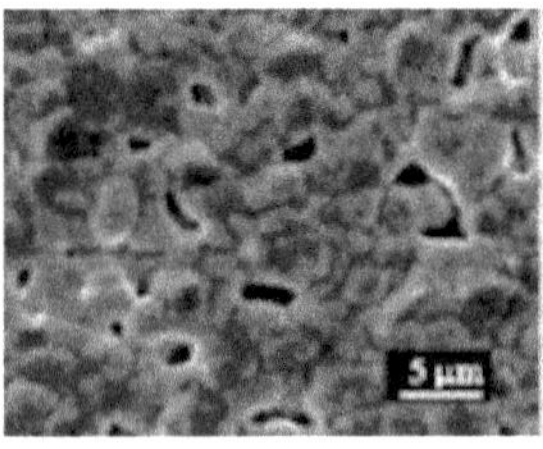

Fig. 15.8. SEM picture of LCL14, film on LAO, sintered at 1300°C in air for 1 h.

Fig. 15.9. SEM picture of LCL11, film on LAO, sintered at 1200°C in O_2 flow for 1 h.

Thick Films on YSZ Substrates

Figures 15.10-15.15 show the SEM pictures of the films on single crystal (100) YSZ substrates. Like the films on LAO, thick films sintered at 1400°C are densely packed. Also, the grain growth is relatively larger in the nitrogen atmosphere at this temperature. The lower sintering temperature (e.g. 1200°C) results a relatively smaller grain growth and porous structure.

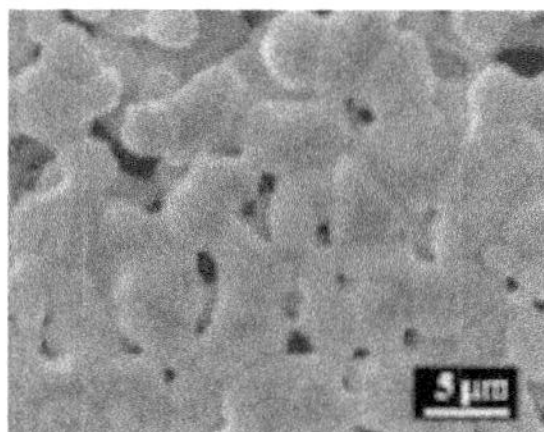

Fig. 15.10. SEM picture of LCY7, film on YSZ, sintered at 1400°C in air for 5 h.

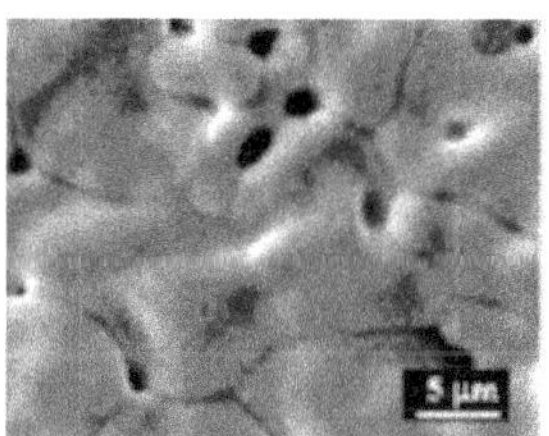

Fig. 15.11. SEM picture of LCY8, film on YSZ, sintered at 1400°C in N_2 flow for 1 h.

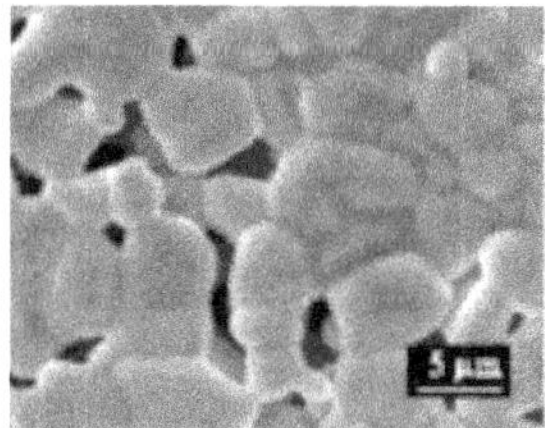

Fig. 15.12. SEM picture of LCY13, film on YSZ, sintered at 1400°C in O_2 flow for 1 h.

Fig. 15.13. SEM picture of LCY12, film on YSZ, sintered at 1300°C in O_2 flow for 1 h.

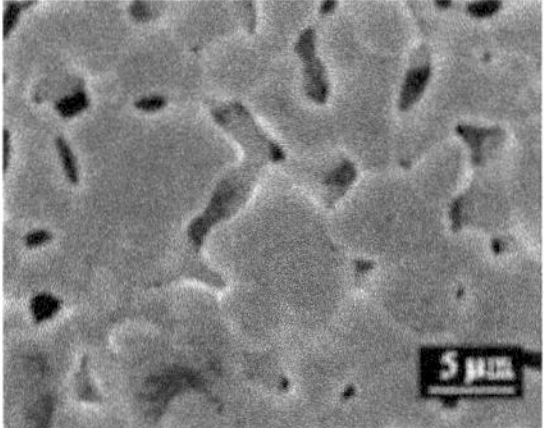

Fig. 15.14. SEM picture of LCY14, film on YSZ, sintered at 1300°C in air for 1 h

Fig. 15.15. SEM picture of LCY11, film on YSZ, sintered at 1200°C in O_2 flow for 1 h.

Thick Films on Polycrystalline Al_2O_3 Substrates

SEM pictures of the films sintered at 1200°C in air and oxygen are shown in Figs. 15.16 and 15.17. The microstructures are very similar to the films on single crystalline (100) LAO and YSZ substrates sintered at similar temperatures. It was observed that grain growth is promoted by longer dwell times in sintering.

Fig. 15.16. SEM picture of AHCA2, film on Al_2O_3, sintered at 1200°C in air for 24 h.

Fig. 15.17. SEM picture of AHCA2, film on Al_2O_3, sintered at 1200°C in O_2 flow for 1 h.

15.2.5 XRD of $La_{0.67}Ca_{0.33}MnO_3$ Thick Films on Different Substrates

X-ray analysis was carried out on some of the representative thick films and also the powder from which the thick films were fabricated. It was found that all samples are single phase within the sensitivity of the X-ray measurements. The unit cell parameters and the volume of each sample are tabulated in Table 15.2.

The $La_{0.67}Ca_{0.33}MnO_3$ powder from which the ink of thick films were prepared has a pseudo-cubic structure with lattice parameter 7.7316±0.0025. From the Table 2, it was observed that the cubic lattice parameter is the smallest for the films sintered in an oxygen flow. This is because the oxygenation of thick films creates more Mn^{4+}, which has a smaller ionic size compared to Mn^{3+} [4].

Table 15.2. Lattice parameters and unit cell volume of the $La_{0.67}Ca_{0.33}MnO_3$ thick films on various substrates sintered at various temperatures and atmospheres. The samples were sintered at the temperature indicated for 1 h.

Substrates	Sintering temperatures (°C)	Atm.	Pseudo-cubic lattice parameters (Å)	Unit cell volume (Å^3)
LAO	1400	Air	7.7227±0.0009	460.5781±0.1674
		O_2	7.7092±0.0018	458.1760±0.3171
		N_2	7.7185±0.0014	459.8271±0.2479
YSZ	1400	Air	7.7237±0.0008	460.7664±0.1472
		O_2	7.7221±0.0014	460.4707±0.2512
		N_2	7.7734±0.0007	469.7081±0.1359
LAO	1300	Air	7.7135±0.0037	458.9368±0.6647
		O_2	7.7140±0.0014	459.0297±0.2545
YSZ	1300	Air	7.7142±0.0029	459.0669±0.5201
Al_2O_3	1200	Air	7.7202±0.0013	460.1392±0.2271
		O_2	7.7080±0.0027	457.9561±0.4847

This result agrees well with those reported earlier [5,6]. The thick film on YSZ substrate which was sintered in a nitrogen flow has the largest lattice parameters, which is the consequence of a drastic decrease of Mn^{4+} content.

15.2.6 Physical Properties of $La_{0.67}Ca_{0.33}MnO_3$ Thick Films

DC Electrical Resistivity

The DC electrical resistance of the thick films were measured by the standard four-point probe technique. The electrical contacts were made on contact pads of the films using conductive silver paint. The temperature dependence of the normalized resistivity, r(T)/r(RT), at zero field for the films deposited on single crystal (100) LAO substrates are shown in Fig. 15.7. In the inset of Fig. 15.18, we show the resistivity of the films.

In this study, all thick films on LAO substrates show a M-I transition at different peak temperatures, T_{p1}. The T_{p1} values are given in Table 15.3. The thick film sintered at 1400°C in oxygen shows two peaks in the r-T curves like the peaks observed in the bulk samples prepared from the powder from which the ink of the thick films were made. Fig. 15.19 shows the normalized r-T curves of a bulk sample prepared from the thick film powder, a thick film and another bulk sample from ref. [7]. The bulk sample prepared from the thick film powder was sintered at 1400°C in air for 5 h. The thick film on LAO substrate was sintered at 1400°C in an oxygen flow and the bulk sample from previous chapter was also sintered at 1400°C in air. The r-T behaviour of the bulk sample prepared from the thick film powder exhibits a very different behaviour, discussed previously [7].

The bulk sample which was prepared from the thick film powder has a very broad M-I transition and double peaks in the r-T curve like the films sintered in

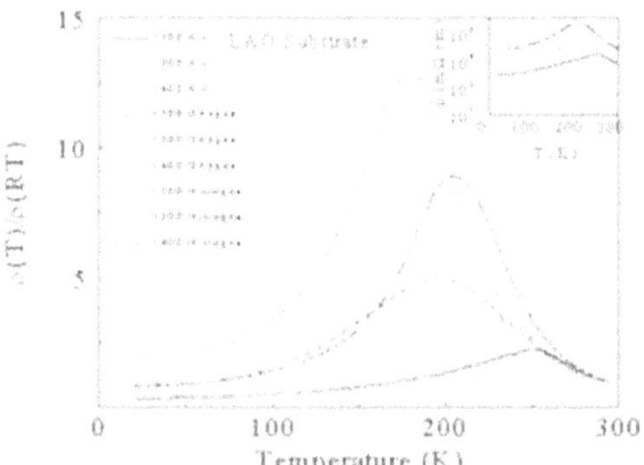

Fig. 15.18. Normalized zero field resistance ratio as a function of temperature for the $La_{0.67}Ca_{0.33}MnO_3$ thick films fabricated on single crystal (100) LAO substrates. Inset shows the resistivity as a function of temperature.

Table 15.3. The T_{p1}, T_c, activation energy E_A, thickness and the $r(T_{p1})/r(20K)$ of $La_{0.67}Ca_{0.33}MnO_3$ thick films on different substrates, where T_{p1} is the temperature corresponding to the resistivity peak. Tc is the Curie-Weiss temperature and is taken from minimum in dM/dT. The M(T) curves were recorded in the presence of 1 T applied field (ZFC). Results of a bulk sample prepared from this powder is also presented in this table.

Substrates	Sintering Temperature	Atmosphere	Thickness (nm)	T_{p1}(K)	$r(T_{p1})$/ r(20K)	E_A (meV)	T_c(K)
LAO	1400	Air	63.4	249.8	7.0	127	252
		O_2	46.2	257.6	9.0	108	264
		N_2	76.5	204.9	5.1	167	
YSZ	1400	Air	42.9	263.1	8.6	116	263
		O_2	54.5	186.8	47.1	133	175
		N_2	58.7	-	-	110	169
LAO	1300	Air	39.6	179.6	6.8	124	-
		O_2	60.2	226.1	9.0	152	
		N_2	31.4	257.0	5.1	120	
YSZ	1300	Air	46.9	243.9	19.7	126	-
		O_2	72.1	264.3	7.0	113	
		N_2	21.1	189.2	9.0	131	
LAO	1200	Air	27.7	251.7	6.1	142	-
		O_2	72.7	197.6	6.2	117	
		N_2	34.2	218.5	5.3	130	
YSZ	1200	Air	30.7	254.8	5.8	130	-
		O_2	30.3	214.0	6.1	114	
		N_2	32.0	243.2	5.8	115	
Al_2O_3	1200	Air	37.4	258.5	5.4	119	-
		O_2	44.5	253.3	4.5	109	
		N_2	41.0	263.3	4.3	110	
Bulk sample	1400	Air	-	254.8	4.5	119	-

the presence of oxygen flow. The origin of different behaviour of r-T in the present composition might be due to the fact that it was prepared using a slightly deviated method. We pre-heated the La_2O_3 powder before mixing with the other ingredients. Usually, $La(OH)_3$ and $La_2O_2CO_3$ coexist within the commercial La_2O_3. To eliminate $La(OH)_3$ and $La_2O_2CO_3$, we heat treated the commercial La_2O_3 at 1000°C for 6 h and then cooled in a desiccator. Recently, Sun et al. [8] reported that nearly stoichiometric composition can be obtained when commercial La_2O_3 is used and purified La_2O_3 (heat treated) results in oxygen deficiency. It possible that the increase in r, in the present composition of

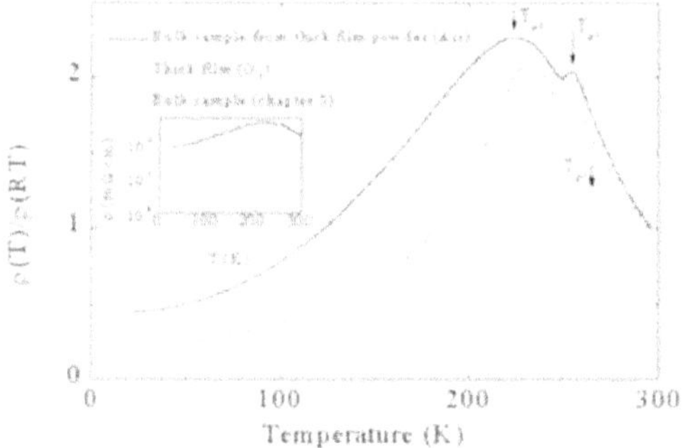

Fig. 15.19. Normalized zero field resistivities as a function of temperatures for a bulk sample prepared from the thick film powder, thick film on single crystal (100) LAO substrate and an as prepared bulk sample from ref.[7]. The nominal chemical composition for all samples are $La_{0.67}Ca_{0.33}MnO_3$.

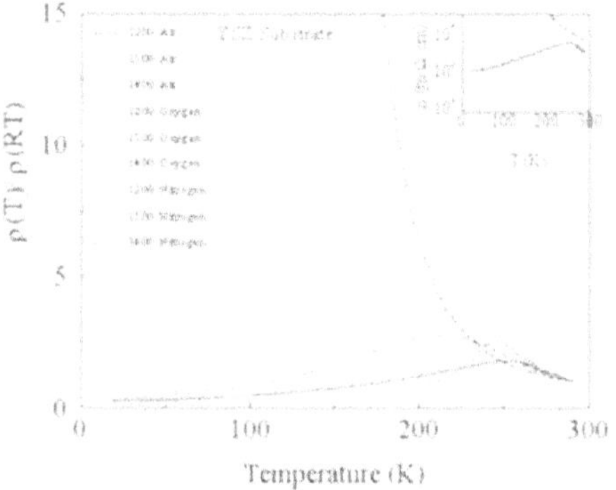

Fig. 15.20. Normalized zero field resistivity as a function of temperature for the $La_{0.67}Ca_{0.33}MnO_3$ thick films fabricated on single crystal (100) YSZ substrates. Inset shows the resistivity as a function of temperature.

$La_{0.67}Ca_{0.33}MnO_3$ (both bulk sample and thick films) is due to the oxygen deficiency as a result of processing route. Most of the thick films on LAO substrates have very broad M-I transitions at T_{p1} similar to the bulk samples prepared from the thick film powder. The T_{p1} varies from 180 to 258 K for the thick films on LAO substrates. The film sintered at 1300°C in air showed the lowest T_{p1} and the highest $r(T_{p1})/r(RT)$, $r(T_{p1})/r(20K)$. The highest T_{p1} was obtained for the film sintered at 1400°C in oxygen and this sample also has the lowest resistivity throughout the whole temperature range (295-20 K). The r above T_{p1} follows an activated behaviour, r(T) exp (EA/kBT) for all samples. The activation energies (EA) are given in Table 15.3. Fig. 15.20 shows the normalized r-T curves of the films fabricated on the single crystal (100) YSZ substrates.

The inset of Fig. 15.20 shows the corresponding r-T curves. These films are also sintered in similar conditions (Table 15.1) to those on LAO substrates. All films except the film sintered at 1400°C in nitrogen flow show a M-I transition at T_{p1}. The T_{p1} varies from 187-265 K for the thick films on YSZ substrates. The film sintered at 1400°C in oxygen has the lowest T_{p1} and the highest $r(T_{p1})/r(RT)$, $r(T_{p1})/r(20\,K)$. The film sintered at 1300°C in oxygen has the highest T_{p1}. The film sintered at 1400°C in air has the lowest resistivity in 295-190 K and the sample sintered at 1300°C in air has the lowest resistivity 190-20 K. The r above T_{p1} follows an activated behaviour with a slight variation of activation energies. The T_{p1} values and the activation energies are given Table 15.3.

The normalized r-T curves for the films fabricated on the polycrystalline Al_2O_3 substrates are shown in Fig. 15.21. These films are sintered at 1200°C in air, oxygen and nitrogen flow. The films sintered at 1300°C and 1400°C reacted with the substrates and were therefore not suitable for further investigations. The measured films showed a M-I transition at T_{p1}. The air and nitrogen sintered samples showed a peak with an additional shoulder. The oxygen sintered sample shows a broad peak only. The T_{p1} varies from 253-264 K. The film sintered in oxygen has the lowest T_{p1} and the sample sintered in nitrogen has the highest T_{p1}. The nitrogen and oxygen sintered samples show the lowest resistivity values. The resistivity above T_{p1} for these films also follow an activated behaviour. The T_{p1} and the activation energies are given in Table 15.3. In this study we have observed a wide variation of T_{p1} for thick films on different substrates sintered at different atmospheres. Choice of substrates and sintering conditions has a great influence on the properties of manganite materials [9,10]. Recently, Balcells and collaborators [11–13] reported magnetoresistive properties of $La_{2/3}Sr_{1/3}MnO_3$ thick films, prepared by either spray printing or screen printing techniques on polycrystalline Al_2O_3 substrates, sintered at 1400°C in oxygen. However, our $La_{0.67}Ca_{0.33}MnO_3$ thick films on Al_2O_3 substrates, sintered at 1400°C in air,

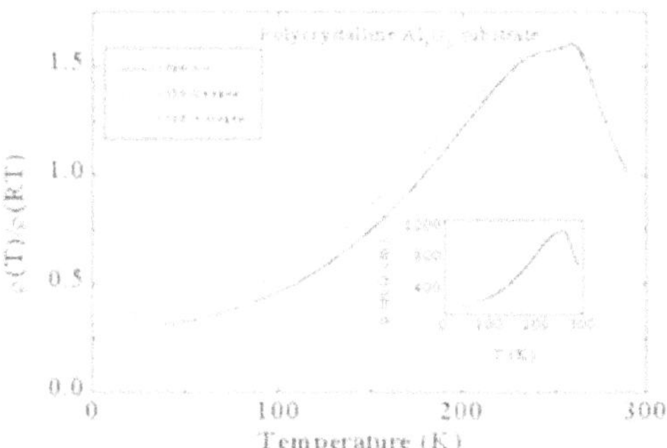

Fig. 15.21. Normalized zero field resistivity as a function of temperature for the $La_{0.67}Ca_{0.33}MnO_3$ thick films fabricated on polycrystalline Al_2O_3 substrates. Inset shows the resistivity as a function of temperature.

oxygen and nitrogen reacted heavily with the substrates. This suggests that the chemical composition also has an influence on the properties of manganite thick films.

Magnetoresistance

The MR(8T) as a function of temperature for some representative thick films on LAO are shown in Fig. 15.22(a). We observed that in addition to MR peaks around T_{p1} all samples have low-temperature MR. The maximum MR, 80 %, was obtained for the sample which was sintered at 1300°C in air at T_{m1}=180 K. The salient feature of the MR of our present thick films is that MR is not limited to a small temperature window, as observed in thin epitaxial films [14,16]. The MR peaks are either (i) very broad, or (ii) have a slightly smaller magnitude of low-temperature MR which is essentially temperature independent.

The MR isotherm as a function of applied field for some representative thick films on LAO substrates are shown in Fig. 15.22(b). This graph indicates the MR sensitivities as a function of temperature and field for the films. MR has four different types of sensitivity as a function of temperature and field (i) Highly sensitive low-temperature ($T \ll T_{p1}$) and low-field (<1T) MR. (ii) Weakly sensitive low-temperature MR with high applied field (1 T-3 T). (iii) MR around T_{p1}. (iv) High temperature ($T>T_{p1}$) MR.

The MR(8T) as a function of temperature for several thick films on YSZ substrates are shown in Fig. 15.23(a). The maximum MR, 94 %, at T_{m1} (184 K) was obtained for the sample sintered at 1400°C in oxygen atmosphere. Figure 12:

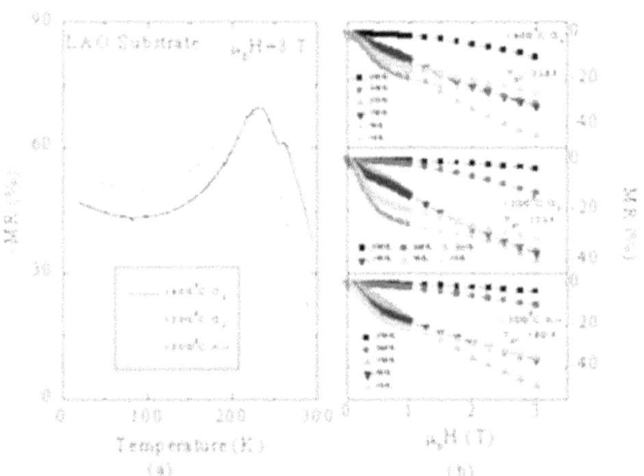

Fig. 15.22. (a) The MR(8T) as a function of temperature. (b) The MR(T, H) for thick films of $La_{0.67}Ca_{0.33}MnO_3$ on single crystal (100) LAO substrates sintered at various temperatures and atmospheres

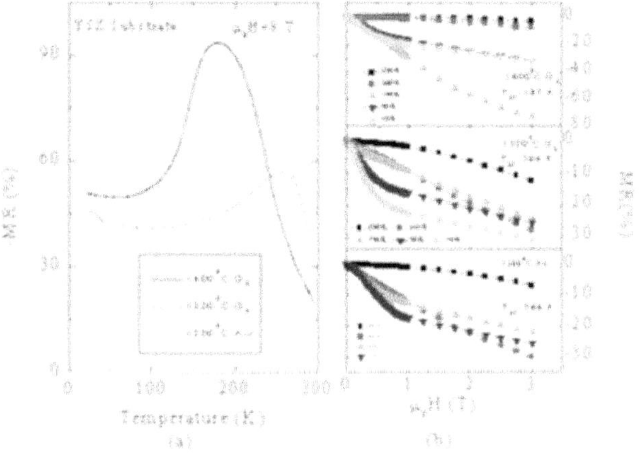

Fig. 15.23. (a)The MR(8T) as a function of temperature. (b) The MR(T, H) for thick films of $La_{0.67}Ca_{0.33}MnO_3$ on single crystal (100) YSZ substrate sintered at various temperatures and atmospheres.

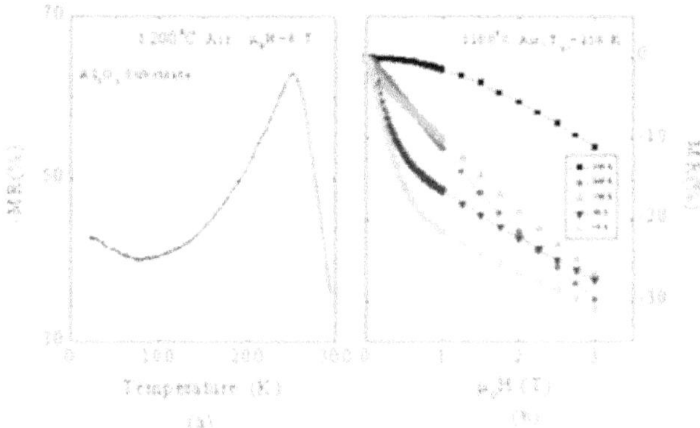

Fig. 15.24. (a) The MR(8T) as a function of temperature. (b) The MR(T, H) for thick films of $La_{0.67}Ca_{0.33}MnO_3$ on polycrystalline Al_2O_3 substrate sintered at 1200°C in Air.

(a)The MR(8T) as a function of temperature. (b) The MR(T, H) for thick films of $La_{0.67}Ca_{0.33}MnO_3$ on single crystal (100) YSZ substrate sintered at various temperatures and atmospheres.

The MR isotherms for thick films on YSZ substrates are shown in Fig. 15.23(b). MR(8T) for the film on alumina substrate is shown in Fig. 15.24(a).

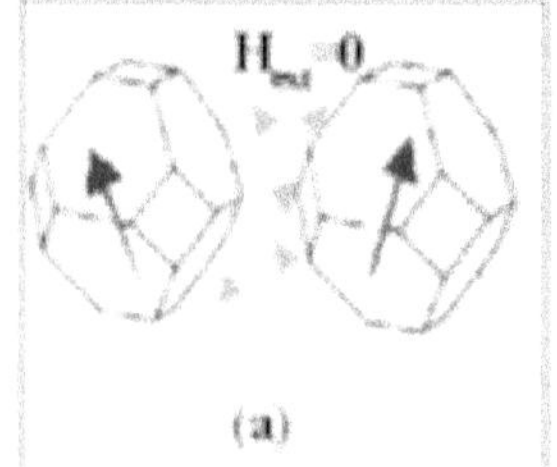
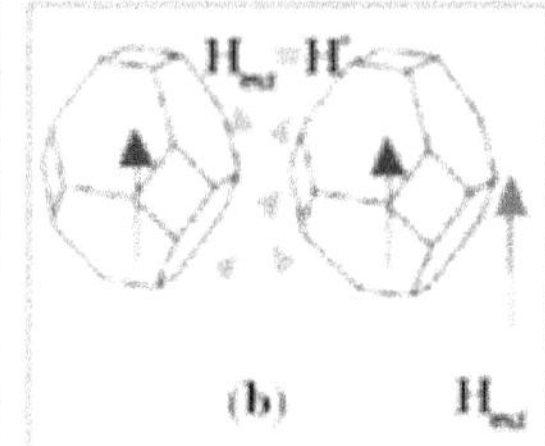
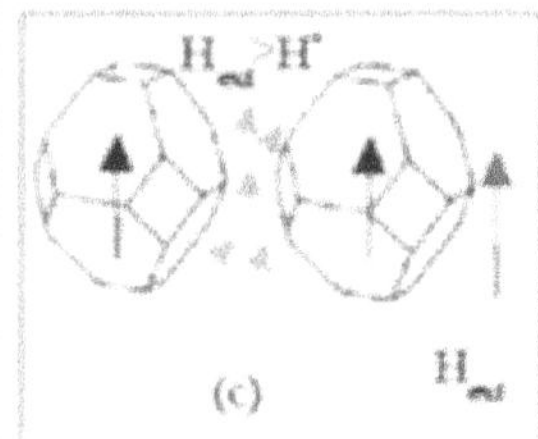

Fig. 15.25. Explanation of the two slopes MR at low-temperature (T≪Tc) with the help of grain and grain boundary magnetisation (a) zero applied field H_{ext}=0, (b) H_{ext}=H* and (c) H_{ext} >H*. We assumed tetrakaidecahedral (truncated octahedral) grain shape.

The maximum MR, 65 %, is observed at T_{m1}=252 K. This sample also shows a low-temperature MR, 43 %, at 20 K. Fig. 15.24(b) shows the MR(T, H) at various temperatures from 0-3 T field.

The low-temperature MR(H) show two slopes. The strong field dependence of MR exists for an applied field below H*. It was observed previously that H* is a function of temperature and H*(T)∼H_s(T) determined magnetically, where H_s is the saturation magnetising field. The low-temperature, low-field (below H*) MR has been attributed to either spin- polarised tunneling [17], spin dependent scattering [18–20] or micromagnetic behaviour associated with alignment of magnetic domain at the grain boundaries [21]. The origin of low-temperature, high field (above H*) MR might be due to the disordered or canted spin in the grain boundary region. Fig. 15.25 represents one possible explanation of both the strong and weak field dependence of MR at low-temperatures. At $T \ll T_c$ (where T_c is the paramagnetic to ferromagnetic transition temperature, which is close to the T_p) the material is in the ferromagnetic regime. However, in the absence of field the magnetisation of the grain of the polycrystalline material will be like that in Fig. 15.25a. Also, the individual spins at the grain boundary region are randomly oriented. In the absence of field, a carrier will be suffer scattering from the unaligned magnetic domain, as well as disordered spin at the grain boundary region. By applying a low magnetic field, the magnetisation of each grain starts to align towards the direction of the external magnetic field direction as shown in Fig. 15.25b. However, a large magnetic field is required to align the spins of the grain boundaries as shown in Fig. 15.25c.

Fig. 15.26 shows a comparison of low-temperature (40 K) MR sintered at different atmospheres (e.g. air and oxygen) on different substrates (LAO and YSZ) at 1300°C. It is obvious from this figure that low-temperature and low-field MR sensitivity is higher in the sample sintered in oxygen atmosphere.

15.2.7 Magnetic Properties $La_{0.67}Ca_{0.33}MnO_3$ Thick Films

Fig. 15.27 shows the normalized M-T curves for a few representative thick films on different substrates sintered only at 1400°C in different atmospheres for an

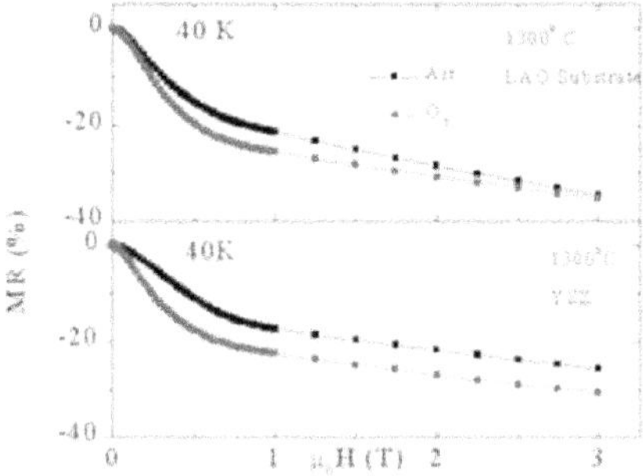

Fig. 15.26. Low temperature MR of $La_{0.67}Ca_{0.33}MnO_3$ as a function of field on different substrates sintered at different atmospheres.

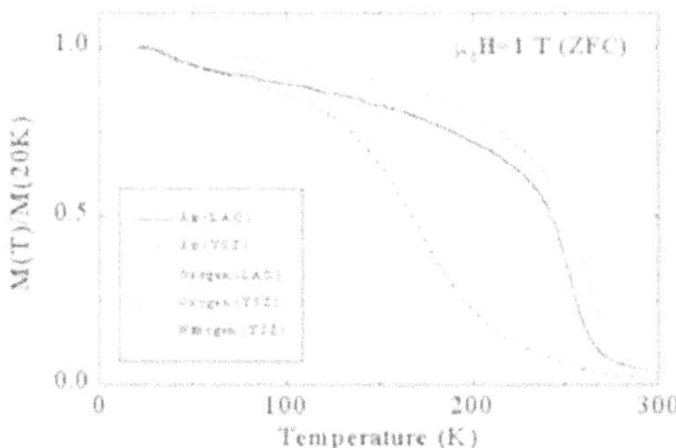

Fig. 15.27. Normalized magnetisation as a function of temperature for a few thick films of $La_{0.67}Ca_{0.33}MnO_3$ on different substrates sintered at 1400°C in different atmospheres. The data were taken in presence of 1 T applied field (ZFC).

hour. This data was taken in presence of 1 T applied field. The samples were cooled down to 20 K in the absence of field and then warmed up with 1 T applied field. All samples show a paramagnetic to ferromagnetic transition at T_c. The Tcs are defined by the minima in dM/dT are given in Table 15.3. It was observed that the Tcs are very close to T_{p1}. However, the thick film on YSZ sintered at 1400°C in nitrogen atmosphere has a T_c ~169 K, but no T_{p1} was present in the r-T curve.

15.2.8 MR as a Function of Magnetisation for $La_{0.67}Ca_{0.33}MnO_3$ Thick Films

We have investigated the relationship between MR and magnetisation for a thick film of nominal composition $La_{0.67}Ca_{0.33}MnO_3$ on polycrystalline Al_2O_3 substrate, sintered at 1200°C in an oxygen flow. Fig. 15.28 shows the MR as a

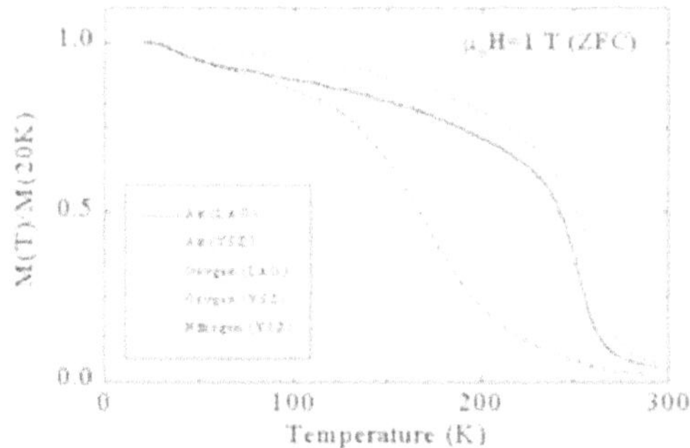

Fig. 15.28. The MR as a function of magnetisation for the $La_{0.67}Ca_{0.33}MnO_3$ film on Al_2O_3 sintered at 1200°C in the presence of oxygen flow. The Ms is taken as M(20 K,8 T). Inset shows the variation of C with various temperatures. The points show the experimental data and the lines are the fit of the theoretical model of Inoue and Maekawa.

function of magnetisation for this film at various temperatures. The points show the experimental data and the lines are the fit of the theoretical model of Inoue and Maekawa [22]. In fact, for the fit, we have used MR(%)=C(M/Ms)2 and the C parameters are shown in the inset as a function of temperatures.

The experimental data fitted well with the theoretical model at $T \gg T_{p1}$ corresponding to $M \ll 0.4(M/Ms)$. The saturation magnetisation Ms is taken as M(20 K,8 T). This result agrees well with the result obtained for the bulk samples of same nominal composition ref [7] and Urushibara et al. [23] in the $La_{1-x}Sr_xMnO_3$ single crystals.

15.2.9 Post-annealing of $La_{0.67}Ca_{0.33}MnO_3$ Thick Films

The thick films sintered at 1400°C in air were post-annealed in a controlled oxygen atmosphere. The film on LAO was annealed at 600°C and film on YSZ was annealed at 900°C. Post-annealing conditions are given in Table 15.4.

Table 15.4. Post-annealing conditions of $La_{0.67}Ca_{0.33}MnO_3$ thick films on various substrates.

Sample No	Subst-rates	Heating rate °C/h	T_{max} °C	Time at T_{max} (h)	Atmos-phere	Cooling rate °C/h
LCL5	LAO	100	600	6	0.8 PO_2	100
LCY5	YSZ	100	900	6	0.8 PO_2	100

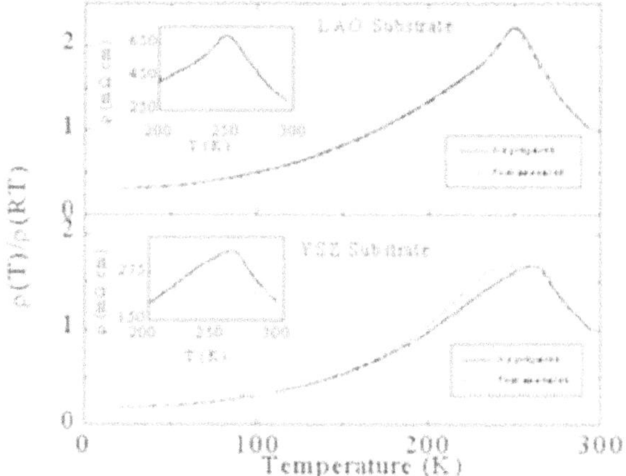

Fig. 15.29. The normalized zero field resistivity as a function of temperature for the as-prepared (see text) and the post-annealed thick films of $La_{0.67}Ca_{0.33}MnO_3$ on LAO and YSZ substrates. Post-annealing conditions were given in Table 15.4. Insets show the r(T).

The normalized r as a function of temperature for the as-prepared and the post-annealed films are shown in Fig. 15.29. The inset gives the r values of the samples.

There is not any significant change in the shape of the r-T curve for the film on LAO substrates; however, the film on YSZ exhibits a broad peak with an additional shoulder. There is a decrease of the r(T) as a result of post-annealing.

15.2.10 Reproducibility of $La_{0.67}Ca_{0.33}MnO_3$ Thick Films

To check the reproducibility of the $La_{0.67}Ca_{0.33}MnO_3$ thick films, we have repeated the 1400°C oxygen sintered sample (LCL13 sample in Table 15.1). We have used thick films on LAO substrate only for the above mentioned repeat sintering. We have also repeated 1200°C oxygen sintered $La_{0.67}Ca_{0.33}MnO_3$ thick film on polycrystalline Al_2O_3 substrate. The transport results of the first and repeat sintered thick films on LAO and Al_2O_3 are shown in Fig. 15.19.

From Fig. 15.30, we found that the form of r(T) is very similar for both first and repeat sintered thick films. From this random reproducibility check, and assuming that R(H, T) is also similar, it is concluded that the thick films are pretty much reproducible.

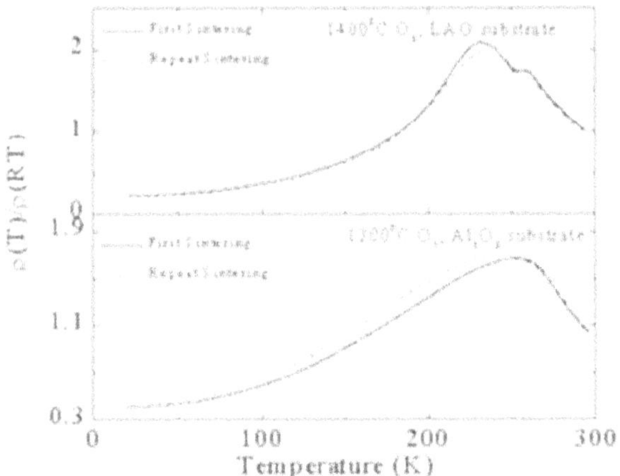

Fig. 15.30. The normalized zero field resistivity as a function of temperature for the first and repeat sintered thick films of $La_{0.67}Ca_{0.33}MnO_3$ thick films on various substrates.

15.3 Processing of $La_{0.63}Y_{0.07}Ca_{0.30}MnO_3$ Thick Films

15.3.1 Calcination of the Composition and Fabrication of Thick Films

The $La_{0.63}Y_{0.07}Ca_{0.30}MnO_3$ powders were synthesised using a conventional solid state reaction technique using the appropriate proportions of ingredients. We followed the same calcination schedule as $La_{0.67}Ca_{0.33}MnO_3$ which was described in Section 15.2. Calcined powders were ultrasonically dispersed in water media for fine particle size. Thick film ink was made from these fine powders by mixing with Blythe Vehicle. Thick films were fabricated by screen printing onto single crystal (100) LAO and polycrystalline Al_2O_3 substrates. The organic vehicle was then burnt out from the thick films.

15.3.2 Sintering of $La_{0.63}Y_{0.07}Ca_{0.30}MnO_3$ Thick Films

We sintered $La_{0.63}Y_{0.07}Ca_{0.30}MnO_3$ thick films only at 1200°C in air and oxygen atmospheres. Table 15.5 gives the sintering conditions of these thick films.

15.3.3 Microstructures of $La_{0.63}Y_{0.07}Ca_{0.30}MnO_3$ Thick Films on Different Substrates

SEM pictures of $La_{0.63}Y_{0.07}Ca_{0.30}MnO_3$ films on both substrates, sintered in oxygen flow are shown in Figs. 15.31 and 15.32.

Table 15.5. Annealing conditions of $La_{0.63}Y_{0.07}Ca_{0.30}MnO_3$ thick films on single crystal (100) LAO and polycrystalline Al_2O_3 substrates.

Sample No	Substrates	Heating rate °C/h	T_{max} °C	Time at T_{max} (h)	Atmosphere	Cooling rate (1) °C/h	Cooling rate (2) °C/h
AHYL1 AHYA1	LAO Al_2O_3	100	1200	1	Air	100 down to 600°C and then halt 12.h	100
AHYL2 AHYA2	LAO Al_2O_3	100	1200	1	O_2	100 down to 600°C and then halt 12.h	100

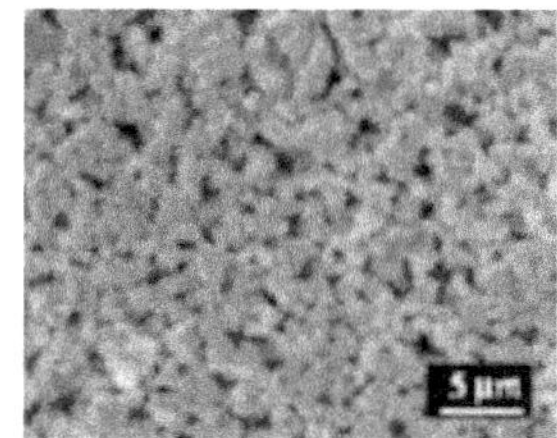

Fig. 15.31. SEM picture of AHYL2, film on LAO, sintered at 1200°C in O_2 flow.

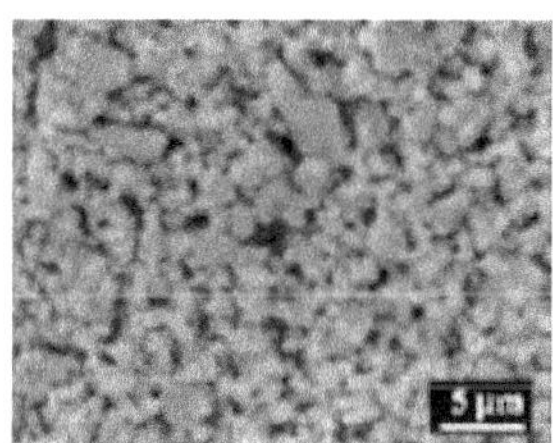

Fig. 15.32. SEM picture of AHYA2, film on Al_2O_3, sintered at 1200°C in O_2 flow.

15.3.4 Physical Properties of $La_{0.03}Y_{0.07}Ca_{0.30}MnO_3$ Thick Films

DC Resistance

The normalized resistance as a function of temperature for films on different substrates and sintered in different atmospheres are shown in Fig. 15.33. The results of a bulk sample prepared from this powder, which was sintered at 1400°C for 5 h in air is also shown in Fig. 15.28.

The transport results are summarised in Table 15.6.

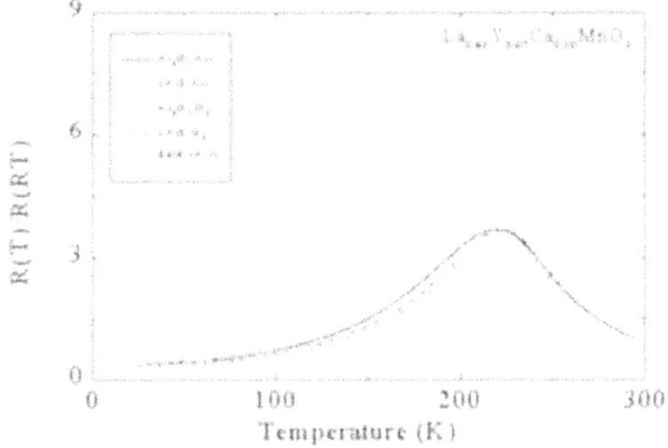

Fig. 15.33. The normalized zero field resistance as a function of temperature for the thick films of nominal composition $La_{0.63}Y_{0.07}Ca_{0.30}MnO_3$ on various substrates sintered at various atmospheres.

Table 15.6. The T_{p1}, the $R(T_{p1})/R(20\,K)$ and the EA for the thick films of $La_{0.63}Y_{0.07}Ca_{0.30}MnO_3$ on different substrates sintered at different temperatures and the bulk sample prepared from thick film powder.

Substrates	Sintering Tempera ture (°C)	Atmos-phere	Thick-ness (nm)	T_{p1}(K)	$r(T_{p1})/r(20\,K)$	E_A (meV)
LAO	1200	Air	-	221.9	10.2	125.9
		O_2	48.98	222.9	10.2	124.2
Al_2O_3	1200	Air	-	219.2	8.9	125.1
		O_2	48.43	224.5	8.2	129.3
Bulk	1400	Air	-	200.5	35.1	125.7

All thick films show a peak in the resistance versus temperature curves at T_{p1} (Table 15.6). There is a slight variation of T_{p1} (2-5 K) among the thick films on different substrates sintered at different temperatures. However, the bulk sample prepared from the thick film powder has a significantly lower T_{p1}. The quantity $R(T_{p1})/R(20\,K)$, which measures the sharpness of the peak, is highest for the bulk samples. This is because the bulk sample was sintered at a higher temperature and for a longer duration of time than the thick films.

Magnetoresistance of $La_{0.63}Y_{0.07}Ca_{0.30}MnO_3$ Thick Films

Fig. 15.34(a) shows the MR as a function of temperature for the films on Al_2O_3 sintered in oxygen flow. This film show 71 % MR near T_{p1} (220 K) in 8 T applied field and 46 % MR at 20 K.

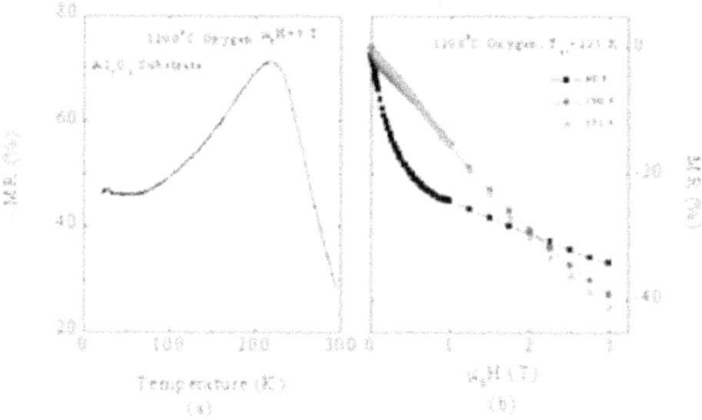

Fig. 15.34. (a) The MR(8T) as a function of temperature (b) MR(T, H) for the thick film of $La_{0.63}Y_{0.07}Ca_{0.30}MnO_3$ on polycrystalline substrates, sintered at 1200°C in oxygen.

The MR isotherms as a function of applied fields are shown in Fig. 34(b). Like the thick films of $La_{0.67}Ca_{0.33}MnO_3$ this film also shows two MR slopes at a temperature much below T_{p1}. The low-temperature and low-field MR sensitivity is much sharper than the bulk or thick films of nominal composition $La_{0.67}Ca_{0.33}MnO_3$.

15.3.5 Reproducibility of $La_{0.63}Y_{0.07}Ca_{0.30}MnO_3$ Thick Films

We repeated the 1200°C oxygen sintering for thick films of $La_{0.63}Y_{0.07}Ca_{0.30}MnO_3$ on polycrystalline Al_2O_3 substrate, to check the reproducibility of the thick films. The electrical transport results for the first and repeat sintered films are shown in Fig. 15.35. Results of two thick films (represented by a and b) of repeated sintering are presented in Fig. 15.23.

It was observed that T_{p1} decreases from 225 K to 217 K, and R/R(RT) at T_{p1} increases from 3.4 to 4.1, for the repeat sintered films compared to first sintered film. It is possible that this difference is due to the fact that repeat sintering was done in a different furnace, and the temperature calibration has slightly deviated from the first furnace. However, the two repeat sintered samples have identical r(T) as shown in Fig. 15.23.

15.4 Conclusions

We have fabricated manganite thick films on various substrates using a screen printing technique. The films were sintered at various temperatures and atmospheres to find the optimum sintering conditions. The magneto-transport properties of these films were investigated. We have observed a wide variation of T_{p1}

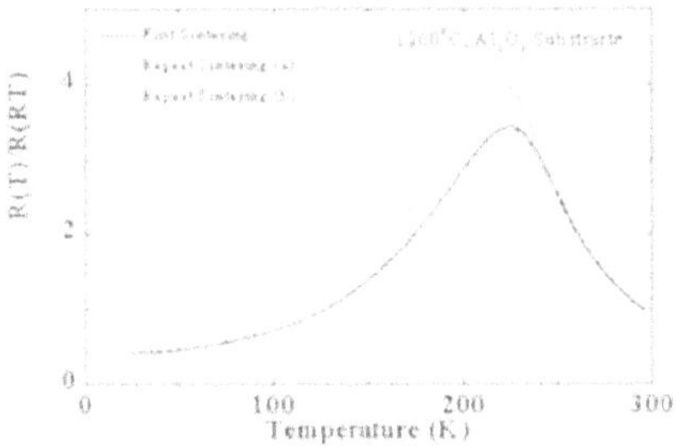

Fig. 15.35. The normalized zero field resistance as a function of temperature for the first and repeat sintered thick films of $La_{0.63}Y_{0.07}Ca_{0.30}MnO_3$.

for the thick films on different substrates sintered at different temperatures and atmospheres. This variation is due to the influence of substrates and different oxygen partial pressures in sintering as the growth conditions are important for the manganite thick films. The physical properties of the thick films are either similar or much better than the bulk samples prepared from the thick film powders. The highest MR obtained for the present thick film is 94 % in 8 T applied field near T_{p1}. Besides this, a low-temperature MR of 51 % has been obtained at 20 K, and for some films this low-temperature MR is temperature independent. The MR as a function of field and temperature for the thick films have characteristics of polycrystalline bulk samples. The low-temperature and low-field MR slope is much sharper in $La_{0.63}Y_{0.07}Ca_{0.30}MnO_3$ and $La_{0.55}Sr_{0.45}MnO_3$ thick films compared to $La_{0.67}Ca_{0.33}MnO_3$ thick films. The ferromagnetic transition temperature, T_c, is near to the T_{p1} observed in the transport measurements. The MR above T_{p1} is found to be proportional to (M/Ms)2 and the magnitude of the proportionality constant decreases with decreasing temperature. It is concluded from our experiment that the highest sintering temperature for the thick films on polycrystalline Al_2O_3 is 1200°C in all atmospheres (e.g. air, O_2 and N_2). The highest sintering temperature for the single crystal substrates was 1400°C. However, the thick films on YSZ, sintered at 1400°C in nitrogen atmosphere depressed the M-I transition below 70 K. The thick film on YSZ sintered at 1400°C in oxygen and film on LAO sintered at 1300°C in air have shown the highest MR and the highest $r(T_{p1})/r(20\,K)$. The screen printing has proven to be a successful technique for the preparation of high quality reproducible thick films.

Acknowledgements

Author would like to thank Association of Commonwealth Universities for the scholarship tenable at Imperial College, London. Author also would like to thank Dr. L. F. Cohen and Dr. J. Driscoll of Imperial College, London, allowing him to use their Laboratory Facilities.

References

1. N. McN. Alford, S. J. Penn, and T. W. Button, Superconducting Science and Technology **10**, 169 (1997)
2. N. M. White and J. D. Turner, Measurement Science and Technology **8**, 1 (1997)
3. R. Valenzuela, Magnetic ceramics (Cambridge University Press, 1994).
4. R. D. Shannon, Acta Crystallography **A32**, 751 (1976)
5. L. Ghivelder, I. Abrego Castillo, N. McN. Alford et al., Journal of Magnetism and Magnetic Materials **189**, 274 (1998)
6. J. Pierre, F. Robaut, S. Misat et al., Physica B **225**, 214 (1996)
7. A. K. M. Akther Hossain, L. F. Cohen, F. Damay, A. Berenov, J. MacManus Driscoll, N. McN. Alford, N. D. Mathur, M. G. Blamire, and J. E. Evetts., Journal of Magnetism and Magnetic Materials **192**, 263 (1999)
8. J. R. Sun, G. H. Rao, and Y. Z. Zhang., Applied Physics Letters **72**, 3208 (1998)
9. E. S. Vlakhov, R. A. Chakalov, K. A. Nenkov et al., Journal of Applied Physics **83**, 2152 (1998)
10. T. Y. Koo, S. H. Park, K.-B. Lee et al., Applied Physics Letters **71**, 977 (1997)
11. Ll. Balcells, R. Enrich, J. Mora et al., Applied Physics Letters **69**, 1486 (1996)
12. Ll. Balcells, R. Enrich, A. Calleja et al., Journal of Applied Physics **81**, 4298 (1997)
13. L. Durand, Ll. Balcells, A. Calleja et al., Journal of Material Research **13**, 2623 (1998)
14. S. Jin, M. McCormack, T. H. Tiefel et al., Journal of Applied Physics **76**, 6929 (1994)
15. S. Jin, T. H. Tiefel, M. McCormack et al., Science **264**, 413 (1994)
16. S. Jin, H. M. O'Bryan, T. H. Tiefel et al., Aplied Physics Letters **66**, 382 (1995)
17. H. Y. Hwang, S-W. Cheong, N. P. Ong et al., Physical Review Letters **77**, 2041 (1996)
18. A. Gupta, G. Q. Gong, Gang Xiao et al., Physical Review **B54**, R15629 (1996)
19. R. Mahesh, R. Mahendiran, A. K. Raychaudhuri et al., Applied Physics Letters **68**, 2291 (1996)
20. X. W. Li, A. Gupta, Gang Xiao et al., Applied Physics Letters **71**, 1124 (1997)
21. J. E. Evetts, M. G. Blamire, N. D. Mathur et al., Phil. Trans. R. Soc. Lond. **A356**, 1593 (1998)
22. J. Inoue and S. Maekawa, Physical Review Letters **74**, 3407 (1995)
23. A. Urushibara, Y. Moritomo, T. Arima et al., Physical Review **B51**, 14103 (1995)

16 Device Applications Using Spin Dependent Tunneling and Nanostructured Materials

Mark Tondra, Dexin Wang, and Zhenghong Qian

NVE Corporation, 11409 Valley View Road, Eden Prairie, MN 55344, USA

Abstract. This paper describes recent developments in magnetoresistive materials and their applications. Also, it will focus on a few magnetoresistive materials and applications that are under active development. Two applications are chosen, in part, due to their connection to nano-scale structures. These are 1) magnetoresistive isolators whose speed will be enhanced by nano-crystalline ferromagnetic layers, and 2) detection and manipulation of nano-magnetic particles using magnetoresistive sensors and integrated straps.

16.1 Introduction

16.1.1 Scope

This paper is intended to be a useful reference for a graduate student who is looking for information about possible research projects in the field of nanostructured magnetic materials and their applications and in understanding the broader technological implications of their work. More experienced researchers may also find it useful as a presentation of a few particular applications that are relatively new in the field of magnetoresistive sensors. No attempt will be made to cover all applications of magnetoresistive devices. Rather, the focus will be on materials and devices that are under active development at NVE. Two major applications will also not be covered, though they are under development at NVE. Specifically, read heads for disk drives are discussed in other papers from the workshop, and Magnetoresistive Random Access Memory (MRAM) requires a separate discussion altogether.

NVE is actively working on two kinds of magnetoresistive materials, Giant Magnetoresistive (GMR) and Spin Dependent Tunneling (SDT). Products using GMR are available, while those using SDT are still under development. Products using GMR include simple resistor bridge field sensors and gradiometers, and digital signal isolators.

Discussion will be limited to magnetoresistive effects related to thin ferromagnetic films of Ni, Fe, Co, and alloys. Specifically excluded are Hall effect devices and Colossal Magnetoresistance effect devices.

16.1.2 Why Magnetoresistance?

Magnetoresistance is the change in the electrical resistance of materials as a function of an externally applied magnetic field. There are many forms of magnetoresistance, but the most useful ones are observed in ferromagnetic thin films, usually alloys of Ni, Fe, and Co. These effects are useful because magnetoresistive thin films can be manufacturable in an integrated circuit process. They can be deposited on semiconductor wafers (with or without integrated circuits) and patterned into many forms of devices using standard photolithography and etching techniques. Resulting magnetoresistive products, then, benefit from the same commercial forces that have made integrated circuits so valuable and successful. They are extremely small, very low cost, and are easily combined with many electronic functions.

The easy combination with electronics is the key to what makes magnetoresistive thin films so important. The same underlying technology can be applied to many different applications. Examples are: generic linear magnetic field sensors; special function magnetic field sensors; read heads for computer disk drives, magnetic tapes, credit cards, currency, etc.; current sensing, digital data transmission isolation, and digital data storage. There are many details that must be addressed in order to have a successful device in any of these categories. The size of the magnetotransducer will vary, as will the circuit complexity, packaging, power requirements, cost, etc. But the basic magnetoresistive effect is common to all of them.

16.1.3 Physics Background

AMR. The understanding of the physics of magnetoresistance in ferromagnetic materials goes back as far as 1857 when the Anisotropic Magnetoresistance (AMR) effect was first discovered in macroscopic ferromagnetic wires [1]. The AMR effect is observed as a relationship between the resistance of the material and the angle between the magnetization vector and the electrical current. The resistance is lowest when the current and magnetization are perpendicular and maximum when they are parallel/antiparallel. This effect is observable in bulk materials and in thin films. The magnitiude of the effect can be as large as 8% at room temperature. However, the AMR effect in the most commonly used materials is between 2 % and 4 %. The main tradeoffs involved are the desire to have low saturation fields, low hysteresis, low magnetostriction, and high linearity; while still maintaining high magnetoresistance. The most popular material for this compromise is the $Ni_{65}Fe_{15}Co_{20}$ (atomic percentage).

The physical origins of the AMR effect are "spin-orbit" interactions, and the imbalance of spin-up and spin-down conduction (Fermi energy) electrons. The spin orbit interactions cause different directional scattering of spin up and spin down electrons. The net spin imbalance at the Fermi surface creates a net imbalance of the scattered charge. These two effects together result in a resistivity that varies with the orientation of magnetization with respect to electrical current. Detailed explanations of observed resistivity variations have been largely worked out by Smit [2] and others.

GMR. The development of thin film deposition technology led to AMR sensors being the dominant technology for the read heads in hard disk drives. This dominance remained until the mid 1990's when Giant Magnetoresistive (GMR) devices took over. The GMR effect is fundamentally a thin film phenomenon. It is empirically observed to be a change in resistance as a function of the angle between the magnetizations of adjacent thin ferromagnetic thin films. Consider the simplest case of two thin NiFeCo ferromagnetic films separated by a thin Cu "spacer" layer (this structure is referred to as a "sandwich"). These films must not be in direct contact with each other or else their magnetizations will be directly coupled, and will not be able to rotate with respect to each other. However, they may not be separated by too much distance, or the electrons flowing in the sandwich will not be able to pass from one ferromagnetic layer to the other without loosing their "spin information". Typical dimensions for such a GMR sandwich are NiFeCo 5 / Cu 3.5 / NiFeCo 5 (thicknesses in nm). The thicknesses must be small compared to the effective scattering lengths of conduction electrons. The GMR effect can be observed with current flowing either in the plane of the thin films or perpendicular to it. Practically speaking, however, devices are made using the Current In the Plane (CIP) configuration because the Current Perpendicular to the Plane (CPP) configuration has a very low resistance unless the lateral dimensions can be made extremely small (~10 nm).

From a developmental standpoint, the discovery of GMR was not possible before vacuum deposition technology advanced to the point where layer thicknesses and roughnesses could be controlled to sub-nanometer dimensions. The usually cited discoverers of GMR were pushing on this technology in their search for magnetoresistive effects [3] and exchange coupling in magnetic multilayers [4].

Since these discoveries in the late 1980's, GMR has rapidly evolved to surpass AMR as the dominant magnetoresistive effect used in hard drive read heads in the mid 1990's. The basic read head structure is called a "spin valve." It is a sandwich as described above with one of the two magnetic layers "pinned" by an antiferromagnetic layer such as FeMn, or CrPtMn. A simple example of a spin valve structure is NiFeCo 5 / Cu 3.5 / CoFe 5 / CrPtMn 30.

The thin layer structure is critical because the physical effect that is being exploited is that the spin up and spin down electrons in the sandwich have different mean free paths that depend on the orientation of the magnetizations. These mean free paths are on the order of 10 nm.

The magnetic operation of a spin valve is such that the resistance is maximized when the magnetic layers are antiparallel and minimized when they are parallel.

The total magnitude of the spin valve GMR effect is 5 % to 25 % depending upon layer thickness, quality, and configuration.

SDT. Spin Dependent Tunneling (SDT) devices are again related to the relative orientations of two adjacent ferromagnetic films. However, the tunneling effect takes place when electrons pass across a nominally insulating barrier separating the two magnetic layers. Thus it is strictly a CPP effect. The first useful SDT devices were demonstrated in 1995 [5,6], and the first one using lithographical fabrication techniques in 1996 [7]. The tunnel barrier must be extremely thin in order to have a measurable tunneling current. A typical SDT structure is NiFeCo 12.5 / Al_2O_3 1.5 / CoFe 50 / CrPtMn 30. Here, the Al_2O_3 tunnel barrier

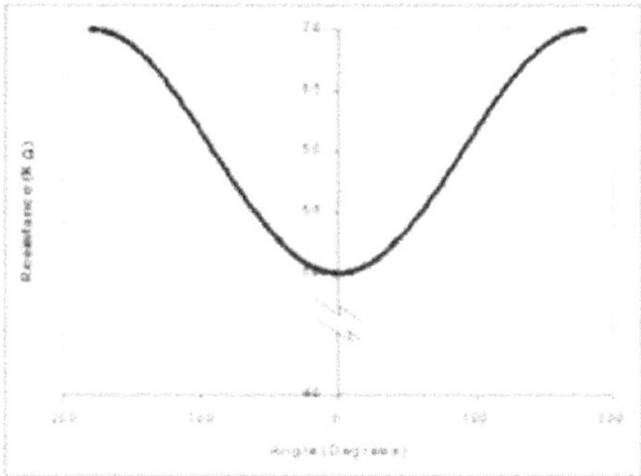

Fig. 16.1. The resistance of a spin valve as a function of the angle between magnetizations of top and bottom magnetic layers. The high resistance values shown are more likely to be observed in a tunnel junction. The typical sheet resistance of a spin valve is about 25 ohms/square, so this data would be representative of a GMR spin valve with 2000. I.e. its length to width ratio is 2000.

has replaced the Cu as the separating film in the spin valve. Magnetically, the operation of a pinned SDT device is identical to that of a GMR spin valve. The magnetoresistance, however, is larger. Total resistance changes on the order of 45 % are common.

SDT construction. Tunnel junction devices are more difficult to fabricate than GMR devices. Two main challenges are the creation of the very thin tunnel barrier with no pinholes, and the subsequent formation of usable structures out of the initial blank tunnel sandwich layers.

At NVE, the SDT devices are made in a Perkin Elmer 2400 with a baseline pressure in the 10^{-8} torr range. The layers are deposited using RF diode sputtering in an Ar plasma using an Ar pressure of 10's of mTorr. The Al_2O_3 barrier is formed by first depositing a 1.2 nm layer of Al, and subsequently oxidizing it by introducing some oxygen into the chamber while a plasma is still present over the sample wafer. As it oxidizes, the Al film expands about 30 % to its final thickness of about 1.5 nm. One or more annealing steps are required to optimize the SDT devices' resistive and magnetic properties.

Patterning of the junctions takes place in two steps. The first step is to etch away portions of the top electrode material while leaving the entire bottom electrode intact. The second step is to etch away undesired bottom electrode material. Both etches are performed using an ion mill for material removal, and a photolithographically defined etch mask. The finished result is a stack where the edges of top electrodes are entirely within the edges of the bottom electrode. The bottom electrode can support one or two tunnel junctions depending upon the intended finished structure.

The two junction SDT structure allows a tighter packing of junctions in a given area, and makes fabrication slightly easier as all connections to junctions are to the top electrodes.

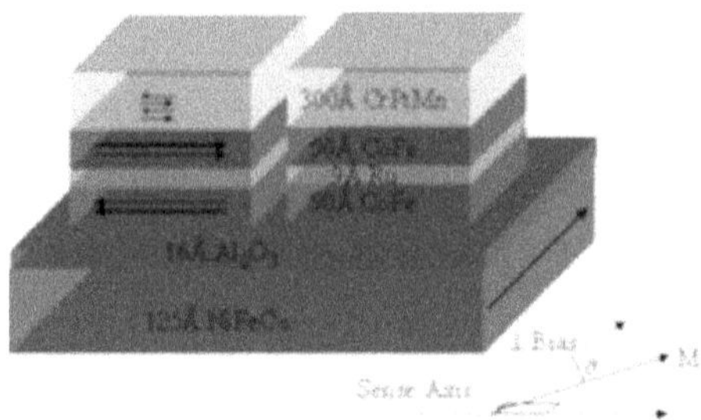

Fig. 16.2. SDT material stack and magnetic biasing mode.

16.2 Applications

16.2.1 Generic Magnetic Field Sensor

In order for the junctions to be useful as magnetic field sensors, more work is required. Such details are addressed here in the context of the SDT sensors. However, similar technology is also applied to GMR and AMR sensors.

Bridge. Typically, a resistance bridge is desirable rather than a single resistor or string of resistors. A bridge allows for null readings, helps compensate for temperature effects, and makes amplifier design much easier. A typical bridge starts with four identical resistors on each of the four bridge legs. For the resistance bridge to do its job, however, features must be added to make some legs behave differently than others. If this is not done, all the legs' resistances will increase and decrease at the same rate, and no net output will be observed. The tool NVE uses most commonly to achieve a non-zero bridge output are flux concentrators.

Field biasing. The SDT device response must be linearized. That is, the "square loop" output of an unbiased device must be turned into a linear, non-hysteretic output. The square loop shape occurs when the free layer magnetization of the SDT junction flips back and forth to be parallel and antiparallel to the pinned layer. This output becomes much smoother when a perpendicular bias field is applied so that the soft layer magnetization rotates rather than flips. The magnitude of the perpendicular bias field is just over the anisotropy field of the soft layer. This is about 18 Oe in NiFeCo, and less than 5 Oe in regular permalloy ($Ni_{80}Fe_{20}$). In the finished devices, the perpendicular field is generated by an on-chip planar coil. A second coil is also provided as a sensitive axis bias. The difference in biased and unbiased SDT response can be seen by comparing Figs. 16.3 and 16.4.

Flux concentrators. These are usually 15 μm thick plated NiFe. Two flux concentrators on a sensor chip are arranged so that they are separated by a long narrow "gap". The sensitive axis of the sensor is parallel to the short dimension of the gap. Two opposite resistor legs (gap legs) are placed in the gap, while the other two are placed underneath the flux concentrators (shield legs). The field in the gap is "amplified" by the flux concentrators such that the gap field is [flux concentrator length] / [gap] x [external field]. The field under the flux

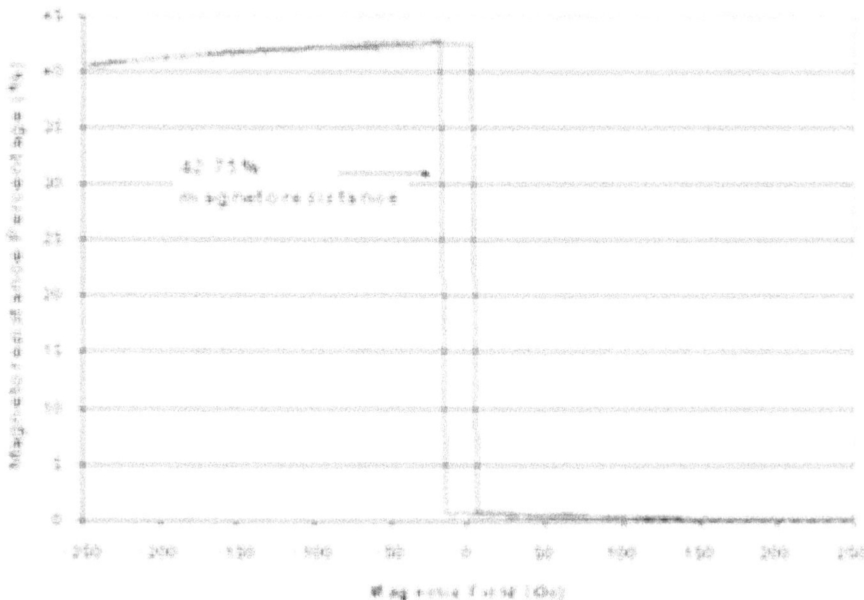

Fig. 16.3. Unbiased SDT response. The change in resistance occurs when the magnetization of the free layer flips 180 degrees. The left side of the plot shows higher resistance where the two magnetizations are antiparallel. The lower resistance on the right side is where the magnetizations are parallel. This tunnel junction is 120 x 320 microns and has a resistance of about 50 Ohms.

concentrators reduced by a similar fraction. The net result is that the gap legs see a much greater field than the shield legs, and a net bridge offset is realized. The flux concentrators also have the effect of attenuating off-axis fields, so that the sensor becomes nearly a vector sensor (outputting the magnitude of the sense-axis field).

Sensor output. Once all of the pieces are together, the sensor output can be observed. Typical sensor output is shown in Fig. 16.5.

16.2.2 Isolators

Digital signal isolators are a relatively new application for magnetoresistive materials. This section describes how they are used, and how adapting a nanocrystalline magnetic material to the SDT stack will enhance the performance of future versions of the magnetoresistive isolator.

The purpose of signal isolators is to allow information to pass from one electrical circuit to another without having a direct conduction path between the two. This helps reduce noise in data transmission, and protects sensitive

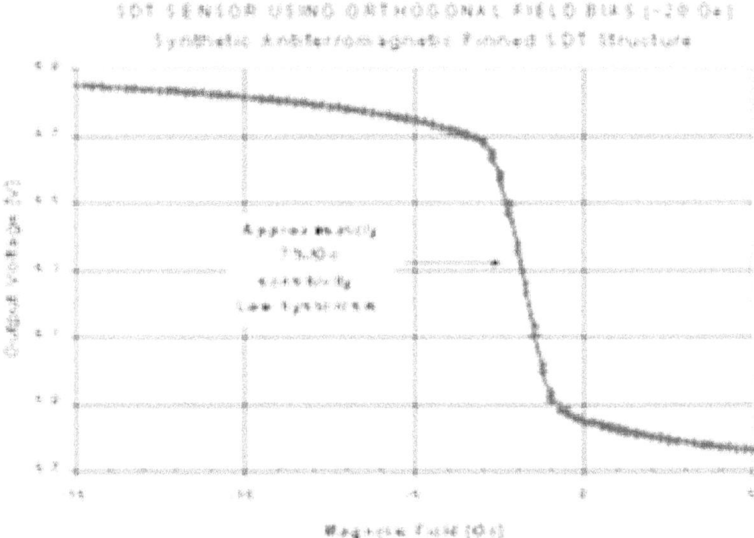

Fig. 16.4. When the SDT device has a proper magnetic bias, the magnetization of the free layer rotates smoothly.

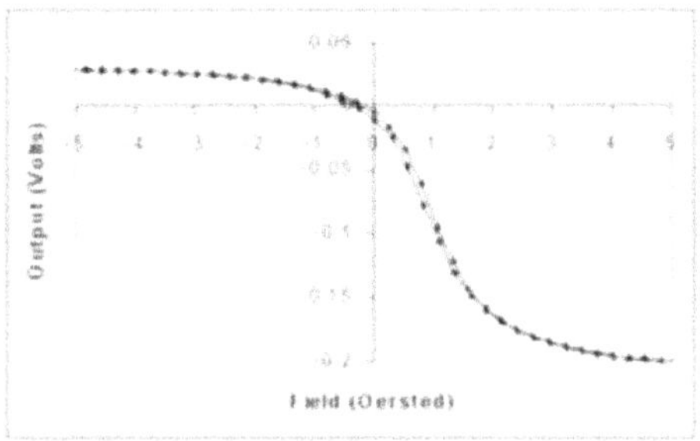

Fig. 16.5. Actual SDT sensor output vs. magnetic field. This bridge is being supplied with 2 Volts, so the output range of 200 mV represents about 10% of the total supply voltage. The most sensitive part of the output is shifted about 1 Oe due to coupling between the pinned and free layers. The shift in voltage (vertical) is due to imperfectly matched SDT resistor legs.

circuitry from potentially damaging voltages. In practice, it is challenging to do this data transmission at a high rate of speed without propagation delays. Isolators are used to protect the integrity of both digital and analog data. This paper will focus on the digital applications because that is what NVE has on the market.

As a way of describing basic isolator functions while also explaining why magnetorestive materials are so well suited for the application, the operation of optical and magnetic based isolators will be contrasted.

Optical. The dominant form of low cost integrated digital signal isolator is optically based. Current flowing into the input side of the isolation barrier goes through a Light Emitting Diode (LED). The light from this LED is directed across the isolation barrier to a semiconductor photodetector. The photodetector registers a signal once sufficient photons have been collected from the beam from the LED.

Magnetoresistive. The newest and fastest form of integrated digital signal isolator is based on integrated GMR devices. Current flowing into the input side of the isolation barrier goes through an on-chip planar coil that generates a magnetic field. The coil is situated so that the GMR spin valve devices are on the other side of the isolation barrier but in the field generated by the coil. The GMR is configured in a bridge so that the voltage out switches high or low depending upon which direction the coil current is flowing. This contrast is shown below in Figure 16.6.

Delays. At first, it is not obvious why the GMR isolator is faster than the optical isolator. Photons travelling from the LED to the photodetector should be just as fast as the magnetic field propogation from the coil to the GMR detector. The difference, though, is in the generation and detection of the signal rather than its propogation. The LED requires some time, typically 10's of ns, to generate enough photons to drive the photodetector beyond the switching threshhold of the detection electronics. This speed can be increased in proportion to the amount of current passing through the LED, but is limited by practical power requirements. The current in the GMR isolator, however, generates a magnetic field instantaneaously. The speed limit for the GMR, then, is how fast the magnetic sensing film's magnetization can rotate in the applied field. Thus, the underlying physics questions that must be asked are: 1) how fast can the sensing layer rotate, and 2) how much current (how large a field) is needed to cause rotation. Mathematically, this boils down to solving the Gilbert equation. Empirically, it is found that for reasonable amounts of current, the switching takes place in 1 ns or so.

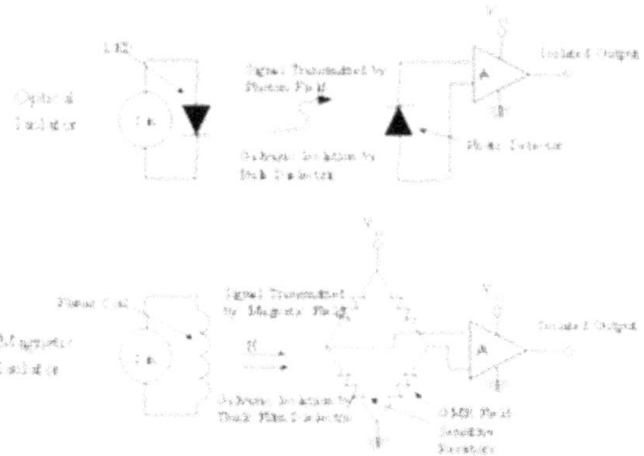

Fig. 16.6. Contrasting GMR isolator with optical isolator concepts.

Since the fundamental limitation for the GMR isolator switching time is related to the dynamics of the soft film's magnetization, research has been done to develop soft films with higher "speed limits". This is where the nanocrystalline ferromagnetic material becomes relevant.

16.2.3 Nanocrystalline Ferromagnetic Devices

CoFeHfO has been developed for use in high frequency transformers. Its high anisotropy and saturation magnetization give it a very high ferromagnetic resonance frequency (FMR) of about 2.8 GHz, while its high and adjustable resistivity reduces the energy lost due to eddy currents. It's properties enable a projection of GMR digital signal isolator operation above 100 GHz, assuming the associated integrated circuitry can keep up. The magnetoresistive and magnetic properties of CoFeHfO are shown below in Figs. 16.7 and 16.8.

The resisitivity has been adjusted over a range from 50 to 2000 micro-Ohm-cm. The saturation magnetization is relatively constant over this range. By making the resistivity as high as possible, one reduces the energy losses due to eddy currents in the films. These induced eddy currents are a main limitation for high frequency transformer applications.

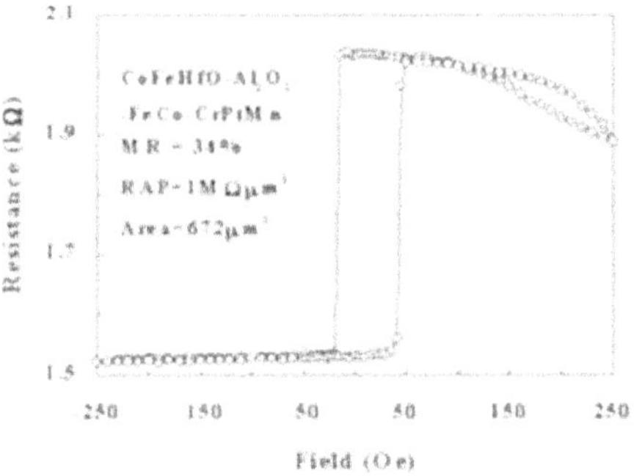

Fig. 16.7. Magnetoresistance of an SDT device with CoFeHfO as a free layer.

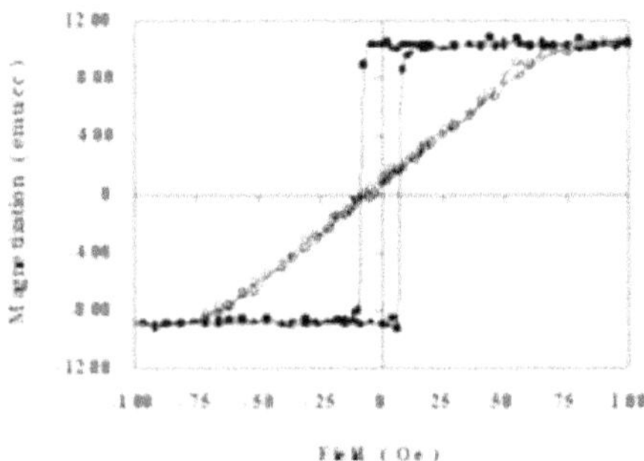

Fig. 16.8. Magnetization vs. field of a CoFeHfO film. The two curves are hard and easy axis magnetization loops.

16.2.4 Magnetic Nano-particle Detection

Magnetoresistive sensors can be used to detect magnetic nanoparticles. Such nanoparticles find uses in a wide range of applications. On one end of the spectrum is as the magnetic material in ferrofluids. On the other is as "labels" in biochemical assays. In between are other useful methods of sorting, heating and cooling, sealing, etc.

This section will focus on the use of GMR sensors as the detection scheme in biological assays. These assays become magnetic in nature when the "label" scheme uses nano-magnetic particles. Labeling is the term used for the attachment of a detectable object to the biochemical species of interest. As an example, a magnetic particle treated with a certain protein becomes "sticky" to certain other proteins but not others. This is because their physiochemical structure is designed to allow these two proteins to bind specifically to each other, but not to other proteins. Such "lock and key" interactions are omnipresent in the biological world. Perhaps the best known example is the DNA chain, where the left and right halves of the rungs of the "ladder" only connect if they match. Other labels are more common. The most common are light-based labels. They emit light due to excitation from a laser, a chemical interaction, or simply reflection. Optical detection, however, is not easy or convenient. Bulky and expensive lenses, mirrors, and other gear are required to achieve high degrees of resolution. Detection of magnetic labels, however, can be done with fully integrated magnetoresistive detectors. That is the basis of interest in this technique.

BARC. It is not the purpose of this paper to explain in detail how biochemical assays work. Rather, it is assumed that such assays generate samples where the number and / or concentration of nano-magnetic particles is an important value to measure. One such assay coats the surface of a GMR chip with an array of different biochemically active materials. Some of these are "sticky" to the specially prepared magnetic nano-particles while others are less so. The areas and degree of magnetic particle adhesion is then a quantifiable assay result. The question is then how to detect and quantify these particles. This idea came from the Naval Research Laboratory (NRL) which has developed the technology into a Bead ARray Counter (BARC) using custom GMR sensor arrays fabricated by NVE [8].

Magnetic Design. There are several magnetic modes that these assays can use. The BARC uses an out-of-plane field to magnetize the particles for steady-state detection. It is also possible to use an in-plane field. In all cases, the sensitive axis of the GMR detector is in the plane. Nominally, the magnetic nano-particles are paramagnetic or superparamagnetic, so their magnetizations and resulting stray fields are linearly proportional to the applied field. The stray fields from the magnetized particle are dipolar in shape. The field that the GMR sensor actually sees from the particle are orthogonal to or in the opposite direction from the applied field. So the total field at the GMR sensor is less than the applied field by an amount determined by the size and separation from magnetic particles. This difference in fields can be detected by comparing the output of the sensing resistor with a reference resistor known to have no particles above it. Signal to noise ratio is enhanced greatly by using a time varying field.

16.3 Conclusions

Magnetoresistive technology is being used in many important and interesting applications. This technology does best when the devices are compatible with high volume semiconductor-type manufacturing. This manufacturing method provides a way to make enormous numbers of sensors at a very low price per part.

The most ubiquitous sensors are the GMR read heads in hard disk drives. Many other sensing applications are possible depending upon the technical and commercial constraints. The most common sensing applications are in automated manufacturing and automotive environments where the position and speed of many components must be monitored. Ultra high performance magnetoresistive digital logic may become a commercial reality in the form of MRAM, currently under development by NVE, Motorola, Honeywell, IBM, and others. NVE has developed magnetoresistive isolators that will ultimately replace many optical isolators.

The current technological thrust in magnetoresistive devices is toward smaller, cheaper, and faster items. Research in nanomagnetics, then, is a clear enabler for these trends. Of special interest are making magnetic devices with stable properties at smaller sizes. Also, improving the properties of nanomagnetic particles (both in magnetic and dimensional terms) will enable many new applications in biochemical sensing and detection.

Acknowledgements

The authors aknowledge the many people at NVE who have participated in the development efforts described here. Jim Daughton, the founder of NVE, has provided not only the ideas behind a lot of the development, but also the energy and drive to bring ideas to reality. John Taylor and Seraphin Akou have done excellent work testing devices. Erik Lange and Dan Reed have helped with electronics development. Cathy Nordman, John Anderson, Bob Sinclair, Carl Smith, and Bob Schnieder have all worked towards improving the underlying technology through various research efforts. And, we must thank our sponsors for their strong and continued financial support. They include NASA, Army, Navy, Air Force, DARPA, NSF, and DOC.

References

1. W. Thomson, "On the electro-dynamic qualities of metals; Effects of magnetization on the electric conductivity of nickel and iron," Proc. Roy. Soc., **8**, pp. 546-550, (1857)
2. J. Smit, "Magnetoresistance of ferromagnetic metals and alloys at low temperatures," Physica **16**, pp. 612-617, (1951)
3. M.N. Baibich, J.M. Broto, A. Fert, F. Nguyen Van Dau, F. Petroff, P. Etienne, G. Creuzet, A. Freiderich, and J. Chazelax: Giant magnetoresistance of (001)Fe/(001)Cr magnetic superlattices, Phys. Rev. Lett. **61**, pp. 2472-2475, (1988)

4. P. Grunberg, R. Schreiber, Y. Pang, M.B. Brodsky, and H. Sowers: Layered magnetic structures: Evidence for antiferromagnetic coupling of Fe layers across Cr interlayers, Phys. Rev. Lett. **57**, pp. 2442-2445, (1986)
5. J.S. Moodera, L. R. Kinder, T.M. Wong, and R. Meservey, Phys. Rev. Lett. **74**, p. 3273, (1995)
6. T. Miyazaki and N. Tezuka, J. Magn. Magn. Mat. **139**, L231 (1995)
7. R.S. Beech, J. Anderson, J. Daughton, B.A. Everitt, and D. Wang, IEEE Trans. Magn. **32**, p. 4713 (1996)
8. D. R. Baselt, G. U. Lee, M. Natesan, S. W. Metzger, Paul E. Sheehan, and R. J. Colton, "A biosensor based on magneotoresistance technology," *Biosensors & Bioelectronics*, **13**, 731-739 (1998)
9. Jacob Fraden, AIP Handbook of Modern Sensors: Physics Designs and Applications, (American Institute of Physics, New York, 1993), 243-262

Lecture Notes in Physics

For information about Vols. 1–555 please contact your bookseller or Springer-Verlag

Vol. 556: D. Page, J. G. Hirsch (Eds.), From the Sun to the Great Attractor. X, 330 pages. 2000.

Vol. 557: J. A. Freund, T. Pöschel (Eds.), Stochastic Processes in Physics, Chemistry, and Biology. X, 330 pages. 2000.

Vol. 558: P. Breitenlohner, D. Maison (Eds.), Quantum Field Theory. Proceedings, 1998. VIII, 323 pages. 2000

Vol. 559: H.-P. Breuer, F. Petruccione (Eds.), Relativistic Quantum Measurement and Decoherence. Proceedings, 1999. X, 140 pages. 2000.

Vol. 560: S. Abe, Y. Okamoto (Eds.), Nonextensive Statistical Mechanics and Its Applications. IX, 272 pages. 2001.

Vol. 561: H. J. Carmichael, R. J. Glauber, M. O. Scully (Eds.), Directions in Quantum Optics. XVII, 369 pages. 2001.

Vol. 562: C. Lämmerzahl, C. W. F. Everitt, F. W. Hehl (Eds.), Gyros, Clocks, Interferometers...: Testing Relativistic Gravity in Space. XVII,507 pages. 2001.

Vol. 563: F. C. Lázaro, M. J. Arévalo (Eds.), Binary Stars. Selected Topics on Observations and Physical Processes. 1999.IX, 327 pages. 2001.

Vol. 564: T. Pöschel, S. Luding (Eds.), Granular Gases. VIII, 457 pages. 2001.

Vol. 565: E. Beaurepaire, F. Scheurer, G. Krill, J.-P. Kappler (Eds.), Magnetism and Synchrotron Radiation. XIV, 388 pages. 2001.

Vol. 566: J. L. Lumley (Ed.), Fluid Mechanics and the Environment: Dynamical Approaches. VIII, 412 pages. 2001.

Vol. 567: D. Reguera, L. L. Bonilla, J. M. Rubí (Eds.), Coherent Structures in Complex Systems. IX, 465 pages. 2001.

Vol. 568: P. A. Vermeer, S. Diebels, W. Ehlers, H. J. Herrmann, S. Luding, E. Ramm (Eds.), Continuous and Discontinuous Modelling of Cohesive-Frictional Materials. XIV, 307 pages. 2001.

Vol. 569: M. Ziese, M. J. Thornton (Eds.), Spin Electronics. XVII, 493 pages. 2001.

Vol. 570: S. G. Karshenboim, F. S. Pavone, F. Bassani, M. Inguscio, T. W. Hänsch (Eds.), The Hydrogen Atom: Precision Physics of Simple Atomic Systems. XXIII, 293 pages. 2001.

Vol. 571: C. F. Barenghi, R. J. Donnelly, W. F. Vinen (Eds.), Quantized Vortex Dynamics and Superfluid Turbulence. XXII, 455 pages. 2001.

Vol. 572: H. Latal, W. Schweiger (Eds.), Methods of Quantization. XI, 224 pages. 2001.

Vol. 573: H. M. J. Boffin, D. Steeghs, J. Cuypers (Eds.), Astrotomography. XX, 434 pages. 2001.

Vol. 574: J. Bricmont, D. Dürr, M. C. Galavotti, G. Ghirardi, F. Petruccione, N. Zanghi (Eds.), Chance in Physics. XI, 288 pages. 2001.

Vol. 575: M. Orszag, J. C. Retamal (Eds.), Modern Challenges in Quantum Optics. XXIII, 405 pages. 2001.

Vol. 576: M. Lemoine, G. Sigl (Eds.), Physics and Astrophysics of Ultra-High-Energy Cosmic Rays. X, 327 pages. 2001.

Vol. 577: I. P. Williams, N. Thomas (Eds.), Solar and Extra-Solar Planetary Systems. XVIII, 255 pages. 2001.

Vol. 578: D. Blaschke, N. K. Glendenning, A. Sedrakian (Eds.), Physics of Neutron Star Interiors. XI, 509 pages. 2001.

Vol. 579: R. Haug, H. Schoeller (Eds.), Interacting Electrons in Nanostructures. X, 227 pages. 2001.

Vol. 580: K. Baberschke, M. Donath, W. Nolting (Eds.), Band-Ferromagnetism: Ground-State and Finite-Temperature Phenomena. IX, 394 pages. 2001.

Vol.581: J. M. Arias, M. Lozano (Eds.), An Advanced Course in Modern Nuclear Physics. XI, 346 pages. 2001.

Vol.582: N. J. Balmforth, A. Provenzale (Eds.), Geomorphological Fluid Mechanics. X, 579 pages. 2001.

Vol.583: W. Plessas, L. Mathelitsch (Eds.), Lectures on Quark Matter, XIII, 334 pages. 2002.

Vol.584: W. Köhler, S. Wiegand (Eds.), Thermal Nonequilibrium Phenomena in Fluid Mixtures. XVII, 470 pages. 2002.

Vol.585: M. Lässig, A. Valleriani (Eds.), Biological Evolution and Statistical Physics. XI, 337 pages. 2002.

Vol.586: Y. Auregan, A. Maurel, V. Pagneux, J.-F. Pinton (Eds.), Sound–Flow Interactions. XVI, 286 pages. 2002

Vol.587: D. Heiss (Ed.), Fundamentals of Quantum Information. Quantum Computation, Communication, Decoherence and All That. XIII, 265 pages. 2002.

Vol.588: Y. Watanabe, S. Heun, G. Salviati, N. Yamamoto (Eds.), Nanoscale Spectroscopy and Its Applications to Semiconductor Research. XV, 306 pages. 2002.

Vol.589: A. W. Guthmann, M. Georganopoulos, A. Marcowith, K. Manolakou (Eds.), Relativistic Flows in Astrophysics. XII, 241 pages. 2002

Vol.590: D. Benest, C. Froeschlé (Eds.), Singularities in Gravitational Systems. Applications to Chaotic Transport in the Solar System. XI, 215 pages. 2002

Vol.591: M. Beyer (Ed.), CP Violation in Particle, Nuclear and Astrophysics. XI, 334 pages. 2002

Vol.592: S. Cotsakis, L. Papantonopoulos (Eds.), Cosmological Crossroads. An Advanced Course in Mathematical, Physical and String Cosmology. XVI, 477 pages. 2002

Vol.593: D. Shi, B. Aktaş, L. Pust, F. Mikhailov (Eds.), Nanostructured Magnetic Materials and Their Applications. XII, 289 pages. 2002

Vol.594: S. Odenbach (Ed.),Ferrofluids. Magnetical Controllable Fluids and Their Applications. X, 257 pages. 2002

Vol.595: C. Berthier, L. P. Lévy, G. Martinez (Eds.), High Magnetic Fields. Applications in Condensed Matter Physics and Spectroscopy. X, 493 pages. 2002

Monographs

For information about Vols. 1–29
please contact your bookseller or Springer-Verlag

Vol. m 30: A. J. Greer, W. J. Kossler, Low Magnetic Fields in Anisotropic Superconductors. VII, 161 pages. 1995.

Vol. m 31 (Corr. Second Printing): P. Busch, M. Grabowski, P.J. Lahti, Operational Quantum Physics. XII, 230 pages. 1997.

Vol. m 32: L. de Broglie, Diverses questions de mécanique et de thermodynamique classiques et relativistes. XII, 198 pages. 1995.

Vol. m 33: R. Alkofer, H. Reinhardt, Chiral Quark Dynamics. VIII, 115 pages. 1995.

Vol. m 34: R. Jost, Das Märchen vom Elfenbeinernen Turm. VIII, 286 pages. 1995.

Vol. m 35: E. Elizalde, Ten Physical Applications of Spectral Zeta Functions. XIV, 224 pages. 1995.

Vol. m 36: G. Dunne, Self-Dual Chern-Simons Theories. X, 217 pages. 1995.

Vol. m 37: S. Childress, A.D. Gilbert, Stretch, Twist, Fold: The Fast Dynamo. XI, 406 pages. 1995.

Vol. m 38: J. González, M. A. Martín-Delgado, G. Sierra, A. H. Vozmediano, Quantum Electron Liquids and High-Tc Superconductivity. X, 299 pages. 1995.

Vol. m 39: L. Pittner, Algebraic Foundations of Non-Com-mutative Differential Geometry and Quantum Groups. XII, 469 pages. 1996.

Vol. m 40: H.-J. Borchers, Translation Group and Particle Representations in Quantum Field Theory. VII, 131 pages. 1996.

Vol. m 41: B. K. Chakrabarti, A. Dutta, P. Sen, Quantum Ising Phases and Transitions in Transverse Ising Models. X, 204 pages. 1996.

Vol. m 42: P. Bouwknegt, J. McCarthy, K. Pilch, The W3 Algebra. Modules, Semi-infinite Cohomology and BV Algebras. XI, 204 pages. 1996.

Vol. m 43: M. Schottenloher, A Mathematical Introduction to Conformal Field Theory. VIII, 142 pages. 1997.

Vol. m 44: A. Bach, Indistinguishable Classical Particles. VIII, 157 pages. 1997.

Vol. m 45: M. Ferrari, V. T. Granik, A. Imam, J. C. Nadeau (Eds.), Advances in Doublet Mechanics. XVI, 214 pages. 1997.

Vol. m 46: M. Camenzind, Les noyaux actifs de galaxies. XVIII, 218 pages. 1997.

Vol. m 47: L. M. Zubov, Nonlinear Theory of Dislocations and Disclinations in Elastic Body. VI, 205 pages. 1997.

Vol. m 48: P. Kopietz, Bosonization of Interacting Fermions in Arbitrary Dimensions. XII, 259 pages. 1997.

Vol. m 49: M. Zak, J. B. Zbilut, R. E. Meyers, From Instability to Intelligence. Complexity and Predictability in Nonlinear Dynamics. XIV, 552 pages. 1997.

Vol. m 50: J. Ambjørn, M. Carfora, A. Marzuoli, The Geometry of Dynamical Triangulations. VI, 197 pages. 1997.

Vol. m 51: G. Landi, An Introduction to Noncommutative Spaces and Their Geometries. XI, 200 pages. 1997.

Vol. m 52: M. Hénon, Generating Families in the Restricted Three-Body Problem. XI, 278 pages. 1997.

Vol. m 53: M. Gad-el-Hak, A. Pollard, J.-P. Bonnet (Eds.), Flow Control. Fundamentals and Practices. XII, 527 pages. 1998.

Vol. m 54: Y. Suzuki, K. Varga, Stochastic Variational Approach to Quantum-Mechanical Few-Body Problems. XIV, 324 pages. 1998.

Vol. m 55: F. Busse, S. C. Müller, Evolution of Spontaneous Structures in Dissipative Continuous Systems. X, 559 pages. 1998.

Vol. m 56: R. Haussmann, Self-consistent Quantum Field Theory and Bosonization for Strongly Correlated Electron Systems. VIII, 173 pages. 1999.

Vol. m 57: G. Cicogna, G. Gaeta, Symmetry and Perturbation Theory in Nonlinear Dynamics. XI, 208 pages. 1999.

Vol. m 58: J. Daillant, A. Gibaud (Eds.), X-Ray and Neutron Reflectivity: Principles and Applications. XVIII, 331 pages. 1999.

Vol. m 59: M. Kriele, Spacetime. Foundations of General Relativity and Differential Geometry. XV, 432 pages. 1999.

Vol. m 60: J. T. Londergan, J. P. Carini, D. P. Murdock, Binding and Scattering in Two-Dimensional Systems. Applications to Quantum Wires, Waveguides and Photonic Crystals. X, 222 pages. 1999.

Vol. m 61: V. Perlick, Ray Optics, Fermat's Principle, and Applications to General Relativity. X, 220 pages. 2000.

Vol. m 62: J. Berger, J. Rubinstein, Connectivity and Superconductivity. XI, 246 pages. 2000.

Vol. m 63: R. J. Szabo, Ray Optics, Equivariant Cohomology and Localization of Path Integrals. XII, 315 pages. 2000.

Vol. m 64: I. G. Avramidi, Heat Kernel and Quantum Gravity. X, 143 pages. 2000.

Vol. m 65: M. Hénon, Generating Families in the Restricted Three-Body Problem. Quantitative Study of Bifurcations. XII, 301 pages. 2001.

Vol. m 66: F. Calogero, Classical Many-Body Problems Amenable to Exact Treatments. XIX, 749 pages. 2001.

Vol. m 67: A. S. Holevo, Statistical Structure of Quantum Theory. IX, 159 pages. 2001.

Vol. m 68: N. Polonsky, Supersymmetry: Structure and Phenomena. Extensions of the Standard Model. XV, 169 pages. 2001.

Vol. m 69: W. Staude, Laser-Strophometry. High-Resolution Techniques for Velocity Gradient Measurements in Fluid Flows. XV, 178 pages. 2001.

Vol. m 70: P. T. Chruściel, J. Jezierski, J. Kijowski, Hamiltonian Field Theory in the Radiating Regime. VI, 172 pages. 2002.

Vol. m 71: S. Odenbach, Magnetoviscous Effects in Ferrofluids. X, 151 pages. 2002.

Vol. m 72: J. G. Muga, R. Sala Mayato, I. L. Egusquiza (Eds.), Time in Quantum Mechanics. XII, 419 pages. 2002.

GPSR Compliance
The European Union's (EU) General Product Safety Regulation (GPSR) is a set of rules that requires consumer products to be safe and our obligations to ensure this.

If you have any concerns about our products, you can contact us on

ProductSafety@springernature.com

In case Publisher is established outside the EU, the EU authorized representative is:

Springer Nature Customer Service Center GmbH
Europaplatz 3
69115 Heidelberg, Germany

www.ingramcontent.com/pod-product-compliance
Ingram Content Group UK Ltd.
Pitfield, Milton Keynes, MK11 3LW, UK
UKHW021333070726
13610UKWH00011B/54

* 9 7 8 3 6 6 2 1 4 3 3 8 4 *